BIOIMPEDANCE
AND
BIOELECTRICITY BASICS

BIOIMPEDANCE
AND
BIOELECTRICITY BASICS

Sverre Grimnes, MSc PhD

Department of Physics, University of Oslo and
Department of Biomedical and Clinical Engineering, Rikshospitalet, Oslo

and

Ørjan Grøttem Martinsen, MSc PhD

Department of Physics, University of Oslo

ACADEMIC PRESS

A Harcourt Science and Technology Company

San Diego   San Francisco   New York
Boston   London   Sydney   Tokyo

Academic Press
A Harcourt Science and Technology Company
Harcourt Place, 32 Jamestown Road, London NW1 7BY, UK
http://www.academicpress.com

Academic Press
A Harcourt Science and Technology Company
525 B Street, Suite 1900, San Diego, California 92101-4495, USA
http://www.academicpress.com

ISBN 0-12-303260-1

A catalogue for this book is available from the British Library

Bioimpedance and Bioelectricity Basics Website: www.fys.uio.no/publ/bbb

Transferred to digital printing 2005.

Typeset by Paston PrePress Ltd, Beccles, Suffolk

00 01 02 03 04 05 MP 9 8 7 6 5 4 3 2 1

Contents

Preface   xi

1. Introduction   1

2. Electrolytics   3
   2.1.   Ionic and electronic dc conduction   3
       2.1.1.   Ionisation   4
       2.1.2.   Molecular bonds   6
   2.2.   The basic electrolytic experiment   7
   2.3.   Bulk electrolytic dc conductance   11
   2.4.   Interphase phenomena   22
       2.4.1.   Faraday's law   22
       2.4.2.   Migration and diffusion   23
       2.4.3.   The electric double layer, perpendicular fields   25
       2.4.4.   The electric double layer, lateral fields   29
       2.4.5.   The net charge of a particle   31
       2.4.6.   Electrokinesis   31
   2.5.   Electrodics and ac phenomena   35
       2.5.1.   Electrode equilibrium dc potential, zero external
              current   36
       2.5.2.   The monopolar basic experiment   39
       2.5.3.   Dc/ac equivalent circuit for electrode processes   44
       2.5.4.   Non-linear properties of electrolytics   48

3. Dielectrics   51
   3.1.   Polarisation in a uniform dielectric   52
       3.1.1.   Coulomb's law and static electric fields   53
       3.1.2.   Permanent and induced dipole moments   53
       3.1.3.   Charge-to-dipole and dipole-to-dipole interactions   56
   3.2.   The basic capacitor experiment   57
   3.3.   Complex permittivity   59
   3.4.   Ac polarisation and relaxation in a uniform dielectric   63
       3.4.1.   Relaxation and dispersion   63
       3.4.2.   Debye relaxation model   64
       3.4.3.   Displacement current density and dielectric loss   70
       3.4.4.   Joule effect and temperature rise   71
   3.5.   Interfacial polarisation   73
       3.5.1.   Maxwell–Wagner effects, MHz region   74
       3.5.2.   Adsorbed counterions and ion diffusion (mHz–kHz region)   77

3.6.   The basic membrane experiment   79
3.7.   The basic suspension experiment   81
3.8.   Dispersion and dielectric spectroscopy   83

4. Electrical properties of tissue   **87**
    4.1.   Basic biomaterials   87
        4.1.1.   Water and body fluids   87
        4.1.2.   Proteins   88
        4.1.3.   Carbohydrates (saccharides)   93
        4.1.4.   Lipids and the passive cell membrane   93
    4.2.   The cell   94
    4.3.   Tissue and organs   99
        4.3.1.   Muscle tissue   101
        4.3.2.   Nerve tissue   101
        4.3.3.   Adipose and bone tissue   103
        4.3.4.   Blood   104
        4.3.5.   Human skin and keratinised tissue   105
        4.3.6.   Whole body   114
        4.3.7.   Post-excision changes, the death process   115
        4.3.8.   Tabulated data for measured conductivity and phase
                 angle, mammalian tissue   118
        4.3.9.   Plant tissue   118
    4.4.   Special electrical properties   119
        4.4.1.   Tissue anisotropy   120
        4.4.2.   Tissue dc properties   120
        4.4.3.   Nerves and muscles excited   122
        4.4.4.   Non-linear tissue parameters, breakdown   122
        4.4.5.   Piezoelectric and triboelectric effects   124

5. Geometrical analysis   **127**
    5.1.   The forward and inverse problems   129
    5.2.   Monopole and dipole sources   129
        5.2.1.   Sphere monopoles   129
        5.2.2.   Spheroid and disk monopoles   132
        5.2.3.   Dipole   136
    5.3.   Recording leads   138
        5.3.1.   Potentials recorded in the monopolar field   138
        5.3.2.   Potentials recorded in a dipolar field   138
        5.3.3.   Lead sensitivity and the lead vector   140
    5.4.   Three- and four-electrode systems   143
        5.4.1.   Three-electrode system, monopolar recording   143
        5.4.2.   Four-electrode (tetrapolar) system   145
    5.5.   Finite-element method   147
    5.6.   Tomography and plethysmography   149

6. Instrumentation and measurement   **153**
6.1. General network theory, the black box   153
6.1.1. Admittance, impedance and immittance   153
6.1.2. The two-port network and the reciprocity theorem   155
6.1.3. Extended immittance concepts   156
6.1.4. The non-linear black box   157
6.1.5. The time constant   157
6.1.6. Kramers–Kronig transforms   158
6.2. Signals and measurement   159
6.2.1. Dc, static values and ac   159
6.2.2. Periodic waveforms, Fourier series of sine waves   160
6.2.3. Aperiodic waveforms   167
6.2.4. Spectrum analysis, Fourier transforms   170
6.2.5. Signal generators   174
6.2.6. The basic measuring circuit   175
6.2.7. Operational amplifiers and filters   176
6.2.8. Ground, reference and common mode voltage   178
6.2.9. Chaos theory and fractals   182
6.2.10. Neural networks   183
6.2.11. Wavelet analysis   184
6.3. Bridges, impedance analysers, lock-in amplifiers   187
6.3.1. Bridges   187
6.3.2. Digital lock-in amplifiers   188
6.3.3. Analogue lock-in amplifiers   190
6.3.4. Current mode lock-in amplifiers   191
6.3.5. Impedance analysers and LCR-meters   192

7. Data and models   **195**
7.1. Models, descriptive and explanatory   195
7.2. Equations and equivalent circuits   201
7.2.1. Maxwell's equations   201
7.2.2. Two-component equivalent circuits, ideal components   203
7.2.3. Constant phase elements (CPEs)   209
7.2.4. Augmented Fricke CPE_F 212
7.2.5. Three- and four-component equivalent circuits, ideal components   214
7.2.6. Cole equations   216
7.2.7. Cole–Cole equations   221
7.2.8. Control of Fricke compatibility   222
7.2.9. The α parameter   223
7.2.10. Symmetrical DRT (distribution of relaxation times)   224
7.2.11. Non-symmetrical DRT   226
7.2.12. Multiple Cole systems   227

7.3.	Data calculation and presentation	232
	7.3.1. Measured data, model data	232
	7.3.2. Indexes	233
	7.3.3. Presentation of measured data, example	233

8. Selected applications **241**
8.1.	Electrodes, design and properties	242
	8.1.1. Coupling without galvanic tissue contact	242
	8.1.2. The electronic conductor of the electrode	250
	8.1.3. The contact electrolyte	252
	8.1.4. Dc voltage and noise generation	254
	8.1.5. Polarisation immittance	257
	8.1.6. Electrode design	262
8.2.	ECG	268
8.3.	Impedance plethysmography	270
	8.3.1. Impedance cardiography (ICG)	274
8.4.	EEG, ENG/ERG/EOG	275
	8.4.1. ENG/ERG/EOG	276
8.5.	Electrogastrography (EGG)	276
8.6.	EMG and neurography	276
8.7.	Electrotherapy	277
	8.7.1. Electrotherapy with dc	278
	8.7.2. Electrotherapy of muscles	280
8.8.	Body composition	282
8.9.	Cardiac pacing	284
8.10.	Defibrillation and electroshock	285
	8.10.1. Defibrillation	285
	8.10.2. Electroshock (brain electroconvulsion)	286
8.11.	Electrosurgery	287
8.12.	Cell suspensions	290
	8.12.1. Electroporation and electrofusion	290
	8.12.2. Cell sorting and characterisation by electrorotation and dielectrophoresis	292
	8.12.3. Cell-surface attachment and micromotion detection	293
	8.12.4. Coulter counter	294
8.13.	Skin and keratinised tissues	294
	8.13.1. Electrical assessment of stratum corneum hydration	294
	8.13.2. Electrical assessment of skin irritation, dermatitis and fibrosis	296
	8.13.3. Electrodermal response (EDR)	297
	8.13.4. Iontophoretic treatment of hyperhidrosis	299
	8.13.5. Iontophoresis and transdermal drug delivery	300
8.14.	Threshold of perception, hazards	301
	8.14.1. Threshold of perception	301
	8.14.2. Electrical hazards	304

8.14.3. Lightning and electrocution 308
8.14.4. Electric fences 309
8.14.5. Electrical safety of electromedical equipment 309

9. History of bioimpedance and bioelectricity 313

10. Appendix 321
10.1. Vectors and scalars, complex numbers 321
10.2. Equivalent circuit equations 325
10.2.1. Equations for two resistors + one capacitor circuits 325
10.2.2. Equations for two capacitors + one resistor circuits 328
10.2.3. Equations for four-component *series* circuit (simple Maxwell–Wagner model) 330
10.3. Global symbols 331
10.4. Physical dimensions 331

References and further reading 335
Recommended books 335
References 336

Index 347

Preface

This book has grown out of a certain frustration of having used unnecessarily much time ourselves learning some of the theory and practice of bioimpedance, and out of a certain hope that a new book could pave an easier and more efficient way for people seeking basic knowledge about our discipline. Bioimpedance and bioelectricity must perhaps be considered as rather specialised fields, but obviously based upon an extract from scientific basic disciplines. All these disciplines cannot be taught in their full extensions, but with this book it should be possible to gather many of them into one single subject and one course. For the newcomer it is also an advantage to be presented a unified set of terminology and symbols, to avoid the start with the silent terminology of the paradigms of each area, bewildering traditions illustrated for instance by the different use of the term "polarisation" and such symbols as m and α.

The starting point for us has been the duality of the electrical properties of tissue. In the lower (<1 MHz) frequency range, the immittance of most tissues are predominantly *electrolytic*. Therefore we start the book with electrolytics, covering also the electric double layer being so sensitive to surface properties. With high resolution techniques it is possible to extract the important capacitive, that is dielectric, properties also in that low frequency range. At higher frequencies the *dielectric* properties of tissue may dominate, and at the highest frequencies tissue properties become more and more equal to that of water, pure water has a characteristic relaxation frequency of approx. 18 GHz.

Another side of the duality is the very early recognised fact that tissue and electrodes in electrolyte solutions have important and special properties in common: a more or less constant phase character. It has become more and more clear that structural Maxwell-Wagner interfacial polarisation and the Debye theory for molecules cannot explain the large measured dielectric decrements found e.g. with small spheres in suspension. Interfacial counterion models have been introduced, implying different electrolytic processes such as the lateral diffusion of counterions in the double layer.

Bioimpedance and bioelectricity is about biomaterials in a broad sense: materials that are living, have lived, or are potential building blocks for living tissue. The basic building block is the living cell, and a prerequisite is that it is surrounded by an electrolyte solution. Great caution must be imposed on the state of the biomaterial sample. A material may change completely from the living, wetted state with large contributions from interfacial counterion mechanisms, via a denaturation or death process to a more or less dead and dry sample, and further out to the extreme: placed in a vacuum chamber. It is important to remember this when e.g. ionic versus electronic/semiconductive properties are discussed. As life is so diversified the situation is of course often somewhat complex: bacteria may be in dry surroundings and encapsulated in a sleeping state, and it is difficult to give them a clear living status.

This book emphasises model thinking. Science is very much about the use of models, in order to *describe* and therefore predict, and in order to *explain* and therefore understand. But models have their shortcomings. Important models for bioimpedance are empirical and can therefore only describe. Because tissue immittance is predominantly electrolytic, a model's treatment of dc conductivity is for instance important. But as shown, the Cole models are specialised models not treating dc conductance necessarily as an independent variable. Also, at sufficiently high energy levels the real world is not linear. It must not be ignored that most models extensively treated by textbooks are limited to linear cases. Many applications such as defibrillation or electroporation are clearly in the non-linear range; a sine wave excitation does not lead to a sine wave response. The principle of superposition is not valid, and the different contributions cannot simply be added. Also the *choice* of the best model is important. Many researchers have been led astray by using a "wrong" model; e.g. a series model for processes actually physically occurring in parallel. Also a dispersion model presupposes that the measured volume is independent of frequency, and this is not always the case in a measuring set-up. This is a part of a very general problem in bioimpedance: how to select or limit the measured volume.

Although this book has been written primarily for graduate and post-graduate students in biomedical engineering and biophysics, we hope it will be useful also for other researchers coming in touch with our area, e.g. from biotechnology in general, electrophysiology, odontology, pharmacy and plant biology. Some devoted medical doctors in the field of neurology, cardiology, dermatology, clinical chemistry and microbiology have not been forgotten. We have on certain subjects reverted to an almost "Adam and Eve" approach. In addition, the number of illustrations have been kept high. We have not renounced on mathematical equations, but usually tried to include an often extended discussion on their implications. To keep the book within the "basic" framework, we have imposed certain boundaries. We have excluded magnetism, which is already well covered by Malmuvio and Plonsey (1995). We have limited the book to sine wave and step function variables, omitting a more general treatment by the theory of Laplace transforms. And we have limited the number of application examples.

Our background in the fields of biomedical engineering, physics and instrumentation is of course discernible. All the same we have found it necessary to cover a much larger range of topics. The emphasis is on systems with galvanic contact to the tissue, not so much on the interaction between tissue and electromagnetic fields and waves. A large part has been dedicated to model thinking: simple equivalent circuit models, the constant phase element (CPE) model and Cole system models. The importance of the measuring system *geometry* can not be overemphasised. We hope that the balance between the descriptive and quantitative/theoretical text parts will be appreciated.

Tips to the reader

Vector symbols are in **bold**. A non-bold symbol is either a scalar, or a magnitude or real part of a vector. In the literature, an intelligent guess sometimes has to be made.

A phase angle is denoted by φ, a loss angle by δ. In the literature the loss angle is often called a phase angle, which it of course also is.

Φ is used for a potential in space and V for a voltage in a circuit. Φ may designate not only the potential at a defined position (e.g. $\Phi(x,y,z)$), but also a potential (scalar) *field*.

Global symbols used all over the book are tabulated in Section 10.2, and are not explained locally in the text.

Impedance variables such as Z, R, X, ρ and C_s are preferably used when components are connected in series. *Admittance* variables such as Y, G, B, σ and C_p are preferably used when components are connected in parallel. *Immittance* is the combined term for both impedance and admittance. It is often used in order to force the reader to be sensitive to the choice: there is no such thing as an immittance equation.

Ideal capacitor and resistor components are drawn as usual:

Electrolytic components with frequency dependent values are drawn:

A Wessel diagram is the same as an Argand diagram: a diagram of the complex plane.

The International System of units (SI) is used in this book. Note that the choice of system also influences the formulas. For instance, Coulomb's law differs by the factor $4\pi\varepsilon_0$ between the old cgs system ($F = q_1q_2/\varepsilon_r r^2$) and the SI system ($F = q_1q_2/4\pi\varepsilon r^2$) used here. Or in cgs: $\mathbf{D} = \varepsilon_r\mathbf{E}$ and in SI: $\mathbf{D} = \varepsilon\mathbf{E}$.

This book has its own home page on the Internet – www.fys.uio.no/publ/bbb. On that page you can find updated information about topics covered in the book and you can also find useful links to other bioimpedance web-sites.

Acknowledgements

We are greatly indebted to the many colleagues and friends who have contributed to this book by reading and commenting on selected chapters. Walter Lund, Tom Henning Johansen and Jakob Sandstad at the University of Oslo and Stig Ollmar at the Karolinska Institute in Stockholm should especially be mentioned for their invaluable help. We are in particular indebted to Herman P. Schwan at the University of Pennsylvania for the long discussions, which have had a significant influence on this book. It has been a pleasure to work with Serena Bureau, Roopa Baliga and Manjula Goonawardena at Academic Press and we are truly grateful for all their professional help and positive spirit. Last, but not least, we appreciate the great support of our wives Kari and Kjersti.

CHAPTER 1

Introduction

Bioimpedance, bioelectricity and the *electrical properties of tissue* are much about the same things. If we apply electricity from an external source outside the living organism under study (*exogenic* current) we measure *bioimpedance*, or perform *electrotherapy*. *Bioelectricity* is a broader concept, covering also the electric currents associated with the life processes, and their biopotentials. Such electricity is called *endogenic* which means that it is internally generated in the tissue. Many of the applications are well known: Recording bioelectric signals from the heart (ECG) was introduced by Waller in 1887 and Einthoven in 1908, and is still a most important examination in hospitals worldwide. Electrosurgery came in the same position from the '30s. Recording bioelectric signals from the brain (EEG) was introduced in the '40s, pacemakers and defibrillators were taken into use in the '60s.

We have chosen the terms bioimpedance and bioelectricity for the title of this book. Immittance is the combined term for impedance and admittance, so a better and more generic term would have been *bioimmittance*. Immittance is related to the *passive* properties of tissue, and less emphasis has been given to the active properties: the excitation of nerve and muscle tissue.

Under linear conditions and for the same tissue, unity cell impedance $\mathbf{Z}$, admittance $\mathbf{Y} = 1/\mathbf{Z}$, complex permittivity ε and complex conductivity $\sigma = j\omega\varepsilon$ all contain the same information, but differently presented. Tissue can be characterised as a dielectric or an electrolytic biomaterial and by a relaxation or immittance model. It may be examined with sine, step or other waveform signals. As long as linear conditions prevail, the information gathered is the same. At high voltage and current levels biomaterials are non-linear, and models and parameters must be chosen with care. The interpretation of the measured data is extremely dependent on the angle of view and the choice of model. One of the most fundamental choices to be made is between impedance and admittance, or series and parallel models.

In tissue and the living cell there is an inseparable alliance between *electricity* and *chemistry*. Electrolytic theory and electrochemistry therefore form an important basis for our field, it is not possible to understand what is going on in tissue during electric current flow without knowing some electrolytic theory. The tissue of interest may be a living plant, fruit, egg, fish, animal, or a human. But it may also be dead biological material such as hair or nail, or excised material such as beef or a piece of stratum corneum. Bioimpedance measurements give information about electrochemical processes in the tissue and can hence be used for characterising the tissue or for monitoring physiological changes. Bioimpedance and its frequency dependence differs greatly between different cell suspensions and tissues and significant changes in the electrical properties are also found when cells or tissues go from one physiological state to another, e.g. living to dead, dry to moist or normal to pathological. Knowing the absolute value or the range of typical values for the impedance is also of great importance e.g. when assessing safety aspects of electromedical equipment or designing the front end of an ECG monitor.

The potential benefits from impedance measurements in medicine and biology are obviously great, and new applications are continuously under investigation and development. Interesting new instruments and methods have the last years been introduced e.g. in areas like electrical impedance tomography, body composition, cell micromotion, organ viability, skin hydration and skin pathology.

Parallel to finding new applications, basic research is going on trying to give us a better understanding of both tissue electric properties and the techniques to be used. Examples of such basic questions still to be answered are whether living tissue is a purely ionic conductor – are there also local electronic or semiconductor mechanisms, what is the conduction mechanism e.g. in keratinised tissue or DNA? Furthermore, is there really a specific constant phase mechanism for the immitance of biological materials, which are the different mechanisms of the α-dispersion and what is the difference between the effect of dc and ac? Dc implies tissue polarisation caused by a transport of matter to an extent not found with ac.

The list of unsolved problems is long. In order to do understand the phenomena of interest, a certain basic knowledge of electrochemistry, electronic engineering, physics, physiology, mathematics and model thinking is needed. And that is what is to be found in the chapters of this book.

CHAPTER 2

Electrolytics

2.1. Ionic and Electronic dc Conduction

An *electrolyte* is a substance with ionic dc conductivity. Living tissue is electrically and macroscopically predominantly an electrolytic conductor. Both intracellular and extracellular liquids contain *ions free to migrate*. In pure electrolytes the charge carriers are ions, and there is no separate flow of electrons, they are all bound to their respective atoms. Tissue dc currents are therefore *ionic* currents, in contrast to the *electronic* current in metals. This does not exclude a possible local electronic conductance due to free electrons, e.g. in the intracellular DNA molecules. New solid materials like organic polymers and glasses may contain an appreciable amount of free ions with considerable mobility, so the materials of an electrolytic measuring cell are not limited to liquid media. Some of these solid media show a mixture of ionic and electronic conductivity.

Two *current carrying electrodes* in an electrolyte are the source and sink of electrons: from electrons of the metal, to ions or uncharged species of the electrolyte. *The electrode is the site of a charge carrier shift, a charge exchange between electrons and ions.*

In a metal the conductance electrons are free to move; they are like an electron gas not linked to particular metal atoms, but with a probability of being at a certain location at a certain time. The metal atoms can be considered bound but ionised—they have lost electrons. Electron transport in a metal involves no transport of metal ions and not even transport of electrons all the way through the bulk. When we supply an electron into a wire end, "another" electron is coming out of the other end. Current flow, which seems to be so fast, is so only because it is not the same electron entering and leaving. Actually, the migration velocity of *electrons* in a metal is very slow, of the order of 0.3 mm/s at rather high current densities. The migration velocity of *ions* in solution is also very slow. As studied by electrophoresis, the ion migration velocity is of the order of 10 mm/s.

At the very low migration velocities there are no collision phenomena in which charge carriers are stopped. The electronic conduction in the vacuum of a cathode ray tube (CRT) is very different. Friction is low and electron velocity is very high, of the order of thousands of meters per second (but with many fewer electrons involved). When these fast electrons are stopped, there is a collision, for instance with the phosphor plate that lights up in a CRT or with the anode of an x-ray tube, which emits x-rays.

Electric current flow in an ionic solution is a more complex event than in a metal. Electron current implies no transport of substance. Therefore, an externally applied dc current can flow forever without changing the conductor. However, ion current implies a transport of substance. Therefore, an externally applied dc current can not flow forever without changing the conductor. At first, changes will occur near the electrodes, but in a closed electrolytic cell with sufficiently long time the change will spread to the bulk of the electrolyte. Accordingly, *electrolytic long duration dc conductivity* is a difficult concept in a closed system.

The transfer of electric charge across the solution–electrode interphase is accompanied by an electrochemical reaction at each electrode (electrolysis). *We must keep the phenomena in the bulk of the solution separate from the phenomena at the electrodes.*

2.1.1. Ionisation

Since the charge carriers of interest are ions, the *ionisation* of atoms is of particular interest. The electrons of an atom are arranged in shells. The forces acting between atoms are of an electrostatic nature. In electrochemistry, the ionisation of an atom is determined by the electron configuration of the *outermost* shell. If this shell is full, the atom has a noble gas configuration. This is a particularly stable form, implying that a large energy is necessary to remove, or add, an electron and thus ionise such an atom (Table 2.1).

For hydrogen and helium the innermost shell, the K-shell, is also the outermost shell. The K-shell is full with 2 electrons (the noble gas helium). The next L-shell is full with 8 electrons (the noble gas neon), and the next M-shell with 18 electrons (the noble gas krypton). The chemical properties of an atom are determined by the

Table 2.1. Electron shell configuration for the lowest atomic number atoms. Ionisation potential is here the *energy* necessary to remove the first electron from the valence (outermost) shell. Values for radii depend on how they are measured, here in vacuum and not aqueous solution

	Protons in nucleus	Shell			Typical electrovalency	Ionisation potential, (eV)	Atomic radius (nm)	Positive ion radius (nm)	Negative ion radius (nm)
		K	L	M					
H	1	1	0	0	+1	13.6	0.037	0.00001	0.154
He	**2**	**2**	**0**	**0**	**Noble**	**24.6**		**N/A**	**N/A**
Li	3	2	1	0	+1	5.4		0.068	N/A
Be	4	2	2	0	+2	9.3		0.044	N/A
B	5	2	3	0	+3	8.3		0.035	N/A
C	6	2	4	0	±4	11.3	0.077	0.016	0.26
N	7	2	5	0	−3	14.5	0.070	0.025	0.17
O	8	2	6	0	−2	13.6	0.066	0.022	0.176
F	9	2	7	0	−1	17.4		N/A	0.133
Ne	**10**	**2**	**8**	**0**	**Noble**	**21.6**		**N/A**	**N/A**
Na	11	2	2	1	+1	5.1		0.097	N/A

Table 2.2. Pauling's scale of electronegativity for some selected atoms

F	4.0	S	2.5
O	3.5	C	2.5
N	3.0	H	2.1
Cl	3.0	P	2.1
Br	2.8	Fe	1.8
I	2.5	Na	0.9

electron configuration of the outermost shell. These electrons are called *valence* electrons, and their ionisation potential (energy necessary to remove an electron) is for most atoms less than 20 eV. Chemical reactions and bonds are related to the valence electrons in the outermost shell; the electrons in the inner shells (affected by x-rays) and the nuclei (high-energy nuclear processes) are not affected. Ordinary chemical methods therefore involve energy levels <20 eV. The *electrovalency z* of an atom is the number of electrons available for transfer. The valency is thus $z = +1$ for Na, and $z = -1$ for Cl, cf. Table 2.1. A valence electron is a rather broad concept comprising those electrons in the outer shell that may combine with other atoms and form molecules, whether it is by gaining, losing or *sharing* electrons.

The electrochemical properties are determined by the inclination of an atom to attain noble gas configuration of the outer electron shell. Atoms with few electrons in the outer shell (e.g. H, Li, Na) have a tendency to empty the shell, that is to lose electrons and form positive ions. Atoms with a nearly filled shell (e.g. O, F) have a tendency to fill up the shell, that is to gain electrons and form negative ions. "Tendency" here simply means that those configurations are lower energy level forms.

Electronegativity is the relative ability of an atom to gain electrons and become a negative ion. Clearly Na is not very electronegative, but F is highly electronegative. Pauling[1] worked out a scale of electronegativity (Table 2.2).

Electronegativity is not a purely quantitative concept, but it is useful in the prediction of the strengths and polarities of ionic bonds between atoms, and thus possible electrochemical reactions. In electrochemistry, the use of electrode equilibrium potential tables (Section 2.5.1) serves the same purpose. Atoms with small electronegativity (e.g. Na) are not inclined to gain an electron at all: it would move the ion away from noble gas configuration. Sodium's natural state is to lose an electron and become a positive ion. Fluorine is very electronegative with a Pauling scale value of 4; its L-shell is filled with just one extra electron. With a value of 2.5 carbon is in a middle position with the ability of both losing and gaining electrons. *Hydrogen* is in a special position; in principle it should be highly electronegative as

---

[1] Linus Pauling (1901–1994), American chemist. Nobel prize laureate 1954 in Chemistry on the structure of proteins (and 1962 in Peace). Important work also on chemical bonding and electronegativity; invented the paramagnetic oxygen analyser.

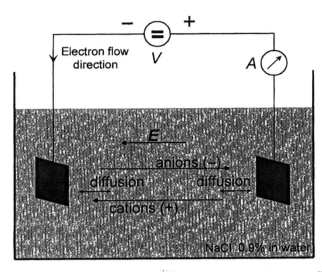

Figure 2.1. The basic electrolytic experiment, shown with material transport directions.

potential difference of one volt or more. But here we presume that the *same* electrode material is used, and that the measured potential difference is small. We will discuss the case for three different electrode materials important in biological work: *platinum, silver coated with silver chloride*, and *carbon*. To the extent that both electrodes are equal we have a symmetrical (bipolar) system, and the voltage–current dependence should not be polarity dependent.

We connect the dc supply to the electrode metal wires and adjust the voltage so that a suitable dc current flows. An electric field E is accordingly set up in the solution between the electrodes. Positive ions (e.g. Na^+) migrate in the same direction as the E-field all the way up to the cathode; they are *cations*. Negative ions (e.g. Cl^-) migrate in the opposite direction, i.e. in the same directions as the electrons in the wires, they are *anions*. Anode and cathode are defined from *current flow direction*, and not necessarily from the polarity of the external voltage source. In the bulk of the electrolyte, no change in composition or concentration occurs during the Na^+ Cl^- migration: the same amount of ions enters and leaves a volume.

We must not forget a second possible transport mechanism different from migration: ionisation of *neutral* species may take place at an electrode. These neutral species can not be transported to the electrode by migration, as they are not charged. The transport is caused by diffusion, that is by the concentration gradient near the electrode.

Findings
Platinum electrodes.   We adjust our dc supply to about 0.5 volt, but no dc current is flowing. We must increase the voltage to about 2 volts to get a dc current, but then the current increases rapidly with voltage. With dc current flowing, gas bubbles are seen on both the anode and cathode metal surfaces.

Carbon electrodes. We must again increase the voltage to about 2 volts to get a dc current flowing. Gas bubbles are seen on both electrodes, but on the anode a process of erosion of the carbon surface seems to take place.

Silver/silver chloride electrodes. Large dc current flows with the voltage supply adjusted to only 1/10 of a volt. Initially no gas bubbles are seen on any of the electrodes. At the anode the colour stays the same, but the cathode loses the silver chloride layer and a pure silver surface appears after some time.

Discussion

With platinum and carbon, an applied dc voltage does not necessarily lead to current flow. There must be energy barriers in the system, and a sufficiently high voltage must be applied to overcome this barrier. It is a non-linear system, not obeying Ohm's law. It can be shown that the bulk solution obeys Ohm's law, and therefore the energy barrier is not in the bulk but near the electrodes. When the voltage is turned on, Na^+ ions migrate to the cathode and Cl^- ions migrate to the anode. But arrival at the electrodes does not lead to an exchange of electrons with the metal, a surface charge is built up opposing the external electric field, and the current stops. An electrode is the interphase at which electronic and ionic conduction meet. Without dc current there is no electron transfer, no chemical reaction and no faradaic current.

At the cathode

With current flowing, anions and cations migrate in opposite directions. The simplest hypothesis dealing with a saline solution would be that Na^+ ions are discharged at the cathode, and Cl^- ions at the anode. It is not that simple: Na^+ ions are not discharged at the cathode. Sodium has a very small electronegativity, which means that it takes a large energy and a large negative voltage on the cathode to impose electrons on Na^+ ions. At much lower voltages, two other processes start: reduction of dissolved neutral oxygen, and decomposition of water molecules. Both processes are linked with *non-charged species*, which are transported to the electron transfer sites by diffusion, not by migration. So in Fig. 2.1 there are two transport mechanisms: migration and diffusion. The reaction of non-charged species at the electrodes must not be overlooked; these species are charged or *ionised* (at least as one step) in the electrode reaction.

The concentration of dissolved oxygen is small, so the dc current from the oxygen reduction is not large. As long as our voltage supply is adjusted for a current lower than this current, the oxygen reduction current is sufficient. If a larger current is wanted, the voltage must be increased so that water is decomposed additionally. The water reaction at the cathode is

$$2H_2O + 2e = H_2 \text{ (gas)} + 2OH^- \text{ (base)}$$

Actually it is more complicated: different versions of the hydrogen ion are active, e.g. the oxonium ion H_3O^+.

In conclusion (cathode): Neutral metals and carbon do not have the ability to be reduced, so electrode material can not be ionised at the cathode and enter the solution. Dissolved oxygen is reduced; at higher currents free hydrogen gas also bubbles up and the solution near the cathode becomes basic. Na^+ need not be considered (but is necessary for the conductivity of the solution, so that the voltage drop in the solution is not too high). The positive silver ions of the silver chloride are neutralised, and little by little the AgCl layer is decomposed and pure silver appears on the surface. The colour changes, but the colour of AgCl is not so easy to define. AgCl is photosensitive, and in films exposed to light there are already grains of pure silver, which are grey or black in colour. Pure AgCl is amber.

At the anode
The electrode reaction at the cathode was not due to the discharge of Na^+. Is the current at the anode due to the discharge of Cl^- ions? Yes. Chloride is highly electronegative, but less energy is necessary for taking electrons from the chloride ions than from water molecules. Neutral Cl_2 gas is formed at the platinum anode. It does not react with platinum, and leaves the area as gas bubbles. It *does* react with carbon and destroys the carbon surface. At the silver chloride surface it reacts with silver oxidised by the anode and forms more silver chloride. Ag^+ will not enter the solution, if it does it will combine with Cl^- and form AgCl. In aqueous solution the solubility of AgCl is very low, only very small amounts will dissolve in the solvent, and it will soon precipitate.

OH$^-$ ions may be discharged, but there are few of them and they do not contribute very much to the dc current. With large currents, water may be decomposed, with oxygen leaving the area as gas bubbles according to

$$2H_2O = O_2 \text{ (gas)} + 4H^+ + 4e^- \text{ (acid)}$$

If oxygen gas is developed, the solution turns acidic near the anode. The importance of this reaction depends on the current level and what current level the Cl^- concentration will take care of alone.

We may therefore conclude that silver chloride behaves rather differently from platinum and carbon. Silver undergoes an electrochemical reaction with one of the ions of the electrolyte (Cl^-): silver may be oxidised or silver ions may be reduced. The transfer of electrons oxidising or reducing species at an electrode is called a *redox* process. *The results indicate that if we are to apply large dc currents to tissue, and we are to use noble metals as electrode material directly on the tissue, the passage of dc current is accompanied by the development of H_2 gas and a basic milieu at the cathode, and Cl_2 gas and perhaps oxygen and an acidic milieu at the anode.* However, in real tissue systems (not the model of Fig. 2.1), organic molecule redox systems will contribute to additional electrode reactions at low current levels.

What happens if we replace the dc voltage with a sinusoidal ac voltage? If the frequency is sufficiently high (e.g. 1 MHz), the migration processes in the bulk electrolyte will take place (back and forth), but no accumulation process or reactions will take place at the electrodes. If the frequency is very low (e.g. 0.1 Hz), the result will depend on the dimensions of the cell and the degree of reversibility of the reactions. If gas has time to bubble away, the process is certainly irreversible.

2.3. Bulk Electrolytic dc Conductance

According to the Arrhenius[4] theory of *dissociation*, molecules of acids, bases and salts react with water molecules to form separate ions. Water *ionises* the substances, and these ions give their solution the property of conducting electricity. Positive and negative ions free to migrate in the electric field contribute separately to the electric current flow, but because of different mobilities they do not carry equal portions of the current.

Environment of ions
In aqueous solutions an ion is not alone. Two zones surround it: the ion attracts ions of opposite sign, and it attracts water molecules. A water molecule has a strong *electric dipole moment*: even though its net charge is zero, water is a polar material. The process of solvent molecules forming a sheath around each electrolyte ion is generally called solvation. When the solvent is water, the process is called *hydration*. Hydration is strong because the water molecules have a large permanent dipole moment. The water molecular sheath stabilises each ion and hinders ions of the opposite sign from approaching so near to each other that they recombine: the substance stays dissociated and ionised. The hydration number is the average number of water molecules forming the sheath. Cations are usually less hydrated, and the hydration sheath less effectively covers large ions. Figure 2.2 shows the hydration process for a sodium ion in water. It is a statistical concept, so that *on average* there are more oriented water molecules (and other ions of opposite sign) near the Na^+ ion.

Hydration is the build-up of a sheath of *dipoles* around a central ion, owing to ion–*dipole* forces. According to Debye[5] and Hückel, the central ion is also surrounded by a slight excess of *ions* of the opposite charge sign, formed by ion–*ion* forces. They called this an *ionic atmosphere*. Both the hydration and the ionic atmosphere will increase the effective dimensions and reduce the apparent charge of the centre ion, and thus retard migration.

The ionic atmosphere is a statistical concept: within the *Debye length* from the central ion, there is an increased probability of finding an ion of opposite charge. A few Debye lengths (of the order of some tenths of nanometres) define a region of space charge in which electroneutrality no longer holds. If the charge of an ion suddenly disappeared, it would take a time of the order of 1 μs for the molecules to rearrange and for the ionic atmosphere to disappear. This is an example of a *relaxation time*.

---

[4] Svante August Arrhenius (1859–1927), Swedish physicist and chemist. 1903 Nobel prize laureate in chemistry on electrolytic dissociation theory.

[5] Peter Joseph Debye (1884–1966), Dutch/American physicist. 1936 Nobel prize laureate in chemistry on molecular structure and dipole moments.

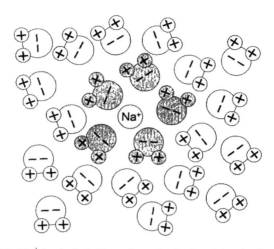

Figure 2.2. Na$^+$ ion hydrated by water molecules forming a sheath around it.

Contributions to ionic conductivity

Kohlrausch[6] showed that conductivity is composed of separate contributions from anions ($-$) and cations ($+$). The current density J (A/m^2) of a single anion–cation pair is (as for all equations in this book, see Section 10.3 for definition of global symbols)

$$\mathbf{J} = (nzev)_+ + (nzev)_- = Fc\gamma\,(\mu_- + \mu_+)\,\mathbf{E} \tag{2.1}$$

$$\mathbf{J} = \sigma\mathbf{E} \tag{2.2}$$

Equation (2.2) is very important and fundamental; it is the version of Ohm's law for volume conductors. It is valid under the assumption of a homogeneous and isotropic medium when the current density and E-field directions coincide. Note that current is not the quantity used in this version of Ohm's law, but current *density J*. As J may vary according to the local E-field strength, current must be found by integrating current density over a cross-sectional area. Current is the sum of charges passing a freely chosen cross section (e.g. of a copper wire) per second (ampere A); current density is the sum of charges passing per *unit area* per second (A/m^2). Current is a scalar sum of charges per second passing some area not entering the equation (*scalar flux*); it has no direction in space. Current density is defined by an area oriented in space, and is therefore itself a vector in space (*vector flux*, also called flux density).

The current density J of eq. (2.2) must be summed up with contributions from each negative and positive ion species. $\mathbf{v}$ is the velocity of the ion (m/s). γ is the activity coefficient: not all the electrolyte may be dissociated, and this is taken care of

[6] Friedrich Wilhelm Georg Kohlrausch (1840–1910), German physicist. Author of *Lehrbuch der Praktischen Physik*.

by the activity coefficient γ, having a value between 0 and 1. Note that it may be difficult to find the activity coefficients of individual ion species because of electroneutrality—an electrolyte cannot consist of only anions or cations.

The contribution to the total conductivity will come from all free ions according to their concentration, activity, charge and mobility. The *transference number* of an ion species is its percentage contribution to the total conductivity.

In the bulk of a solution with free ions there is *electroneutrality*: in a volume L the sum of charges is zero:

$$L\sum(nze)_+ + L\sum(nze)_- = 0 \qquad (2.3)$$

If this were not the case, a space charge would build up driving excess ions out of the volume. During current flow, equal amounts of charge must enter and leave a solution volume. Electroneutrality is valid for a volume much larger than ionic dimensions. Electroneutrality does not prevail at boundaries with space charge regions, cf. Section 2.4.3 on electrical double-layers.

The current density according to eq. (2.1) must be summed for all free ions present, e.g. for NaCl the conductivity of eq. (2.2) may be written

$$\sigma = F(c\gamma)_{\text{NaCl}}(\mu_{\text{Na}} + \mu_{\text{Cl}}) \qquad (2.4)$$

The *molar conductivity (equivalent conductance)* Λ is conductivity per mole of solute per volume:

$$\Lambda = \sigma/c = \gamma F(\mu_+ + \mu_-) \qquad (2.5)$$

The molar conductivity is therefore a parameter directly linked with the mobility and not with concentration. The basic unit is (S/m) per (mol/m^3), or Sm^2/mol, Table 2.4. The mobility μ is related to the random molecular collisions and corresponding frictional force (viscosity η) experienced by the migrating ion. Ideally, the frictional force $\mathbf{f}$ is related to the hydrodynamic radius a of the ion according to Stokes' law

$$\mathbf{f} = 6v\pi\eta a \qquad (2.6)$$

The bulk electrolyte solution obeys the linear Ohm's law (eq. 2.2). The force on a charge q in an E-field is proportional to the electric field strength according to $\mathbf{f} = q\mathbf{E}$. The linear Ohm's law therefore shows that ions are not formed by the external field, they are in existence already without a field.

Equation (2.2) is valid also for dc, under the condition that electrochemical changes occurring at the electrodes do not spread to the bulk. σ is also considered frequency independent up through the whole kHz frequency range.

The diameters of many ions have been determined by x-ray diffraction of ionic crystals (dry) (Table 10.2). From Table 2.4 we see that the number of water molecules bound in the sheath around an ion (the *hydration number*) has a certain correlation with the molar conductivity, and therefore according to eq. (2.5) to the mobility (the Ca^{2+} value must be halved for this comparison). This is so because a stronger hydration increases the effective radius of the ion and therefore the friction,

Table 2.4. Limiting molar conductivity Λ_0 in aqueous solution. Λ_0 in $S \cdot cm^2$ per $mol \cdot z$ at infinite dilution and 25°C. Hydration number is the average number of water molecules in the hydration sheath

Cation	Λ_0	Hydration no.	Anion	Λ_0	Hydration no.
H^+/H_3O^+	350	–	OH^-	198	
Na^+	50	5	Cl^-	76	0
K^+	74	4	HCO_3^-	45	0
Ca^{2+}	119	10	CO_3^{2-}	72	0

and also reduces the effective charge of the ion. Both hydration and the ionic atmosphere reduce the molar conductivity.

Electrolytic concentrations are in practice often given by weight, e.g. 0.9% NaCl in water meaning 9 g/1000 g = 9 g/litre. From the relative atomic mass (Table 2.5), the molar concentration can be found. For instance 23.0 + 35.5 = 58.5 g NaCl = 1 mol. 0.9% NaCl is therefore equal to 154 mmol/litre.

The conductivity of a given electrolyte solution can be found from the tables of molar conductivity, at least for low concentrations (Table 2.4). The conductivity of, e.g., 0.9% NaCl at 25°C should therefore according to eq. (2.5) be $\sigma = \Lambda c = (50 + 76)\, 0.154 \times 10^{-1} = 1.94$ (S/m).

Figure 2.3 shows how the actual conductivity may be lower, because the dependence is not linear. Molar conductivity Λ_0 relates to the limiting value at low concentrations. For NaCl the relationship is quite linear up to the physiological concentration of 0.9% by weight. Sweat concentrations are somewhat lower, urine concentrations vary but may be higher, seawater concentrations are around 3.5%. Much higher concentrations are sometimes used for contact electrolytes, cf. Section 8.1.3.

At high concentrations the interaction between ions reduces their mobility. The ions are so tightly packed that their fields interact, so that ions of opposite charge form *ionic pairs* (with lower mobility, they become quasi-dipoles). This effect is dependent on solvent permittivity, because the electrostatic force between charges and the formation of ionic pairs are stronger the lower the permittivity (eq. 3.1). Also, higher ionic thermal energies (kT) give less coherent ionic atmospheres.

Table 2.5. Relative atomic mass = gram weight per 1 mole (gram mole)

	Protons	Gram weight/mol		Protons	Gram weight/mol
H	1	1.0	F	9	19.0
Li	3	6.9	Na	11	23.0
B	5	10.8	Cl	17	35.5
C	6	12.0	K	19	39.1
N	7	14.0	Ca	20	40.1
O	8	16.0			

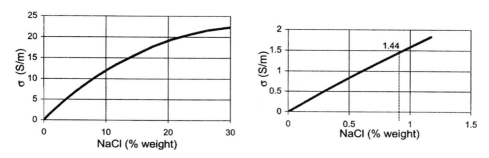

Figure 2.3. Conductivity as a function of NaCl concentration in aqueous solution (20°C).

Higher temperature and permittivity therefore give less friction and higher conductivity. According to the *Onsager*[7] theory, molar conductivity Λ is dependent on $\sqrt{c}$ according to $\Lambda = \Lambda_0 - b\sqrt{c}$, where b is dependent on, e.g., temperature and viscosity.

Conductivity of water itself
Water, hydroxyl ions and hydrogen ions have very special electrical properties. Water is strongly polar, but is also to a small extent an electrolyte in itself, from Table 2.7 we see that the intrinsic conductivity is low, but not zero. The small rest conductivity is due to a protonic self-ionising process: there is a small statistical chance that a water molecule transfers one of its protons to another water molecule in the following way:

$$H_2O + H_2O \rightleftharpoons H_3O^+ + OH^- \tag{2.7}$$

The traditional description of the hydrogen ion H^+ alone is not correct. Actually the proton, the hydrogen ion H^+, cannot exist as such in water. Without the electron it consists only of the nucleus with a radius of perhaps 10^{-14} m, about 4 decades smaller than an ion with electrons in orbit. The electric field near the naked proton is extremely strong, and polar water molecules are immediately attracted; the proton is hydrated and the *oxonium* (also called *hydroxonium* or *hydronium*) ion H_3O^+ is formed. This is also strongly polar and attracts new water molecules forming $H_5O_2^+$ and $H_9O_4^+$ in a dynamic and statistical way. For simplicity they will all be called the *hydrogen ion* in this book when it is not necessary to differentiate.

A small naked proton would have a high mobility and therefore high molar conductivity. The oxonium ion is of more ordinary size, and so the molar conductivity actually measured should be in a normal range. However, hydrogen (and OH^-) ions have irregularly high molar conductivity (Table 2.4). This is due to a special *proton hopping* mechanism (Grotthuss mechanism), illustrated in Fig. 2.4,

---

[7] Lars Onsager (1903–1976), Norwegian/American chemist. 1936 Nobel prize laureate in chemistry for his work on irreversible thermodynamics.

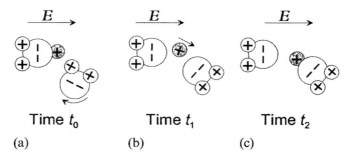

Figure 2.4. Proton hopping conductance, same molecules at three different times: (a) water molecule rotation; (b) hopping; (c) new proton position.

which is different from ordinary migration. The protons are transferred by hopping directly from water molecule to water molecule by a tunnelling effect instead of being slowed by viscous forces (Jonscher 1983). The speed limiting process is believed to be a necessary rotation of a neighbour water molecule before it can accept a proton; a statistical process. The proton can be considered to hop along a string of water molecules with a statistical net flow in the direction of the E-field. Parallel to the hopping conductance there is of course the ordinary viscous conductance.

Because of their high molar conductivity, the hydrogen ions will often dominate the conductance found. Because of electroneutrality there are, for instance, equal numbers of H^+ ions and Cl^- ions when the strong acid HCl dissociates. But the electrolytic conductivity contribution from the hydrogen ions is about 5 times as large as that of Cl^-. From Table 2.4 it is seen that for NaCl the contribution is largest from the Cl^- ions; for KCl the contributions from each ion are approximately equal.

Conductivity of weak acids

The electrolytes do not only determine conductivity; they are also strongly related to the acid–base balance and the pH defined as $pH = -\log[H_3O^+]$ of the tissue, where $[H_3O^+]$ represents the concentration in mol/L.

With salts and strong acids, water dissociates all molecules. With weak acids, the situation is different: some of the acid molecules stay undissociated. An example is the conductivity due to dissolved CO_2 gas in water. Some of the CO_2 physically dissolved is transformed to H_2CO_3, which is a weak acid. The Guldberg–Waage[8,9] law (mass action law) is useful for the calculation of ionic concentrations in such cases. In its original form it states that the rate of a chemical reaction is proportional to the *mathematical product* of the masses of the reacting substances, and that equilibrium can be expressed by an equilibrium constant characterising the chemical

---

[8] Cato Maximilian Guldberg (1836–1902), Norwegian mathematician, particularly interested in physical chemistry.
[9] Peter Waage (1833–1900), Norwegian chemist, particularly interested in practical applications of chemistry.

reaction. In the particular case of a dissociation process, electroneutrality also imposes the condition that the ionic concentrations be equal (for unity valency ions). In the case of dissolved CO_2, the equilibrium dissociation constant can be set to $K_1 = 10^{-6.3}$; the Guldberg–Waage law therefore gives

$$[H_3O^+][HCO_3^-]/[H_2CO_3] = K_1 = 10^{-6.3}$$

For such a weak acid, a doubling of the concentration $[H_2CO_3]$ leads to $\sqrt{2}$ increase in $[H_3O^+]$, because $[HCO_3^-]$ must also increase by $\sqrt{2}$ (electroneutrality) so that the product increases by 2. Because of the high molar conductivity of hydrogen ions, the total conductivity of such a solution is dominated by the hydrogen ions and is therefore roughly proportional to $\sqrt{[CO_2]}$ or the square of the partial pressure of the carbon dioxide gas in equilibrium with the solution.

Accordingly, when water is in equilibrium with air in nature, H_3O^+ and HCO_3^- ions will contribute to an additional dc conductivity as well as a reduction of pH. Actually the reduction in pH is substantial, the pH of pure (non-polluted) water is 5.7 (acid) when in equilibrium with the normal atmospheric CO_2 concentration levels (0.03% by volume).

Water itself is very weakly self-ionised, with a low equilibrium constant K_w:

$$[H_3O^+][OH^-] = K_w = 10^{-14}$$

For electroneutrality $[H_3O^+] = [OH^-]$, which then is equal to 10^{-7}, corresponding to pH 7. The conductivity according to eq. (2.5) is

$$\sigma = (\Lambda_+ + \Lambda_-)c = (350 + 198) \times 10^{-7} \times 10^{-1} = 5.5 \times 10^{-6} \text{ (S/m)}$$

cf. Table 2.7.

Other factors influencing conductivity

The *temperature* dependence of conductivity of most ions in aqueous solution is about $+2.0\%/°C$. This strong temperature dependence results from the decrease in viscosity of water with temperature. The temperature coefficients of H_3O^+ (1.4%/°C) and OH^- (1.6%/°C) are important exceptions, and are the result of the different conduction mechanisms of these ions. Viscosity increases with increasing *pressure*, and the conductivity is reduced.

The conductivity increases at high *frequency* (> 3–30 MHz, Debye–Falkenhagen effect). It takes about 0.1–1 μs to form an ionic atmosphere, the time is dependent on the ion concentration. The literature is not clear as to the conductivity frequency dependence of electrolytes such as NaCl, but Cooper (1946) found no variations in the concentration range 1–4 (weight%) and frequency range 1–13 MHz.

As shown, pure de-ionised water has a low conductivity, but the permittivity is constant up to the lower GHz range with a relaxation frequency around 20 GHz (Section 4.1.1).

Special electrolytes

Some substances are completely ionised in water (strong acids), others are only partly ionised, e.g. weak acids. Water is often necessary for the ionisation or splitting of a molecule; pure HCl liquid, for instance, is an insulator. Many substances dissolved in water are not ionised at all, and therefore do not contribute to electrical conductivity. They are true non-electrolytes, such as sugar/glucose. The molecules of such substances are not ionised or dissociated by water. Some may have a symmetrical distribution of charges, with the centre of positive and negative charge coinciding. However, many molecules have centres not coinciding, forming permanent dipoles with zero net charge, such substances are called *polar*. Water itself is polar, and a substance has to be polar in order to be soluble in water.

NaCl as a dry salt at room temperature is not an electrolyte. The dc conductance is negligible; still the Na and Cl atoms are ionised, but they are "frozen" so that they can not migrate.

NaCl dissolved in water is the true electrolyte, and the Na and Cl ions are split and free. Even if NaCl is the true electrolyte, the whole electrolyte solution is often also termed "the electrolyte".

In a solution with colloidal particles, a charged double layer will surround each particle, and the particle may be regarded as a macro-ion. The colloidal particles free to migrate contribute to the solution's electrical dc conductance, and may be regarded as a *colloidal electrolyte*. Particles are called colloidal when two of the dimensions are in the range 1 nm–1 μm (the third dimension does not have this constraint, e.g. for a very thin string).

A *solid electrolyte* is also possible. In liquids, ions generally are more free to move (have a higher mobility) than in solids. Solid electrolytes at room temperature therefore have a relatively low conductivity.

AgCl is an important molecule for electrode surfaces. The chloride ions Cl^- are bound, but the silver metal ions Ag^+ are genuine charge carriers giving a certain electrolytic ionic conductivity.

The dry glass core in a pH electrode exhibits solid electrolytic conductance, not semiconductivity. The electrode is proton sensitive, but the small conductivity in the glass core stems from ions of Li^+, Na^+ and K^+, which have mobilities 10^3–10^4 times larger than the protons. Water is absorbed in the leached surface layers of the glass, however, and there the proton mobility is high and contributes to local dc conductance in an important way.

NaCl as a solid crystal has atoms in ionised form. But neither the Na^+ nor the Cl^- ions nor any electrons, are free to migrate (at room temperature). There is no dc conductivity: NaCl is an insulator. When solid NaCl is dissolved in water, the water splits the NaCl molecules into free Na^+ and Cl^- ions: NaCl is *dissociated*. NaCl is the electrolyte when dissolved in water, but we can hardly call solid NaCl an electrolyte, even if its atoms are in ionised form. However, if NaCl is warmed to 800°C it melts, the ions are free to migrate with low friction (high mobility), and we have a strong *ionic liquid* or *fused electrolyte*.

There are also *mixed conductors* with both ionic and electronic conductance. The sulphides, selenides and tellurides of silver and lead are examples. New plastic materials with ionic conductance have been discovered in polymer chemistry. Nafion is a polymer with ionic conductivity; it is already much used in multigas analysers in anaesthesia, where the high permeability to water is useful.

Some polymers have mixed electronic and ionic conductivity, and some are purely electronic conductors with free electrons like a metal, or electrons locally linked to centres with electron-donor properties. Carbon, as the most basic element for all biochemistry, is a very special element that deserves attention also as an electrode material. In the form of graphite it is an electronic conductor, but as diamond it is an almost perfect insulator.

Proteins in the body liquids may be considered as a colloidal electrolyte solute in a water solvent. Contact with water is the natural state of a protein. In more or less dry form a protein loses some of its electrolytic character; it loses the charged double layer on the surface, and behaves electrically very differently from protein with water. Such materials may well be mixed conductors: electronic in the dry state, and ionic with water content. Keratin is a more or less dry protein found in the natural state of no longer living biological materials such as hair, nail and stratum corneum. The water content of such materials is dependent on the relative humidity of the ambient air. The question of ionic or electronic conductivity in proteins is important, and an electronic conduction mechanism must be considered in many cases.

Body liquid electrolytes (Table 2.6)
By far the most important ions for extracellular conductance are Na^+ and Cl^-. Note that free protein in plasma provides charge carriers with a negative charge (anions), and in this context can be regarded as consisting of macro-ions and a contributor to conductance. This charge is also the basis of dc electrophoresis as an important analytical tool in clinical chemistry (Section 2.4.6). In order to maintain electroneutrality, increased protein concentration must increase the concentration of cations or reduce the concentration of other anions. The anion HCO_3^-

Table 2.6. Concentration of electrolytes in body liquids (meq/L) is ion concentration in milliequivalents (mmole·valency z) per litre

Cations (meq/L)			Anions (meq/L)		
	Plasma	Intracellular		Plasma	Intracellular
Na^+	142	10	Cl^-	103	4
K^+	4	140	HCO_3^-	24	10
Ca^{2+}	5	10^{-4}	Protein$^-$	16	36
Mg^{2+}	2	30	$HPO_4^{2-} + SO_4^{2-} +$	10	130
H^+ (pH 7.4)	4×10^{-5}	4×10^{-5}	organic acids		
Sum	153	180	Sum	153	180

(bicarbonate) is related to the transport of carbon dioxide (CO_2) in the blood. A change in bicarbonate (anion) concentration will therefore have consequences for the cation concentration.

Ionic and electronic conduction with respect to semiconductor theory

The charge carriers in metals are electrons that are free to migrate in the energy band called the conduction band (Fig. 2.5).

In semiconductors there is a special valence energy band below the conduction band. With pure (intrinsic) semiconductors the energy levels in between are forbidden levels, and at room temperature very few electrons statistically have sufficient energy to cross the forbidden band and reach the conduction band: the conductivity is low. The energy gap is, e.g., 0.7 eV (germanium), 1.1 eV (silicon) or 5.2 eV (diamond, an insulator).

However, with *local impurities* in the material, *local* energy centres may exist in the forbidden gap. Here electrons reside in energy levels that may be characterised as local energy *wells*. With a certain amount of added energy, electrons can come up from the well and reach the conduction band (n-type impurities). This can increase conductivity considerably (extrinsic conduction). With closely spaced energy wells (< 0.15 nm) an energy barrier may also be crossed by *tunnelling*.

The idea of a possible semiconductive mechanism (electrons and holes) in biomaterials is old, as illustrated by the book by Pethig (1979). Takashima (1989) did not mention semiconductivity as a possible mechanism. Indications of an electronic, semiconductive conduction for the DNA molecule have subsequently appeared again (Fink and Schönenberger 1999). Such experiments are performed under non-physiological conditions in vacuum, which implies that every water molecule free to do so has disappeared.

Ions do not obey the laws applying to semiconductors. However, the concept of local energy wells can also be adapted to ionic conduction. Debye (1929) proposed a model in which an ion may be translocated between a pair of neighbouring energy

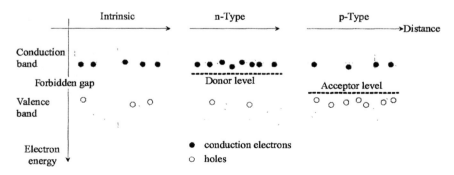

Figure 2.5. Electronic semiconductivity. Energy levels in a semiconductor without (left) and with impurities. Local impurities create local energy levels (energy wells) as local reservoirs in the forbidden energy gap.

wells by an applied electric field. With an applied ac field, an ion can be made to hop between these two wells. However, in solids such local hopping does not contribute to dc conduction, only to ac polarisation (Jonscher 1983). The extension of these theories is discussed in Section 7.2.9.

Materials classified according to conductivity (Table 2.7)
Materials are often classified according to their dc conductivity: conductors, semiconductors (electronic), and insulators. For electronic conductors (charge carriers: electrons) this classification is based upon energy levels and Fermi–Dirac statistics of the free electrons. A division between metals and semiconductors is sometimes based upon the temperature coefficient of the conductivity σ: semiconductors (as for ionic conductors) have a positive $d\sigma/dT$, metals a negative $d\sigma/dT$.

For ionic conductors the classification is a little more problematic, they are in the conductivity range of pure semiconductors, but the nature of the charge carriers and therefore the conduction mechanism is very different. It is often found with liquid or

Table 2.7. Electronic (e) or ionic (i) dc conductivity σ at 20°C if not differently specified

Material	Type of conductivity	Conductivity, σ (S/m)
Superconductors (low temperature)	e	∞
Silver	e	61×10^6
Copper	e	58×10^6
Aluminium	e	35×10^6
Steel (hard)	e	2×10^6
Mercury	e	1×10^6
Coal	e	17×10^3
Graphite	e	100×10^3
NaCl, fused, 800°C	i	300
NaCl in water, 25%	i	22
NaCl in water, 5%	i	7
NaCl in water, 0.9% 37°C	i	2
NaCl in water, 0.9%	i	1.3
Whole blood 37°C	i	0.7
Muscle 37°C	i	0.4
Germanium (pure semiconductor)	e	2
Silicon (pure semiconductor)	e	300×10^{-6}
Bone, living	i	10×10^{-3}
Tooth (human enamel)	i?	5×10^{-3}
Ethyl alcohol	i	330×10^{-6}
AgCl	i	100×10^{-6}
Water, de-ionised	i	4×10^{-6}
Bone, dry	e?	100×10^{-12}
Transformer oil	i?	10×10^{-12}
Mica	e?	10×10^{-15}
PTFE (Teflon)	e	10×10^{-15}
Diamond	e	10×10^{-15}

solid electrolytes that the difference in conductivity is the result of different *mobilities* and not so of much different numbers of charge carriers.

2.4. Interphase Phenomena

We have indicated that the processes in the bulk of the solution (*ionics*) are very different from the processes at the electrodes (*electrodics*). Since tissue is the material of interest, and from a number of perspectives it is electrolytic, the field of bioimpedance and bioelectricity is much concerned with ionics. On the other hand, tissue is full of membranes, and each membrane shows distinct surface phenomena. The smaller the cells, the more important the surface phenomena.

In addition the electrodes also form interphases with the tissue. The electrodes are usually regarded only as tools, and our main interest is to be able to minimise or at least control their influence on the tissue data. With current-carrying electrodes, the polarisation processes represent special interphase phenomena.

2.4.1. Faraday's law

In the solution there is a two-way flow of ions and there are charge carriers of both signs. In the metal wires of the electrodes there is only one type of charge carrier of one sign, the electrons. The charge of one electron is exactly equal to that of one proton (with opposite sign): 1.6×10^{-19} coulomb, the *unit charge*.

In the solution the current spreads out from the current-carrying electrodes, and we must use *current density* (ampere/m^2) as quantity instead of current, cf. eq. (2.2). A *flux* is a rate of flow through a cross section of unit area, so current density is the flux of charges. The total sum of ionic charges passing through the solution is the current density integrated over the whole cross-sectional area, and per second this must equal the electronic current I in the external electrode wire.

A sum of electric charges q is also a *charge* and is also called a *quantity of electricity* Q, with the unit *coulomb*. If one ampere flows through a cross-sectional area in one second, one coulomb of charge has passed. Faraday[10] found that the amount of electrochemical reaction is proportional to the quantity of electricity passed by the electrolyte. The Faraday constant is the charge of 1 mole of electrons. 1 mole is the number 6×10^{23} of the particles under consideration, and the charge of one electron is 1.6×10^{-19} coulomb, so the Faraday constant $F = 96\,472$ coulomb/mol. *Faraday's law of electrolysis* links the amount of chemical change at the electrode to a quantity of electricity:

$$M = Q/Fz \tag{2.8}$$

where M = amount of substance produced (mol) and
z = valency of the element produced

---

[10] Michael Faraday (1791–1867), English physicist and chemist (natural philosopher), found the laws of electromagnetic induction and of electrolysis. A self-educated man who knew very little mathematics.

Faraday's law of electrolysis defines the term *electrolytic*: an electrolytic system is a system that is essentially characterised by Faraday's law. A current creating a reaction at an electrode according to Faraday's law is called a *faradaic current*.[11] The quantity of electricity involved may be obtained with a dc current or a current pulse. A dc current implies faradaic current only. A pulsed current has additional capacitive, non-faradaic current components.

2.4.2. Migration and diffusion

The electrode is the final boundary for ionic migration. Some of the ions originate here, and here they meet a physical hindrance. The metal of an anode may furnish metal ions to the solution, across the interphase.[12] Depletion or accumulation of matter and charge may occur at the electrodes, as well as chemical reactions. An electrode may also exchange electrons with *neutral species*, e.g. the reduction of dissolved oxygen at a cathode. The transport of these species in the bulk of the electrolyte is not by ionic migration but by a diffusion transport process caused by a concentration gradient. *Diffusion may be as important a transport mechanism as migration in an electrolytic cell.*

Equations (2.1) and (2.2) are to be interpreted according to the kinetic molecular theory of the transport properties of liquids. The charge carriers do not move in an orderly, linear fashion through the liquid. *Diffusion* is the process resulting from random motion of molecules by which there is a net flow of matter from a region of high concentration to a region of low concentration. This process is related to the concepts of Brownian motion, molecular collisions and mean free path between collisions. Ions in a solution also experience diffusion, in addition to migration. The *migration* of charge carriers in an electric field may be regarded as a special case of diffusion caused by an external influence, not driven by an internal concentration gradient. The transport of charged molecules/ions may be due to both concentration gradients and electric fields; *electrodiffusion* is the general term for both these transport processes. The migrational part is dependent on the electric field. In order to reduce the migrational effect, an indifferent electrolyte not intervening in the process to be studied may be added to increase conductivity and reduce the electric field. Generally transport generating electricity is classified as *electrogenic*. The sedimentation potential (Section 2.4.6) is for instance electrogenic.

Molecular diffusion is described by *Fick's laws*. Consider a simple system in the form of a compartment with unit width and height dimensions, and a concentration gradient in the infinite-length *x* direction. Fick's first law is

$$M = -D\frac{\partial c}{\partial x} \qquad (2.9)$$

where M is the molar flux (mole per s and per m^2) and D the diffusion coefficient

---

[11] Not to be confused with *Faradic* current which is the pulse current from an induction coil used for muscle stimulation.
[12] *Interface*: a surface that forms the boundary between two materials (sharp transition). Inter*phase*: diffuse transition zone between two phases (e.g. solid/liquid).

(m^2/s). The minus sign indicates that the transport is towards *lower* concentration. During stationary diffusion, e.g. in a tube, eq. (2.9) shows that a linear concentration will be set up in the diffusion zone out from the electrode surface.

Fick's second law is

$$\left(\frac{\partial c}{\partial t}\right)_x = D\frac{\partial^2 c}{\partial x^2} \tag{2.10}$$

Warburg solved this equation in 1899, finding concentration waves extending into the electrolyte at a distance from an ac-polarised electrode surface (cf. Section 2.5.3).

As another example, consider a diffusion process starting in a specified compartment with an infinitely thin (*x*-direction) band of solute of concentration *S* (mole per unit width *y* and depth *z*) released in the middle of the compartment at $x = 0$ and $t = 0$. Solving Fick's second law under these boundary conditions, the concentration as a function of position *x* and time *t* is

$$c(x, t) = \frac{S}{e^{x^2/4Dt}\sqrt{4\pi Dt}} \tag{2.11}$$

The initial conditions correspond to $c(0,0) = \infty$. $\sqrt{Dt}$ has the dimension of metre, and corresponds to the net distance diffused by a molecule in one second. Such time dependence is due to the random walk of the molecule. With time the solute spreads out, and the concentration at the band position ($x = 0$) decreases according to $1/\sqrt{t}$. At a distance *x* the concentration at first is extremely small, but then reaches a maximum and thereafter follows the falling and uniform concentration around the origin (see Fig. 2.6). With $t \rightarrow \infty, c \rightarrow 0$, because the compartment is infinite in the *x*-direction. An electrolytic cell of finite dimensions corresponds to different boundary conditions, and a different equation. This is of importance when considering increasing *t* and the approach towards dc conditions. When the diffusion reaches the boundary of the cell compartment in the *x*-direction, the process no longer follows eq. (2.11). With diffusion-controlled processes, new frequency-dependent parameters enter at very low frequencies with a measuring cell of small dimensions (cf. the Warburg impedance, Section 2.5.3).

Equation (2.11) and Fig. 2.7 illustrate the diffusion process from an electrode surface positioned at $x = 0$, whether it is the export of reaction products or the import of reactants in the other direction. Equation (2.11) also illustrates that

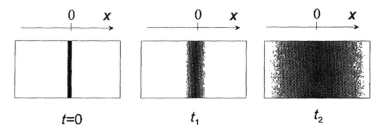

Figure 2.6. Spreading of a thin strip of solute from $x = 0$ at $t = 0$ by pure diffusion.

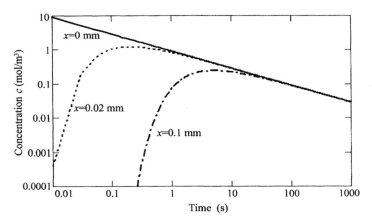

Figure 2.7. Diffusion of a thin strip of solute from position $x = 0$ at $t = 0$, according to eq. (2.11) and Fig. 2.6. Concentration as a function of time at three selected values of x. $D = 10^{-9}$ m^2/s, $S = 10^{-4}$ mol/m^2.

physical processes in electrochemistry do not necessarily follow exponential laws. This is an important reason why it is often difficult to model electrolytic cells with ideal electronic components.

The diffusion constant D in aqueous solution is of the order of 10^{-11} for large molecules to 10^{-9} m^2/s for small molecules. This means that a spread of just 1 mm in liquids will take hours for larger molecules, whereas small molecules will diffuse about 0.1 mm in one second. In gases the values of D are higher. For instance, for water molecules in air $D = 0.24 \times 10^{-4}$ m^2/s, and without any convection water molecules will therefore move about 5 mm in one second. The spreading or mixing of gases in a room across a metre or more is therefore purely convection-controlled.

2.4.3. The electric double layer, perpendicular fields

At the electrode–liquid interface the transformation from *electronic* to *ionic* conduction occurs. The electrode metal is the source or sink of electrons, and *electron transfer* is the key process whereby the electrode exchanges charges with the arriving ions, or ionises neutral substances (a second mechanism of charge transfer is by oxidation of the electrode metal; the metal leaves the surface as charged cations and enters the solution). Without electron transfer there is no chemical electrode reaction, no dc electrode current and no faradaic current. In the solution at the electrode surface an electric *double layer* is formed as soon as the metal is wetted. Electron transfer takes place somewhere in that double layer.

In all interphases, such as the transition zone between the metal of an electrode and the electrolyte, tissue or gel, or at a cell surface, there will be a non-uniform distribution of charges. Hence there will be an electric potential across the interphase according to the Poisson equation (7.6). This effect is particularly pronounced at the interphase between a solid and a polar medium, e.g. water, where the surface charges of the solid will attract counterions (ions of opposite polarity) from the polar medium. When the polar medium is liquid and the ion mobility thus high, the formation of an electric double layer will take place in the liquid phase.

In this section the double layer charge will be treated as a function of the distance *perpendicular* to the surface. In the next section (2.4.4) we will treat counterion movements *laterally* along the surface.

Simple Helmholtz layer
The double layer can be thought of as a molecular capacitor, where one *plate* is represented by the charges in the metal and the other plate by the ions at a minimum distance in the solution. The distance between the "plates" is of the order of only 0.5 nm, so the capacitance values are enormous. This simple model of the electric double layer was introduced by Helmholtz in 1879, and is *valid only for rather high concentration* electrolyte solutions (Fig. 2.8, left). In this simple model the double layer is depleted of charges. In more dilute solutions the transition will not be so abrupt and the thickness of the double layer will *increase*. The thickness is related to the distance from the metal surface at which the ions can escape to the bulk by thermal motion. In that case, the counterion atmosphere will be more like the ionic atmosphere around an individual ion, and this is commonly referred to as the *diffuse electric layer*.

General theory of Gouy–Chapman
In the combined theory of Gouy and Chapman, the exchange of counterions between the double layer and the bulk solution due to thermal motion is taken into account. Both coulombic forces and thermal motion hence influence the equilibrium

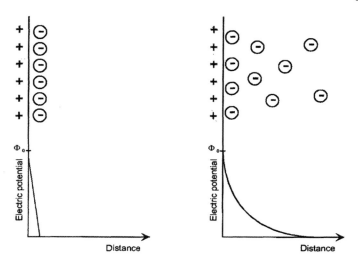

Figure 2.8. Helmholtz' (left) and Gouy–Chapman's (right) model of the electric double layer and electric potential as a function of distance from the electrode surface.

distribution of counterions in their model for the diffuse double layer (Fig. 2.8, right). The assumptions of the theory are, among others:

1. The surface charge is continuous and uniform.
2. The ions in the solution are point charges.

The electric potential in the double layer is given by the Poisson equation:

$$\nabla^2 \Phi = -\frac{q_{vf}}{\varepsilon} \tag{2.12}$$

where q_{vf} is the volume density of free charges. Furthermore, by the Boltzmann equation

$$n_i = n_0 \exp\left(\frac{-W_{\text{el}}}{kT}\right) \tag{2.13}$$

where n_i is the concentration of an ion i at a given point, determined by coulombic forces and thermal motion, n_0 is the concentration in the bulk solution and the electrical work $W_{\text{el}} = z_i e \Phi$, where z_i is the charge per ion. The Poisson equation for the diffuse double layer will then be

$$\nabla^2 \Phi = \frac{2zen_0}{\varepsilon} \sinh(ze\Phi/kT) \tag{2.14}$$

We then introduce

$$\kappa = \sqrt{\frac{2z^2e^2n_0}{\varepsilon kT}} \tag{2.15}$$

where $1/\kappa$ is referred to as the thickness of the double layer (also called the Debye length), and is used in the decision between simplifications of eq. (2.14).

Debye–Hückel approximation
The Debye–Hückel approximation may be used if the surface potential is small:

$$\Phi_0 \ll \frac{kT}{ze} \tag{2.16}$$

(~ 25 mV for a monovalent electrolyte at 25°C). The simplification will then be made by replacing $\sinh(x)$ with x.

For a spherical double layer, the solution using Debye–Hückel approximation will be

$$\Phi = \Phi_0 \frac{a}{r} \exp[-\kappa(r - a)] \tag{2.17}$$

For a flat double layer, i.e. when $a \gg 1/\kappa$, the solution will be

$$\Phi = \Phi_0 \exp(-\kappa x) \tag{2.18}$$

General theory of Stern
The theory of Gouy–Chapman becomes inadequate when κ and/or Φ_0 is large. The theory of Stern takes the finite size of the counterions and their binding properties at the surface, into account. The diffuse layer is divided into an inner layer (the Stern

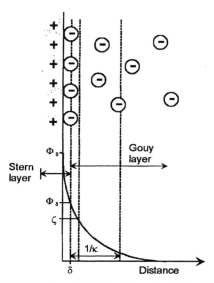

Figure 2.9. Stern's model of the diffuse electric double layer.

layer) and an outer layer (the Gouy layer), as shown in Fig. 2.9. The Stern layer includes any adsorbed layer of ions. It is separated from the Gouy layer at the Stern plane at a distance δ from the surface. This distance corresponds roughly to the radius of a hydrated ion. The ions are adsorbed in the Stern layer according to the Langmuir adsorption isotherm. The theory of Gouy–Chapman is still applicable in the Gouy layer, but Φ_0 is replaced by Φ_δ which is the potential at the Stern plane. In case of strong specific adsorption in the Stern layer, Φ_0 may be smaller than Φ_δ, or they may have opposite polarity. The potential at the Stern plane, Φ_δ, may in most cases be assumed to be equal to the so-called zeta (ζ)-potential, the electrokinetic potential, which can be determined experimentally by means of, e.g., electrophoresis or streaming potential measurements. The zeta-potential is the potential at the shear plane between the charged surface and the liquid, i.e. the potential at the boundary, for example, of a moving particle with its adsorbed ions. The Stern theory is complicated, and will not be treated in any detail in this book. Stern theory should be used when the surface potential is high, or when the solution is concentrated so that a significant part of the potential drop occurs in the Stern layer. Gouy–Chapman theory may be adequate also at higher concentrations if the surface potential is small.

Grahame made a further division of the Stern layer into the inner Helmholtz layer and outer Helmholtz layer. These layers are separated by the inner Helmholtz plane at a distance from the surface corresponding to the radius of non-hydrated specifically adsorbed ions. These ions are smaller than the counterions, and the inner Helmholtz plane is hence located between the Stern plane and the surface. The outer Helmholtz layer is limited by the outer Helmholtz plane, which is identical to the Stern plane.

2.4.4. The electric double layer, lateral fields

Schwartz theory for a suspension of spheres

Schwartz (1962) wanted to use the theories of electric double layers to describe the measured α-dispersion (Section 3.8) of particle suspensions. He considered the case of an electric double layer at the surface of a spherical particle, as shown in Fig. 2.10. The counterions will be electrostatically bound to the surface charges of the sphere, but will be free to move laterally along the surface. When an external field is applied, the positive counterions in Fig. 2.10 will move towards the cathode, but without leaving the surface of the sphere (this polarisation effect is largely exaggerated in Fig. 2.10). The re-establishment of the original counterion atmosphere after the external field is switched off, will be diffusion controlled, and the corresponding time constant according to Schwartz theory is

$$\tau = \frac{a^2}{2D} \tag{2.19}$$

where $D = \mu kT/e$, and a is the particle radius.

Later improvements of Schwartz theory

Schwartz theory provides a practical tool for analysing measured data, but the theory has been criticised for neglecting the diffusion of ions in the bulk solution near the surface. Efforts have been made by, among others, Dukhin, Fixman and Chew and Sen to employ Gouy–Chapman theory on particle suspensions, but the resulting theories are very complex and difficult to utilise on biological materials. Mandel and Odijk (1984) have given a review of this work.

Simplified models that use the Gouy–Chapman theory have been presented by, e.g., Grosse and Foster (1987), but the assumptions made in their theory limit the utility of the model.

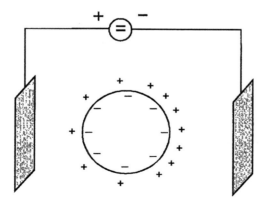

Figure 2.10. Counterion polarisation near the surface of a spherical particle.

Effect of hydration and specific adsorption

Hydration of ions is due to the dipole nature of water. In the case of a cation in water, the negative (oxygen) end of the neighbouring water molecules will be oriented towards the ion, and a sheath of oriented water molecules will be formed around the cation. This sheath is called the *primary hydration sphere*. The water molecules in the primary hydration sphere will attract other water molecules in a *secondary hydration sphere*, which will not be as strongly bound as the primary sphere. Several spheres may likewise be involved until at a certain distance the behaviour of the water molecules will no longer be influenced by the ion.

Specific adsorption (chemisorption) is the process of chemical binding of ions to, e.g., the metal surface of an electrode. These ions are not counterions, and are hence bound to the surface by chemical bonds and not primarily by coulombic (van der Waals) forces. Specifically adsorbed ions will not be hydrated since the layers of water molecules would impede the chemical interaction with the solid surface. Counterions may be hydrated, however, and the corresponding effect will be that the counter charges are moved away from the solid surface. Models have also been presented that include the possibility that the solid surface is also hydrated, thus involving a monolayer of water molecules between the surface and the counterions.

The adsorption or desorption may occur at very sharply defined potentials, and result in a charging current in the external measuring circuit. The process may be quick or slow, and therefore frequency dependent in the case of ac. As long as there is no electron transfer, there is no electrode reaction and no faradaic current, and the current is capacitive (with losses).

The hydration sheath on an electrode surface and around an ion will not stop the electrode processes, but will increase the activation energy of the electron transfer because the average distance of the water molecules from the ion will change when, for example, the ion accepts an electron. The reduced charge of the ion will extend the hydration sheath, and additional energy is required to stretch the sheath to its new position. The increased activation energy will increase the voltage needed to sustain a faradaic current, and also influence the faradaic impedance of the electrode (both the electron transfer resistance and the slow process impedance will be affected – see Section 2.5). It is moreover likely that hydration of the electrode surface and of the counterions will reduce the double layer capacitance.

The effect that specifically adsorbed ions may have on the overall electrode impedance is complex and only a brief discussion will be given here. If the ion is involved in the electron transfer process, a multistep reaction must be taken into account where, for example, an electron is transferred from an ion in the solution to the adsorbed ion, and subsequently transferred from that ion to the electrode surface. This will most likely change the activation energy for the electron transfer, and hence influence the faradaic impedance of the electrode as discussed above. Adsorbed species that do not take part in the electron transfer, e.g. neutral molecules, may block reaction sites on the electrode surface and thereby affect the faradaic current. The double layer capacitance will also be reduced as a result of reduced effective electrode area.

2.4.5. The net charge of a particle

The charged double layer at the surface of colloidal particles (metallic, semiconductive, non-conductive) and living cells in aqueous solution makes them behave as macro-ions and migrate in an electric field. The net charge of a particle is related to the electrokinetic potential (zeta-potential), which was introduced in Section 2.4.3. This is what occurs by *in vitro* electrophoresis. In human body liquids, the range of pH is very narrow: 7.35–7.45. But *in vitro* it is usually possible to find a pH value at which an amino acid, a protein or a cell does not migrate. This pH at which no migration occurs is the *isoelectric point*, or sometimes it is a pH *range*. If the particle is conductive, there is a common charge for the whole particle, with a sheath of counterions around it. When the total charge is zero, no migration occurs. If the particle is not conductive, we are dealing with local charges and counterions, and when the net migrational forces of these are zero, we are at the isoelectric point. The isoelectric point is dependent on the solution used for the examination.

The charge of proteins or cells is the basis for the reciprocal repulsion of particles in a suspension. Loss of charge means losing repulsive forces and implies *coagulation* and *precipitation*. Blood coagulation is the result of such a process.

Osmosis is the transport of a solvent through a semipermeable membrane into a more concentrated solution. In reverse osmosis a sufficiently high pressure is applied on the more concentrated side to reverse the transport direction. *Dialysis* in medicine is the *separation* of suspended colloidal particles from dissolved ions or molecules of small dimensions by means of their unequal rates of diffusion through the pores of semipermeable membranes. If charged carriers are involved, applying an electric field can accelerate the process: *electrodialysis*.

Usually an *ion exchanger* is packed into a tube or column through which a solution is made to flow in order to capture anions or cations. The column arrangement forces the ion-exchange reaction, which is intrinsically reversible, to completion. The solution flowing down the column continually meets fresh exchanger. When the exchangeable ions do start to emerge from the end of the column, the column has become completely saturated. The column may be regenerated by passing through it a solution of the ions that it originally contained.

2.4.6. Electrokinesis

There are four electrokinetic effects due to the electric charge of the double layer at the solid–liquid interphase. In an *E*-field, *electrophoresis* is the migration of charged particles through a liquid, and *electro-osmosis* is bulk liquid flow through a pore caused by a migrating ionic sheath. Furthermore, a current and a potential difference are *generated* by falling charged particles (*sedimentation* potential), and when a liquid is pressed through a pore (*streaming* potential).

The conductivity of the solution involved is of general interest in connection with these effects. High conductivity results in a high current density for a certain *E*-field strength, with a possible problem of rise of temperature. High conductivity also results in small generated potentials.

Electrokinetic effects are not restricted to charged particles, however. Field-induced polarisation will make *uncharged* particles move in inhomogeneous or moving electric fields. The resultant forces increase with the volume of the particle, and are therefore called "body" or "*ponderomotive*" forces. These forces are the basis of phenomena and techniques such as *electrorotation, levitation, dielectrophoresis, pearl chain formation* and *travelling wave dielectrophoresis* (Fuhr *et al.* 1996). Electrorotation and travelling wave dielectrophoresis are techniques in which angular or linear movement of the field relative to the particle produces the desired movement of the particle. Inhomogeneous fields, on the other hand, can cause levitation, dielectrophoresis and pearl chain formation, although pearl chain formation will also occur in homogeneous fields. Ponderomotive effects will be described later in this chapter.

Transport caused by applied electric field

Electrophoresis. Free amino acids, proteins, ions, colloidal particles, bacteria and cells are possible charged particles migrating in an electric field, and can therefore be studied by electrophoresis. As described in Section 2.4.3, some molecules of the solvent are attached to charges on the particle, and hence some solvent will move together with the particle. This is a part of the electrophoretic effect.

Electrophoretic flux can be calculated from the zeta-potential, the permittivity and the viscosity. As these quantities are difficult to estimate, electrophoretic mobility is a more practical quantity (Table 2.8). Migration velocity is simple to measure and the different electrophoretic mobilities are the basis of a very powerful *in vitro* analytical tool for amino acids and proteins in clinical laboratories.

Electro-osmosis. Electro-osmosis is the transport of bulk liquid through a pore under the influence of an electric field. The volume $\dot{V}$ of solution transported per unit time, is

$$\dot{V} = A\zeta\varepsilon E/4\pi\eta L \tag{2.20}$$

where A = effective total area of pores,
   L = pore length,
   ζ = double layer zeta-potential.

Electro-osmosis is a kinetic process that is used for the determination of the ζ-potential of a surface/electrolyte solution interphase. It is also a process found in the sweat pores of the human skin. A dry cathode at a skin surface will fill the ducts with liquid/sweat from deeper skin layers.

Flow-generated potentials

Sedimentation potential. Charged particles falling through a column filled with an electrolyte solution set up a potential difference between column ends. The falling charge carriers constitute an electric current flow, which results in a potential

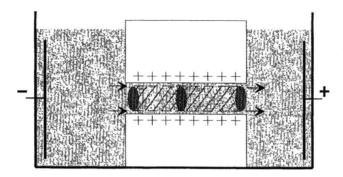

Figure 2.11. Electro-osmosis in a capillary, caused by an externally applied electric field. Double-layer ion sheath migrates and brings the inside bulk capillary volume into motion.

Table 2.8. Electrophoretic mobility (10^8 m^2 V^{-1} s^{-1}) at pH 7.0

Particle	Mobility
Human blood cells	-1
Streptococcus	-1
Methicillin-resistant staphylococcus	-1.5
Proton	$+37$
Cl^-	-7
Colloidal gold	-3.2
Oil droplets	-3.1

difference and hence a flow of charge carriers the opposite direction, opposing the motion of the falling particles. High-conductivity solutions result in small generated potentials.

Streaming potential. When a liquid is forced through a capillary tube, a potential difference $\Delta\Phi$ is generated between the ends of the capillary:

$$\Delta\Phi = \zeta\varepsilon\Delta p/4\pi\eta\sigma \qquad (2.21)$$

where Δp = pressure difference.

The core electrolyte is neutral and no electromotive voltage is generated from the bulk flow, only from the ion sheath migration.

Ponderomotive effects
Forces on particles (bodies) generated by electric fields are called ponderomotoric.

Dielectrophoresis. Dielectrophoresis (DEP) is the movement of non-charged dielectric particles in an inhomogeneous electric field. The force **f** on a charge is **f** = q**E**, so if the net charge is zero, there is no force. A dipole has net charge of zero and will experience a torque in an electric field, but no translational force. However, in an inhomogeneous field, there is also a translational force on a dipole. Consider a spherical non-charged particle in a medium. It is polarised in an electric field and if there is a field gradient ∇E, there is a force $\mathbf{f}$ on the particle (Pethig 1979):

$$\mathbf{f} = (\mathbf{p} \cdot \nabla)\mathbf{E} = 2\pi a^3 \Re[\varepsilon_2(\varepsilon_1 - \varepsilon_2)/(\varepsilon_1 + 2\varepsilon_2)]\nabla E^2 \qquad (2.22)$$

where $\mathbf{p}$ is the induced dipole moment, subscript 1 is for the particle and 2 for the medium. Equation (2.22) is the basic equation for DEP and shows that, depending on the relative polarisability of the particle with respect to the medium, the particle will move either in the direction of the field gradient (positive DEP) or in the opposite direction (negative DEP). The electrophoretic and dielectrophoretic mechanisms can be completely separated and Pohl (1958) adopted the term dielectrophoresis to identify this distinction. Only DEP gives a net force in an ac electric field, the E^2 term has no polarity dependence. One problem with eq. (2.22) is that it does not account for surface and double layers on wet surfaces. The ponderomotoric effects were earlier studied also by Schwan (1982).

Electrorotation.   Particles suspended in a liquid will experience a torque in a rotating E-field (Arnold and Zimmerman 1982). A dipole is induced in the particle. As the polarisation process (redistribution of charges) is not immediate, the induced dipole will lag the external field and a frequency-dependent torque will exist. It can be shown that the torque is dependent on a relaxation time constant identical to the time constant in the theory of β-dispersion (Schwan 1985). Cell rotation is therefore a direct physical manifestation of dispersion. Theory predicts that the torque may have two maxima, usually of opposite sign (co-rotation and anti-rotation) and is given by (Zhou *et al.* 1995):

$$\Gamma = -4\pi a^3 \Im[\varepsilon_2(\varepsilon_1 - \varepsilon_2)/(\varepsilon_1 + 2\varepsilon_2)]\nabla E^2 \qquad (2.23)$$

Rotation also occurs in non-rotating external fields, and onset is often at a sharply well-defined frequency. The rotation is therefore also called *cellular spin resonance*. One possible mechanism is the dipole–dipole interaction between neighbouring cells. The peak rotation frequency is many decades lower than the excitation frequency, the latter being in the β-dispersion range (MHz).

Cellular spin resonance indicates various physiological states of living or dead cells, and also represents a method for cellular manipulations. It should, for example, be possible to rotate intracellular organelles with respect to the whole cell.

Levitation.   Negative dielectrophoresis causes particles and cells to be repelled from regions of high electric field strength. This effect can be used to levitate particles, for example, over a planar array of electrodes. A common geometry is a four-electrode field trap with electrodes forming the sides or corners of a square and with the particle levitated over the centre of the square. The viscous forces from the medium will damp the motion of the particle.

Travelling wave dielectrophoresis.   If the E-field in an electrorotation chamber is made to move in a linear rather than an angular manner, an interesting combination

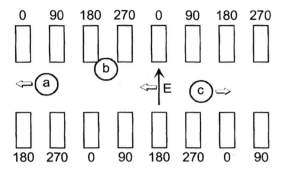

Figure 2.12. Travelling wave dielectrophoresis achieved by using cosine voltages with the indicated phase relationships as excitation on each pair of facing electrodes. The electric field will travel from right to left in this example. Particle **a** could be a non-viable cell, **b** a cell trapped by positive dielectroporesis and **c** a viable cell.

of dielectrophoretic and electrorotation effects will occur. The principles are shown in Fig. 2.12, where phase-shifted cosine voltages are applied to facing electrodes along an array of such electrodes. Only particles experiencing negative dielectrophoresis will move along the inter-electrode channel. They will otherwise be trapped at the electrodes by positive dielectrophoresis.

As shown above, the dielectrophoretic effect depends on the real part of the induced dipole moment and the electrorotation effect on the imaginary part of the induced dipole moment. Both these effects contribute to the motion of the particle in the travelling wave chamber. Viable cells will exhibit negative dielectrophoresis at low frequencies (typically < 10 kHz) and in a frequency window below this frequency, where the imaginary part of the induced dipole moment is positive, the cell will move in the opposite direction to the travelling electric field. Below this frequency window, the cell will be levitated over the electrode array. For non-viable cells, the negative dielectrophoretic effect will typically occur at high frequencies, over 1 MHz. The corresponding imaginary part of the dipole moment will be negative and the cell will hence travel in the direction of the moving field (Wang *et al.* 1993).

Pearl chain formation. A dielectric particle disturbs the local E-field, so even in a homogeneous external field there is a local gradient. Two neighbouring dielectric particles will therefore be attracted to each other. They form a dipole and will be oriented in the direction of the E-field. In a homogeneous field, cells will therefore tend to form pearl chains in the direction of the E-field. In a gradient field the pearl chains will protrude from the electrode surface (Schwan and Sher 1969).

2.5. Electrodics and ac Phenomena

The *half-cell* concept is often used for analysing an electrode with its surrounding solution. However, we do not have access to an isolated half-cell, so a real electrolytic cell consists of two half-cells. Often the second electrode is considered to be a reference electrode.

2.5.1. Electrode equilibrium dc potential, zero external current

Metal/ion and redox systems

According to Nernst, metals have a tendency to release their ions into a solution as solvated particles. The potential difference arises from this transfer of positive metal ions across the double layer, with a resulting negative potential on the metal electrode. The standard potentials of such *metal/ion* half-cells are shown in Table 2.9, where the reference electrode is the hydrogen electrode. Volta's original findings were in rough accordance with this table. The least noble metals have the most negative values. For the noble metals, the number of metal ions released into the solution may be small, and the potential must then be measured under strict zero-current conditions.

With an external dc power supply connected to the electrolytic cell, the applied voltage that gives *no dc current flow* in the external circuit corresponds to the *equilibrium potential* of the half-cell (or actually the cell). It is the same voltage as read by a voltmeter with very high input resistance and virtually no current flow (pH meter). In electrochemistry, *potentiometry* is the measurement of the potential of an electrode at zero current flow, that is when the cell is not externally polarised. In order to understand the equilibrium potential with zero external current, we must introduce the concept of electrode reaction, and link it with an electric current in the external circuit. An electrode reaction going rapidly both ways (ionisation/de-ionisation, reduction–oxidation) is called *reversible*. A pure redox system presup-

Table 2.9. Metal/ion equilibrium potentials

Metal/metal ion	V_0 (volt)
Li/Li$^+$	-3.05
K/K$^+$	-2.93
Na/Na$^+$	-2.71
Al/Al^{3+}	-1.66
Zn/Zn^{2+}	-0.76
Fe/Fe^{2+}	-0.44
Ni/Ni^{2+}	-0.25
Sn/Sn^{2+}	-0.14
Pb/Pb^{2+}	-0.13
H$_2$/H$^+$	**0**
Carbon	dependent on structure
Cu/Cu^{2+}	$+0.34$
Ag/Ag$^+$	$+0.80$
Pt/Pt^{2+}	$+\sim 1.2$
Au/Au^{3+}	$+1.50$

poses an inert electrode metal, e.g. platinum, with no electrode metal ion transfer. At the electrodes the redox system is:

The *anode* is the *electron* sink, where ions or neutral substances lose electrons and are oxidised: red → ox + *ne*.

The *cathode* is the *electron* source where ions or neutral substances gain electrons and are reduced: ox + *ne* → red.

The metal/ion half-cell generates a potential by the exchange of *metal ions* between the metal and the electrolyte solution. In contrast, a *redox* half-cell is based upon an exchange of *electrons* between the metal and the electrolyte solution. So actually there are two sets of standard potential tables, one for metal/ion half-cells (Table 2.9), and one for redox half-cells (Table 2.10). The half-cell potential is of course *independent* of the interphase area, because equilibrium potential is without current flow. As soon as the cell is externally polarised and a current is flowing, electrode area is of interest (current density).

Table 2.10 shows some typical half-cell standard potentials for redox systems.

The Nernst equation
The Nernst equation relates the redox processes and the potential to the concentration/activity of the ions in the solution of an electrolytic cell. It indicates the *redox equilibrium potential V with no dc current flow*:

$$V = V_0 + (RT/nF)\ln(a_{ox}/a_{red}) \qquad \textit{The Nernst equation} \qquad (2.24)$$

Here V_0 is the standard electrode potential of the redox system (with respect to the hydrogen reference electrode at 1 mole concentration), n is the number of electrons in the unit reaction, R is not resistance but the universal gas constant, and F is the Faraday constant. a_{ox} and a_{red} are *activities*, $a = \gamma c$, where c is the concentration and γ is the activity *coefficient*. $\gamma = 1$ for low concentrations (no ion interactions), but < 1 at higher concentrations. The half-cell potentials are referred to standardised conditions, meaning that the other electrode is considered to be the standard hydrogen electrode (implying the condition: pH 0, hydrogen ion activity 1 mol/L).

The Nernst equation presupposes a *reversible* reaction: that the reaction is reasonably fast in both directions. This implies that the surface concentrations of

Table 2.10. Standard equilibrium electrode potentials for some redox systems

Electrode	Electrode reaction	V_0 (volt)
Pt	$2H_2O + 2e^- \rightleftharpoons H_2 + 2OH^-$	-0.83
Pt	$O_2 + 2H_2O + 2e^- \rightleftharpoons H_2O_2 + 2OH^-$	-0.15
Pt	$\mathbf{2H^+ + 2e^- \rightleftharpoons H_2}$	**0**
Carbon	→ ?	?
AgCl	$AgCl + e^- \rightleftharpoons Ag + Cl^-$	$+0.22$
Calomel		$+0.28$

reactants and products are maintained close to their equilibrium values. If the electrode reaction rate is slow in any direction, the concentrations at the electrode surface will not be equilibrium values, and the Nernst equation is not valid. Then the reactions are *irreversible*.

For metal cathodes the activity of the reduced forms a_{red} is 1, and the equation, e.g., for an AgCl electrode is

$$V = V_{AgCl/Ag} - (RT/F) \ln a_{Cl^-} \qquad (2.25)$$

where R is the universal gas constant. Then the dc potential is only dependent on the activity of the Cl^--anion in aqueous solution (and on temperature). The silver–silver chloride electrode is an example of an "electrode of the second kind": the electrode metal (Ag) is in equilibrium with a low-solubility salt of its ions (Ag^+).

Total galvanic cell dc voltage

According to Tables 2.9 and 2.10, two different electrode materials in the same electrolyte solution may generate one volt or more dc. Superimposed on signals in the microvolt range, this may create noise and be a problem for input amplifiers.

If both electrode surfaces are of, say, stainless steel in saline, there are not necessarily any redox reactions at the surface. The voltage is not well defined and may easily attain 100 mV or more, the system is highly polarisable and the output voltage is noisy, cf. Section 8.1. Even strongly polarisable electrode metals like stainless steel, platinum or mercury have an electrode reaction if the applied dc voltage is high enough. This is on both the anodic and cathodic side, but the reactions are irreversible and therefore not redox reactions.

Under zero-current conditions, many metal electrodes are of interest in biological work. The platinum electrode, for instance, becomes an interesting redox potential recording electrode. An inert platinum electrode may actually be used in potentiometry in the bulk of a redox system, during titration or for example in sea water analysis. With zero current the inert platinum will pick up the redox equilibrium potential of the process.

Poisson's law (eq. 7.6) defines the relationship between the potential function (representing a possible emv (electromotive voltage, Section 10.3) source to the external circuit) and the charge distribution in an electrolytic volume. A change in the charge distribution near the electrode surface results in a changed potential function, and consequently a changed half-cell potential. A polarisable electrode therefore implies a current-induced change of charge distribution near the electrode surface, and a changed half-cell potential.

The liquid junction equilibrium dc potential

Between two dissimilar electrolyte solutions a potential difference is created just as between a metal and an electrolytic solution. By Brownian motion the ions randomly walk with a velocity proportional to the Boltzmann factor kT. The corresponding E-field will have a direction that slows down the rapid ions and accelerates the slow ones in the interface zone. The resulting potential difference is called the *liquid junction potential* (Φ_{lj}), and follows a variant of the Nernst equation called the *Henderson* equation:

$$\Phi_{lj} = \frac{\mu^+ - \mu^-}{\mu^+ + \mu^-} \frac{RT}{nF} \ln \frac{c_1}{c_2} \tag{2.26}$$

where R is the universal gas constant and μ^+ and μ^- are the mobilities of cations and anions, respectively.

The liquid junction potential is usually less than 100 mV. For instance for a junction of different concentrations of NaCl and with $c_1 = 10c_2$, the dilute side is 12.2 mV negative with respect to the other side.

The K^+ and Cl^- ions have about the same mobilities, and therefore KCl creates a lower liquid junction potential than, e.g., NaCl. The liquid junction dc potential can be kept small by inserting a *salt bridge* between the solutions, so that there will be two junctions instead of one. By using a concentrated solution of KCl in the bridge, it can be shown that the two junction potentials will tend to be equal and will be dominated by the concentrated salt solution, but with opposite sign so that they more or less cancel.

Membrane equilibrium potentials (Donnan potential difference)
The liquid junction potential was defined with no membrane separating the two media. A membrane separating an electrolyte in two compartments is often selectively permeable, e.g. rather open to water but less permeable to certain ions or larger charge carriers. The selectivity may be due to the mechanical dimensions of pores or to charge-dependent forces.

With different concentrations on each side, such membranes generate an osmotic pressure difference. With different ionic concentration an electrical potential difference is also generated. This is called the *Donnan potential difference*, Φ_D.

$$\Phi_D = (RT/F) \ln a_1/a_2 \tag{2.27}$$

where a_1 is the activity of a specified ion on compartment side one, and R is the universal gas constant.

2.5.2. The monopolar basic experiment

When we are to study electrode reactions a little further, we must be able to differentiate between cathodic and anodic processes. We therefore change the setup shown in Fig. 2.1 a little. Instead of two equal electrodes, we reduce the area of one of them to be, e.g., $< 1/100$ of the other.

With no current through the electrolytic cell, it does not matter whether the electrodes are large or small, the equilibrium potentials are the same. But with current flow, the current density and therefore the voltage drop and the polarisation, will by much higher at the small electrode. An increased potential drop will occur in the constrictional current path near the small electrode, and in general the properties of the small electrode will dominate the results. The small electrode will be the electrode studied, the *working electrode*. It is a *monopolar* system, meaning that the effect is determined by *one* electrode. The other electrode becomes the *indifferent* or *neutral* electrode. Note that this division is not true in potentiometry; electrode area is unimportant under no-current conditions.

We let the external dc voltage change slowly (e.g. ramp voltage from a polarograph), and we record the dc current (Fig. 2.14). In electrochemistry this is called *voltammetry* (volt-am-metry), the application of a varying voltage with the measurement of current. *Ampereometry* is more generally the measurement of current with a constant amplitude voltage. *Polarography* is a form of voltammetry, preferably with the dropping mercury electrode and with a diffusion-controlled current in a monopolar system.

The solution (Fig. 2.13) is as usual NaCl 0.9% in water. With a *platinum* working electrode there will be no dc current over a rather wide voltage range (Fig. 2.14a). The *incremental* dc resistance $R = \Delta V/\Delta I$ is large in this range ($\Delta I \approx 0$). Platinum is therefore a strongly polarisable electrode material.

With a sufficiently large voltage of any polarity, dc faradaic current flows and we have an *electrode reaction* at the platinum surface. R falls to smaller values. Evidently a large *activation energy* is necessary to obtain electron transfer and an electrode reaction. It is not likely that this reaction involves chemical reaction with platinum, or that platinum metal enters the solution as ions. However, in our basic experiment of Fig. 2.1 we did consider the effect of dissolved oxygen. With a suitable negative voltage, the neutral oxygen is reduced at the cathode (Fig. 2.14a, broken curve). This reaction causes a faradaic current. However, when the negative voltage is large enough, we reach a *current plateau*, again with a large incremental R. Then the electron transfer is no longer the rate-limiting factor, but rather the diffusion of

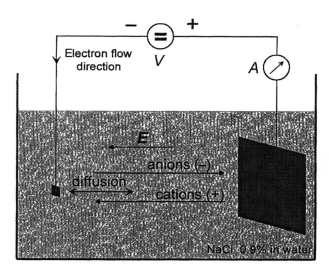

Figure 2.13. The basic electrolytic experiment, *monopolar* electrode system. The bipolar version is shown in Fig. 2.1.

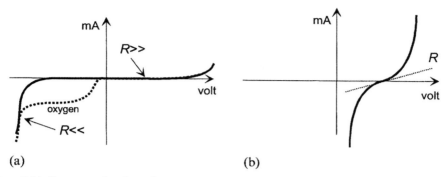

Figure 2.14. Dc monopolar electrode current. Incremental resistance R is the tangent to the curve at a given point. Working electrode (a) platinum, (b) silver–silver chloride.

oxygen to the cathode. At the cathode surface the concentration of oxygen is approximately zero; all available oxygen is reduced immediately. *The current is diffusion controlled.* The oxygen molecules are neutral in the solution and are not migrating in the electric field, they move because of the concentration gradient. At the platinum surface they are ionised (reduced), accepting electrons. This is the principle of the polarographic oxygen electrode. If we increase the voltage to about -1.4 volt, a much larger dc current flows, and the decomposition of water occurs with bubbles of H_2 appearing.

Now let us change to a small *silver/silver chloride* working electrode (Fig. 2.14b). Even with a small deviation ΔV from the equilibrium voltage with zero current, the ΔI will be large, and in either direction. Even at the equilibrium voltage, R is rather small. We have an immediate, large electrode reaction. This is the non-polarisable electrode; even with relatively large dc currents the charge distribution does not change very much, neither does the dc potential. With a positive overvoltage on the electrode, the AgCl layer thickens; with a negative overvoltage it becomes thinner and is at last stripped off. Then we have changed the electrode surface to a pure silver electrode, and with a different equilibrium potential.

Suppose a small ac signal is superimposed on the dc voltage. At equilibrium dc voltage with no dc current, the ac current will flow in both directions because it is a reversible redox electrode process at the AgCl surface. With a dc current, the small ac current will be superimposed on the larger dc current and will thus change the reaction rate by only a small amount. As seen in Fig. 2.15, the resultant *ac current* will depend on the local slope of the dc current curve. The slope defines the *incremental resistance*:

$$R = \Delta V / \Delta I \tag{2.28}$$

The incremental resistance varies according to the dc (*bias*) level. As the dc current curve is not linear, it is clear that with large amplitudes the current response will not be a sine wave even if the superimposed ac voltage is a pure sine wave. Because of the

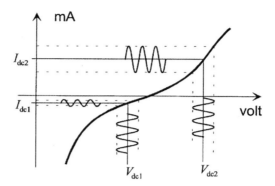

Figure 2.15. Superimposed ac voltage as independent variable, with dc (bias) voltage as parameter. The resulting ac current is dependent on the incremental resistance, that is the slope of the dc current curve.

capacitive properties of the cell, the current will not be in phase with the applied sine voltage.

From the curves in Fig. 2.15 it is clear that by changing the frequency and recording the ac current it is possible to examine electrode processes and find out how rapid they are. By studying the harmonic content of the current waveform as a function of dc voltage and ac amplitude/frequency, non-linearity phenomena can also be studied.

With no redox reactions, e.g. using a platinum electrode in saline, the small ac voltage will result in an ac current dependent on double layer capacitance and other components (see the next subsection).

The voltage deviation ΔV from equilibrium necessary for a certain dc current flow is called the *overvoltage*. With small deviations there is a linear relationship between ΔV and electrode current I, but with larger ΔV it is strongly non-linear. An overvoltage is linked with an external current and therefore the electrode is externally *polarised*.

In *cyclic voltammetry* the voltage is swept as in polarography, but at a predetermined voltage level the sweep is reversed and the cycle ends at the starting voltage. More than one cycle may be used, but usually the recorded current curve changes for each cycle. The single-cycle experiment must therefore not be confused with a steady-state ac condition. Two examples are shown in Fig. 2.16.

As the sweep is linear, the x-axis is both a voltage and a time axis. The charge transferred from the sweep generator to the cell is therefore proportional to the area between the curve and the x-axis ($=0$ mA). If the enclosed areas over and under the x-axis are equal, no net charge is supplied to the system. The currents may be due to double layer charging, sorption at the metal surface, or electrode reactions.

The electrolytic cell voltammograms in Fig. 2.16 are for irreversible (a) and reversible (b) processes. The irreversible process represents a net charge transfer to the cell, because very little reverse current is present. During voltage sweep a non-faradaic charge current will also flow to or from the double layer capacitance. The

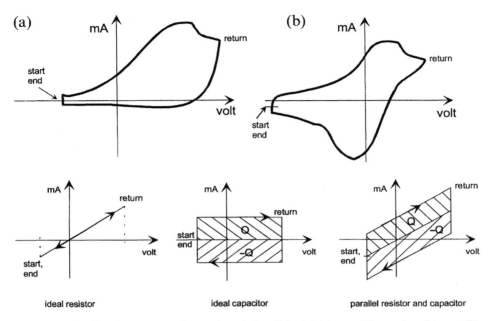

Figure 2.16. Cyclic voltammograms for an electrolytic cell (top): (a) irreversible reaction, (b) reversible. Bottom: with ideal components.

current steps at the sweep ends represent the reversal of such a capacitive charging current. In the reversible process most of the charge is returned—the redox reaction is reversed.

With an ideal resistor instead of the electrolytic cell (Fig. 2.16, bottom), the current curve is the same straight line, and a charge transfer equal to twice the area under the line is transferred and lost. With an ideal capacitor the charging current I is $I = C \, dV/dt$. When dV/dt is constant, I is constant; when dV/dt changes sign, I changes sign. The area under the current curve is Q, and the same charge $= Q$ is returned at the return sweep. No net charge is transferred to the capacitor. The current to the parallel RC components is then the sum of the separate R and C currents.

One type of electrode has no defined dc voltage at all: the ideally *polarisable electrode*. It can attain any potential; it is just a double-layer capacitor that can be charged to any voltage if the necessary charging current is supplied. No electron transfer occurs, no free charge carriers cross the double layer or flow in the solution. With a real electrode this implies virtually zero dc current within a certain range of applied voltage. A platinum electrode in NaCl aqueous solution is a practical example. As a current-carrying dc reference electrode in NaCl electrolyte, the platinum electrode is the worst possible choice. Another example is the dropping mercury electrode in an indifferent electrolyte, with a perfectly smooth surface continually renewed. No reversible charge transfer occurs within a rather broad range of dc voltage (oxygen-free solution).

To obtain a well-defined potential *with a variable dc current flowing*, a reversible electrochemical reaction is necessary at the electrode. Usually this implies that there must be redox reactions, with easy electron transfer in both directions. Such an electrode with easy electron transfer is called a *non-polarisable electrode*.

A redox reaction is dependent not only on the electrode material but also on the electrolyte solution. As we have seen, platinum is highly polarisable in NaCl solution. However, if the surface is saturated with dissolved hydrogen gas, a redox system is created (H/H^+), and then the platinum electrode becomes a *non-polarisable reference* electrode. Surface oxidation, adsorption processes and organic redox processes may reduce the polarisability and increase the applicability of a platinum electrode in tissue media.

2.5.3. Dc/ac equivalent circuit for electrode processes

The faradaic polarity-dependent dc currents just described are dependent not only on the applied dc voltage but also on elapsed time. With a small change in excitation voltage, it takes a certain time to reach the new current level. The four processes to be considered are:

1. A faradaic component: the rate of *electron* transfer to the electro-active species of the solution (occurring near/at the electrode surface in the double layer).
2. A faradaic component: the amount of species that can be *transported to* the reaction site from the bulk of the electrolyte, and the amount of reaction products that can be transported away *from* the reaction site.
3. The electric *charging* of the double layer.
4. The *sorption* of species at the electrode surface.

The mechanisms and speed of the *electron transfer process* have long been discussed. Does it take the electron a long time to meet the particle, or is the particle immediately ready to donate or accept an electron? When a chemical reaction is to occur, the species approach one another to a necessary close distance. In an electrode reaction, the ions enter the outer part of the double layer, and the chance increases of gaining or losing electrons. If the electroactive species are ions, what is the effect of the hydration sheath and ionic atmosphere surrounding them? Must a part of it be stripped off? The reacting partners must possess sufficient energy (translational, vibrational, rotational) to obtain reaction. The concept of necessary *activation* energy is therefore important. Sodium is very electropositive (cf. Section 2.1.1), thus tending to get rid of electrons. The most probable electrode reactions can be predicted from the electronegativity scales.

A plausible but rough electrical dc/ac equivalent circuit for the electrode processes is shown in Fig. 2.17. The electrodic part consists of three principal current paths in parallel. The elements are Cole-like as discussed in Section 7.2, and some of the component symbols used indicate that their values are non-ideal and frequency dependent.

As for the electrode processes, dc parameters are contraindicated, because a dc study should be performed with virtually zero applied overvoltage in order to be in

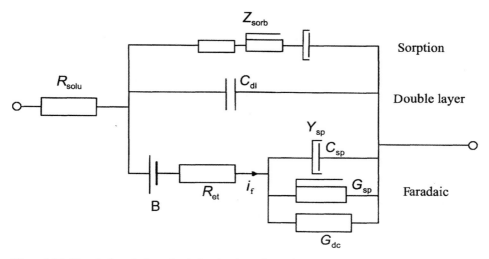

Figure 2.17. Electrical equivalent circuit for the three electrode processes. The subscript sp means slow processes.

the linear region (Section 6.2.1). Using a superimposed ac signal of sufficiently small amplitude with $f > 0$, the model may be a linear model, and the values of the components of the equivalent circuit may vary according to the applied dc voltage or current.

The series resistance

The bulk of the electrolyte obeys Ohm's law (eq. 2.2). Accordingly, the bulk electrolyte is modelled as an ideal resistor R_{solu} in series with the electrode components. This is to indicate that bulk electrolytic conductance is considered frequency independent, but dependent on the geometry and possible current constrictional effects (see Fig. 5.2). If the bulk electrolyte is replaced by tissue, a more complicated equivalent circuit must replace R_{solu}, and we are confronted with the basic problem of division between tissue and electrode contributions.

The series resistance causes an overvoltage due to a simple IR-drop ($\Delta V = IR_{solu}$) in the solution. Actually this is not due to a polarisation process. However, it is often practical to include it in the total overvoltage and the electrode polarisation concept. The IR drop is proportional to current, and can be reduced by the addition of a suitable strong electrolyte not intervening with the processes of interest (*indifferent electrolyte*). It can also be reduced by introducing a reference electrode reading the potential very near to the working electrode, and connecting all three electrodes to a potentiostat (see Section 5.4.1).

1. Double layer path with leakage. With a dripping mercury electrode the surface is ideal and the double layer is modelled as a pure, frequency-independent capacitor, somewhat voltage dependent. The capacitance values are very high because of the small double layer thickness: C_{dl} is about 20 $\mu F/cm^2$. With solid electrode materials the surface is of a more fractal nature, with a distribution of capacitive and resistive properties. The actual values are dependent on the type of metal, the surface conditions, the type of electrolyte and the applied voltage. The capacitance increases with higher electrolyte concentration (cf. Section 2.4.3). The double layer capacitor is inevitable, it is there as long as the metal is wetted. C_{dl} may dominate the circuit if there are no sorption or electrode reaction processes.

2. Surface sorption ac path. The additional capacitive impedance Z_{sorb} is due to the adsorption and desorption of species at the electrode surface. These species do not exchange electrons with the electrode surface but they change the surface charge density and therefore cause a pure ac current path. A current-limiting resistor in series may model the processes with a Cole-like element without dc conductance (drawn as a series constant phase element because it is in series with a resistor, cf. Section 7.2.3). Z_{sorb} may dominate the circuit in the middle and higher Hz and lower kHz frequency ranges. Sorption currents are ac currents, and as the adsorption/desorption may occur rather abruptly at a certain dc voltage, these currents may be very dependent on the applied dc voltage. Sorption currents may dominate noble metal electrodes, so the measurement of their polarisation impedance must be performed under controlled dc voltage.

3. Electrode reaction dc/ac path (faradaic). The metal and the electrolyte also determine the dc half-cell potential, modelled by the battery B. If there is no electron transfer, R_{et} is very large and the battery B is decoupled, the electrode is then polarisable with a poorly defined dc potential. But if there is an electrode reaction, R_{et} has a lower value and connects an additional admittance in parallel with the double layer admittance. This current path is through the *faradaic impedance* Z_f, and the current is the faradaic current i_f. Faradaic current is related to electrode reactions according to Faraday's law (Section 2.4.1). The faradaic impedance may dominate the equivalent circuit in the lower Hz and sub-Hz frequency ranges and at dc. The faradaic impedance is modelled by a complete Cole-like series system. It consists of the resistor R_{et} modelling the electron transfer, in series with a Cole-like parallel element of *admittance* Y_{sp} related to slow processes (mass transport and slow electrode reactions) and therefore introducing time delays necessitating an equivalent admittance and not just a conductance.

The electron transfer resistance R_{et} is related to the activation energy and to the extent to which electroactive species have reached sufficiently near the electrode surface so that acceptance or donation of electrons can occur. If the electrode voltage is not sufficient to create electron transfer and an electrode reaction, R_{et} will be very large. R_{et} is purely resistive, which means that there is little transfer time delay. The process is almost immediate, but energy dependent and therefore probability dependent. R_{et} is clearly current dependent, and for non-polarisable electrodes it is small for all dc current values. For polarisable electrodes it is very large at zero dc current. R_{et} is clearly a non-linear element. It only dominates to the extent that electron transfer is the current-limiting process. Under that condition, the concentration of active species at the electrode is independent of current flow. Then the relationship between the *overvoltage* ΔV and the electrode current I is given by the *Butler–Volmer* equation:

$$I = I_0\{\exp(\psi k\,\Delta V) - \exp[-(1 - \psi)k\,\Delta V]\} \qquad (2.29)$$

where I_0 is the exchange current present when the external current is zero, the exponent ψ is called the *transfer coefficient*, and $k = zF/RT$. At equilibrium voltage with no external current, there may actually still be large reducing and oxidising local currents that cancel externally, this is I_0.

The slow process *admittance* Y_{sp} is related to reactions that are rate limited by the necessary time of transport to or from the electrode active sites, and the time of accompanying chemical reactions. If the faradaic current is limited by *diffusion alone*, the immittance is called *Warburg* immittance. R_{et} is then negligible, and the current is determined by Y_{sp}. The electron transfer flux and the chemical reaction rate are so high that the process is controlled by diffusion alone. Diffusion is a transport process dependent only on concentration gradients, cf. Section 2.4.2. The diffusion may be of reactants to the electrode, or of reaction products *away* from the electrode, both influencing Y_{sp}. Warburg (1899) was the first to solve Fick's second law (eq. 2.10) for an electrolytic cell under ac conditions. He presumed a diffusion-controlled process with negligible migrational effects and found that under ideal conditions the concentration at the electrode surface is $+45°$ and the voltage $-45°$ out of phase with the applied current, independently of the applied frequency. Such an ideal and purely diffusion-controlled electrolytic cell is therefore a perfect model for a constant phase element (CPE). Warburg found that the concentration wave spreads out longer in the electrolyte the lower the applied frequency (Fig. 2.18).

The penetration will be delayed and can be described by a damped sine wave, so that concentrations both lower and higher than the bulk concentration may exist. It is analogous to the slow penetration of heat into a housing wall or the ground during a change in cold to warm weather. Warburg found the *diffusion zone*, defined as the distance δ from the electrode at which the concentration wave is reduced to $1/e$ of its value at the electrode, to be

$$\delta = (2D/\omega)^{1/2} \qquad (2.30)$$

where D is the diffusion constant. The diffusion zone length δ may therefore extend all the way to the countercurrent-carrying electrode at low frequencies. With $D = 10^{-9}$ m²/s and at 10 μHz, δ is 14 mm. In order to introduce negligible phase distortion, the cell length should be several centimetres long. Pure 45° properties will be found only if convection effects are negligible during the period. A purely diffusion-controlled electrode reaction has an *infinite-length Warburg admittance* with a constant (frequency-independent) phase angle $\varphi = 45°$. A more general *finite-length Warburg admittance* does not have the same constant phase character (MacDonald 1987).

The thickness of an electric double layer is about 0.1 nm with strong electrolytes,

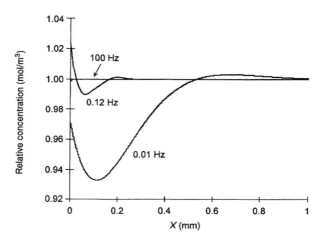

Figure 2.18. Ideal diffusion controlled concentration wave into the electrolyte caused by a sine wave ac current. According to Warburg (1899), his eq. 5, with the frequency of the applied current as parameter. Shown at a given moment; actually the concentration amplitude at the electrode surface will vary as a function of time according to a 45° phase-shifted sine wave. Current density and electrolyte concentration arbitrarily chosen, $D = 10^{-9}$ m^2/s. X is the distance from the electrode surface.

about the same size as the ions, and perhaps 10 nm in dilute solutions (diffuse double layer). The thickness of a diffusion zone in unstirred solutions is enormously larger. The diffusion zone length is not dependent on concentration (as long as D is constant).

The influence of the electron transfer resistor R_{et} and the slow process admittance Y_{sp} on the total faradaic impedance is determined by which factor is reaction rate limiting. It is possible to study this by plotting the faradaic impedance as a function of frequency in a log–log plot, or in the Wessel diagram, and looking for circular arcs (see Section 7.2).

The term *electrode polarisation immittance* is sometimes related to the total immittance of the equivalent circuit of Fig. 2.17, but it is sometimes useful to exclude the series resistance R_{solu} if it has the character of being an access resistance to the electrode processes. Without electrochemical reaction, measured currents are due to double layer components and sorption. The term electrode polarisation immittance should be either avoided or defined. The processes involved are of a very different nature, and in a measuring setup different variables must be controlled depending on what effects are to be studied. The electrode immittance of an electronic conductor in contact with 0.9% NaCl is of special interest to us. Data for such interfaces are found in Chapter 8.1.

2.5.4. Non-linear properties of electrolytics

Electrolyte
The bulk electrolyte solution obeys Ohm's law (eq. 2.2), which is linear. If the E-field changes the viscosity η in eq. (2.6) or the number of ions per volume n in eq. (2.1),

then the system is non-linear and does not obey Ohm's law (Wien 1928, Onsager 1934). This will be the case at very high electric field strengths. According to the Debye–Hückel theory, the ionic atmosphere is symmetrical about the ion in the absence of an external electric field. In an electric field the ion migrates, but its atmosphere (ionic and hydrational) is retarded by friction and is no longer symmetrical about the ion (asymmetry effect). Accordingly, at high electric fields the conductivity increases because the ions move so fast that the retarding ionic atmosphere does not have time to form: it is stripped off (the Wien[13] effect, Wien 1931).

Electrode processes
From Fig. 2.15 and the Butler–Volmer equation (2.29) it is clear that the dc resistance is strongly dependent on the dc current through the electrode. With an ac superimposed on a dc, the resultant ac is dependent on the incremental resistance/conductance of the dc curve. If excitation is sinusoidal, and the measured ac voltage or current also is sinusoidal, then the system is linear with the amplitudes used. On increasing the amplitude, there will be a level at which non-linearity is reached.

A dc or pulsed current polarises the electrode, and from the basic electrolytic experiment described in Section 2.2 it is also clear that faradaic current flow changes the chemical environment at the electrode surface. Current-carrying electrodes are used in such different applications as nerve stimulation, pacemaker catheter stimulation (Jaron *et al.* 1968, 1969) and defibrillation with 50 amperes passing for some milliseconds. Often a square wave pulse is used as stimulation waveform (e.g. in a pacemaker), and the necessary overvoltage is of great interest, cf. Section 8.1. In such applications a clear distinction must be made between tissue non-linearity (Section 4.4.4) and electrode non-linearity. Non-linear network theory is treated in Section 6.1.4. Non-linear behaviour in suspensions has been studied by e.g. Block and Hayes (1970) and Jones (1979); for electroporation see Section 8.12.1.

For the electrode polarisation impedance it has been shown that it was possible to state a frequency-independent *voltage* amplitude limit for linear behaviour (Onaral and Schwan 1982). This limit is about 100 mV (average, corresponding to about 300 mV p-p) ac. The corresponding *current* limit will of course be frequency dependent, and be as low as 5 μA/cm^2 in the lower millihertz range and as high as 100 mA/cm^2 in the higher kilohertz range. A typical current limit for a platinum black electrode in saline is 1 mA/cm^2 at 1 kHz (Schwan 1963). There is reason to believe that as the frequency approaches zero, the current limit of linearity flattens out around 5 μA/cm^2 where the electrode impedance becomes resistive (Onaral and Schwan 1982).

With composite waveforms the electrode may therefore operate in the non-linear region for the low-frequency components, and in the linear region for the high-frequency components.

---

[13] Wilhelm Karl Werner Wien (1864–1928), German physicist, famous for the Wien displacement law. 1911 Nobel prize laureate in physics.

The current density under a surface plate electrode is not uniform, with larger densities at the edge. The fractal properties of the electrode surface also create local areas of high current density. The onset of non-linearity may therefore be gradual, and start very early at very limited areas on the electrode surface. By harmonic analysis (Bauer 1964, Moussavi *et al.* 1994, Section 6.3.2) it has accordingly been found that very weak non-linearity is measurable at much lower voltages than 100 mV.

In a practical case when current-carrying electrodes are used with tissue, it may also be difficult to differentiate between the non-linearity of the electrode processes and the tissue processes, cf. Schwan (1992).

Electrode behaviour in the non-linear region may be studied by electrode polarisation impedance $Z = R + jX$ measured as a function of sinusoidal amplitude. The limit current of linearity i_L may, for instance, be defined as the amplitude when the values of R or X deviate more than 10% from low current density values. Often i_L increases with frequency proportionally to f^m (Schwan's law of non-linearity) (Onaral and Schwan 1982, McAdams and Jossinet 1991a, 1994). m is the constant phase factor (as defined in this book) under the assumption that it is obeying Fricke's law and is frequency independent (Section 7.2.3). When measuring current is kept $< i_L$ they showed that Fricke's law is valid down to 10 mHz. The limit current of linearity will usually be lower for X than for R.

CHAPTER 3

Dielectrics

We have seen that tissue electrolytes are electrolytic *conductors*, with ions free to migrate, and with a considerable dc conductivity. In Chapter 2 a purely electrolytic system was characterised with immittance variables. But the same system may as well be considered as a *dielectric* and characterised with another set of variables. If the dielectric and the electrodes are dry with no free ions, there are no galvanic contacts with the electrodes, and accordingly there is no electrode polarisation impedance in series with the dielectric. Dielectric theory is well adapted to such systems. Biomaterials can, of course, be studied in the dry state, but the structure of complex molecules will usually be very different from that in the natural or living state. With living tissue and the electrolytic conductance of the extracellular fluids, the situation is more complicated and hydration, double layer formation and electrode polarisation must be accounted for. Then both electrolytic and dielectric theories are relevant for the study of biomaterials.

A dielectric may be defined simply as any material placed between the plates of a capacitor for examination by an applied electric field. Etymologically, however, a *dielectric* is a material that the electric field penetrates. This is not true for conductors in a static electric field. Basically, a perfect dielectric is a substance without free charges. Synonyms for a perfect dielectric may then be an insulator or a non-conductor; antonyms may be a conductor or an electrolyte. A more direct definition is a material in which the capacitive (displacement) current is larger than the in-phase current, $\omega C > G$ or $f > \sigma/2\pi\varepsilon$. According to this definition, saline ($\sigma = 1$ S/m, $\varepsilon = 80 \times 8 \times 10^{-12}$) is a conductor for $f < 250$ MHz, and a dielectric for $f > 250$ MHz. At sufficiently high frequencies even a metal becomes a dielectric. The definition is frequency-dependent, and this is not very practical for a general grouping of materials. In any case, a material may be classified as a dielectric if it has an ability to *store energy* capacitively, not just to dissipate it.

Thus tissue may be regarded as a conductor or a dielectric, the choice is ours. An electrolytic conductor is characterised by immittance, a dielectric by permittivity or capacitance. However, as we shall see, the conductivity may be complex and thus take care of a capacitive component as well, and permittivity and capacitance may be complex and also take care of conductance. Muscle tissue is more a conductor with certain capacitive properties, stratum corneum is more a dielectric with certain conductive properties. We have a situation that is confusing for users of bioimpedance data such as medical doctors. Some bioimpedance groups characterise tissue by *immittance* terminology, others by *dielectric* terminology. A dermatologist is confronted with skin characterised by a number of parameters such as capacitance, permittivity, impedivity, impedance, resistivity, resistance, reactance, admittivity, admittance, conductivity, conductance or susceptance. These are different, but possibly correct, ways of describing a biological material.

There have been two mainstream traditions in the development of dielectric theories for biomaterials: the Debye tradition of regarding biomolecules as polar materials, and the even older tradition of regarding them as inhomogeneous materials with important interfacial polarisation contributions. As the dielectric theory for polar biomaterials can not explain many of the experimental findings, interfacial polarisation theory has come increasingly into focus.

3.1. Polarisation in a Uniform Dielectric

Polarisation is a difficult concept, widely but variously used. It is a key concept in understanding the electrical properties of tissue, because it covers some very important characteristic phenomena. We will start with the most general definition of polarisation, and then differentiate.

> Polarisation is the electric field-induced disturbance of the charge distribution in a region.

Generally we have to *apply energy* in order to polarise a system from the outside, for instance by electromagnetic radiation, or from an externally applied electric field. The exogenic energy may be stored or dissipated in the dielectric. The membrane of a living cell is polarised because the energy-consuming sodium–potassium pump transfers ions across the membrane so that the cell interior is negatively charged with respect to the extracellular fluid. This is an *endogenic* polarisation mechanism in living tissue.

All materials are polarisable, but vacuum is not (though it can hardly be called a material). With only bound ions, an electric field displaces the charges only small distances, dipoles are formed, and the material is polarised. However, with free ions there are important additional effects from the migration of these ions in an electric field. With free ions, their migration in the *bulk* of a homogeneous electrolyte implies *no change in the local charge distribution* (equal amounts of charge entering and leaving a volume), and therefore no polarisation strictly speaking according to the definition above.

In electrochemistry an electrolytic cell may be said to be polarised if a dc current flows through the cell. In potentiometry there is zero dc current, and the recording electrodes are not considered to be polarised.

3.1.1. Coulomb's law and static electric fields

Coulomb's law governs the mechanical force **F** between two charges q_1 and q_2 at distance L:

$$F = q_1q_2/4\pi L^2\varepsilon_s, \qquad \mathbf{F} = q_1q_2\mathbf{L}/4\pi L^3\varepsilon_s \tag{3.1}$$

Coulomb's law is a most fundamental law in electrostatics, and it is *empirical*. The force will try to unite charges of opposite sign (even if other forces hinder it at the closest distance in an atom). ε_s is the static *permittivity*, expressing a property of the material surrounding the charges. Note that the *coulomb force values are smaller the larger the permittivity of the medium*. If the medium is water with the large relative permittivity of around 80, the forces are relatively small, and water will tend to break up (dissociate) solute molecules held together by coulomb forces.

From Coulomb's law the concept of the *electric field* is derived. There is an electric field at a location if a charge there is influenced by a mechanical force **f** (space vector) proportional to the charge:

$$\mathbf{E} = \mathbf{f}/q \tag{3.2}$$

A free *positive* charge migrates in the same direction as the **E**-field. The unit for the electric field strength E is V/m; **E** is the *potential gradient*. **E** defines a vector field not linked to a material and is also defined for vacuum. The work (energy) to bring a charge q from infinity to a location is equal to $q\Phi$, where Φ is the potential at that location. It is usually convenient to chose the potential to be zero at infinity when the charges extend over only a finite region. *A potential difference is work per charge*, 1 volt = 1 joule/coulomb. This is the basis for defining potentials in electrolytic systems and in the electric double layer at electrode or cell membrane surfaces.

The equipotential lines will be perpendicular to the **E** and **J** vectors (isotropic medium), but the potential Φ itself is not a vector and has no direction (although the *change* of Φ has a direction, so that Φ has a *gradient* vector, $\mathbf{E} = -\nabla\Phi$). Vector quantities that are perpendicular at all points in space are treated with what mathematicians call *conformal mapping*.

3.1.2. Permanent and induced dipole moments

Without free charges, e.g. in Teflon or dry NaCl, there will be no dc current flow and no local build-up of free charge at dielectric interfaces, and no electrolytic polarisation, either in the bulk or at the electrodes. However, a local disturbance of the distribution of bound charges will occur. The bound ions of the dielectric can only move (translate or rotate) locally and under strong confinement. *No electric charges leave the electrodes and enter the dielectric.*

Suppose that two charges are equal but of opposite sign and are kept at a small distance and thus prevented from recombining. Such an electric doublet is called a *dipole* (cf. Fig. 5.11). An atom with its electrons at a distance from the positive nucleus does not necessarily form a net dipole: the electrical centre of the electrons may coincide with that of the nucleus. However, every such atom is polarisable, because the electrical centres of the charges will be displaced by an *external electric field*. As positive and negative charges move in opposite directions, *dipoles* are formed (*induced*) and the material is *polarised*. In addition, molecules may form *permanent* dipoles, which of course will also be influenced by an externally applied electric field. A polyatomic molecule is non-polar if it fulfils certain symmetry criteria. Water molecules are asymmetrical and therefore polar.

In electrostatic theory a dipole is characterised by its electrical *dipole moment*, the space vector **p**:

$$\mathbf{p} = q\mathbf{L} \qquad (3.3)$$

The SI unit for p is coulomb meter (C m); the *Debye unit* ($D = 3.34 \times 10^{-30}$ C m) is also used. A pair of unit charges $+e$ and $-e$ held at a distance of 0.1 nm has a dipole moment of 4.8 D; a water molecule has a *permanent* dipole moment of about 1.8 D.

The dipole moment **p** may be the resultant dipole moment of a molecule, of many molecules or of a whole region. Polarisation **P** (in C m/m^3 = C/m^2) is the electrical dipole moment *per unit volume* (dipole moment volume density). It is therefore a more macroscopic concept than the dipole moment of molecules or atoms. **P** is a space vector, for isotropic materials it has the same direction as the **E**-vector:

$$\begin{aligned}
\mathbf{P} &= \mathbf{D} - \varepsilon_0\mathbf{E} = (\varepsilon_r - 1)\varepsilon_0\mathbf{E} = \chi\varepsilon_0\mathbf{E} \\
\mathbf{D} &= \varepsilon\mathbf{E} = \varepsilon_0\mathbf{E} + \mathbf{P}
\end{aligned} \qquad (3.4)$$

D (either as a surface charge density q_s (C/m^2), or dipole moment volume density C m/m^3) is called the *electric flux density*, or *displacement*. ε_r is called the *relative permittivity* (*dielectric constant*) of the material.

Equation (3.4) shows the relationship between polarisation and permittivity. High polarisation means high permittivity; $E = 0$ means $P = 0$ and no polarisation. Permittivity is a measure of the amount of dipole moment density created (induced) by an electric field. Bound charges that can be displaced a long distance L cause a larger dipole moment (eq. 3.3), resulting in a higher permittivity. A substance in the liquid state is therefore more strongly polarised than in the solid state: ice has lower permittivity than water. Tissue components such as proteins are characterised by particularly long distances between charges, and can thus have a surprisingly high permittivity. Polarisation effects are actually net effects of a constantly changing charge distribution; the degree of disorder increases with temperature according to the Boltzmann factor kT.

Polarisation P cannot itself be measured. Dielectric theory is therefore invariably linked with the concept of a capacitor formed by two plates with a dielectric between them (Fig. 3.1). The capacitance of this capacitor can be measured, and the polarisation *calculated*.

Some molecules are *polar*, e.g. water and many proteins. The *permanent* dipole moments are oriented at random, but with an externally applied electric field they reorient statistically. *Induced* dipoles have the direction of the applied E-field, but if the medium is polar, the polarising E-field also *rotates the permanent dipoles already in existence* (orientational polarisation). Polar materials have a large permittivity in addition to the permittivity caused by the induced dipoles of the non-polar

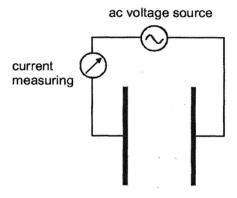

Figure 3.1. The basic capacitor experiment.

molecules present. The permanent dipoles will experience a rotational force, defined by the *torque* τ:

$$\tau = \mathbf{p} \times \mathbf{E} \qquad (3.5)$$

If the charges are bound, the torque will result in a limited rotation. The positions and angles are statistically distributed with less ordering the higher the temperature. As we shall see in Section 8.12.2, in a *rotational* E-field the *induced* dipoles may lag the external field, and thus also give rise to a torque and continuous rotation.

In some polar media the dipole moments are "frozen" in positions that statistically results in a net component in a given direction. They are called *electrets*, and they have a net internal **P**-field in the absence of an externally applied field. An electret is the electrical equivalent of a permanent magnet. In the form of a thin sheet, the resultant equivalent surface charge density may be of the order of 50×10^{-3} C/m^2. The polarisation will slowly diminish, but the process may take many years.

The polarisation vector **P** is composed of all three components:

1. An external field produces new dipoles (induced dipoles).
2. An external field orients the permanent dipoles already there.
3. Without an external field electrets have a permanent net dipole moment.

$\chi = \varepsilon_r - 1$ is called the *electric susceptibility*[14] of a material. It is a macroscopical parameter and is correlated with a microscopical factor called the *polarisability* α (in Cm^2/V) so that $\mathbf{p} = \alpha \mathbf{E}_L$, where $\mathbf{E}_L$ is the local electric field strength. Then macroscopically $\mathbf{P} = N\alpha E$, where N is the volume density of atoms or molecules. In

---

[14] Not to be confused with susceptivity in the relation: admittivity, conductivity and susceptivity.

non-polar media there is a simple relationship between the polarisation and the molecular structure, the *Clausius–Mossotti* equation:

$$\frac{\varepsilon_r - 1}{\varepsilon_r + 2} = \frac{N\alpha}{3\varepsilon_0} \tag{3.6}$$

Equation (3.6) can be extended to comprise also the contribution from polar molecules, the *Debye* equation:

$$\frac{\varepsilon_r - 1}{\varepsilon_r + 2} = \frac{N_A \left(\alpha + \dfrac{p^2}{3kT} \right)}{3v_m} \tag{3.7}$$

where v_m is the molar volume (N_A/N), and N_A is Avogadros constant. The kT factor is due to the statistical distribution of polar molecules causing the orientational polarisation. Equations (3.6) and (3.7) are in best agreement for gases, in less agreement for liquids and least for solids. For the last-mentioned cases, Onsager and Kirkwood have extended the theory.

Polarisation linked with bound charges is usually divided into three components; the total polarisation is the sum of the three.

1. *Electronic polarisation*: Dipole moment density as a result of very small translational displacements of the electronic cloud with respect to the nucleus, whether in single atoms or in molecules. Displacement of electrons is very fast (picosecond), and the dispersion is in the gigahertz region. Induced dipoles will have the direction of the local E-field, and undergo no rotation.
2. *Orientational polarisation*: Rotational movement caused by the torque experienced by *permanent* dipoles in an electric field. The effect is seen only in polar materials.
3. *Ionic polarisation*: This is displacement of positive ions with respect to negative ions. To clarify the division between electronic and ionic polarisation: electronic polarisation is the displacement of the electron cloud with respect to the nucleus; ionic polarisation is the displacement of ions relative to each other. The hydrated sheath around an ion at rest is symmetrical (though the ion is not really at rest: everything is bumping around at room temperature and we are speaking statistically). When current flows, the sheath will lag behind the migrating ion and will no longer be symmetrical to the ion. This is local polarisation of charges bound to each other.

A molecule with a non-zero net charge will experience a translational force and therefore try to migrate in a *homogeneous* electric field. A dipole with net charge of zero will also experience a translational force, but *only* in a *non-homogeneous* field.

3.1.3. Charge-to-dipole and dipole-to-dipole interactions

Equation (3.3) defines the electrostatic dipole moment $\mathbf{p} = q\mathbf{L}$. The potential field Φ from such a dipole is given by

$$\Phi = \mathbf{p} \cdot \mathbf{r}/4\pi\varepsilon_s r^3 \tag{3.8}$$

Table 3.1. Energy dependence of interactions

Interaction	Distance dependence of $\hat{E}$
Ion–ion	$1/r$
Ion–dipole	$1/r^2$
Dipole–dipole	$1/r^3$
Dipole–dipole (rotating randomly)	$1/r^6$

where **r** is the vector distance from the dipole to the point where the potential is to be defined. This is visualised in Fig. 5.11. According to eq. (3.8) the potential falls off as $1/r^2$ in radial direction. Then the E-field falls off as $1/r^3$.

In a molecule there may be local dipoles and single charges in the sense that the countercharge is far away. Electrostatic forces will occur that are basically of a charge-to-dipole or dipole-to-dipole nature. These forces will be intermolecular and intramolecular and will therefore influence the geometrical form of the molecule.

The static energy $\hat{E}$ related to a point charge q at a distance r in perpendicular direction from an ideal dipole is

$$\hat{E} = -pq/4\pi\varepsilon_s r^2 \tag{3.9}$$

Table 3.1 shows the distance dependence of the static energy of the different interactions. Note that the energy falls off more rapidly when two dipoles rotate at random than under stationary non-thermal conditions.

3.2. The Basic Capacitor Experiment

Set up

In Chapter 2 we treated the electrolytic cell. Now we introduce a similar cell: the capacitor with two plates. The substance between the electrode plates was termed an electrolyte (in a solution); the substance between the plates of a capacitor we call a *dielectric* (Fig. 3.1). In one sense the difference between Fig. 2.1 and Fig. 3.1 is not very great. In Fig. 2.1 we considered a liquid between metal plates, and we studied dc conductance. The material between the plates had a *high* volume density of free charges. In Fig. 3.1 we consider a solid material with a low volume density of *free* charges (but many bound charges) between the metal plates. Both dc and ac current are measured, with a phase-sensitive volt/ammeter or displayed on a 2-channel oscilloscope.

Findings

Applying a dc voltage of even 10–20 volts does not lead to a dc current. Applying an ac sinusoidal voltage leads to an ac sinusoidal current, and the current is proportional not only to voltage amplitude but also to the signal frequency. The current leads the voltage by almost 90°. As long as this is the case, even large currents do not heat up the dielectric.

Discussion

A capacitive current is an ac current. It is not an electron current of free charges passing between the plates through the dielectric. In the electrode wires it is a true current of electrons charging the capacitor plates, but without electrons leaving or entering the dielectric. There is a simple relationship between the charge (Q, coulomb) on the plates, the voltage (V, volt) between the plates, and the capacitance (C, farad) of the capacitor:

$$Q = VC \qquad (3.10)$$

The capacitance C (farad) of the capacitor is dependent on the plate area A (m^2), the distance between the plates d (m), and a property of the dielectric between the plates already defined in Section 3.1.1 as static *permittivity* ε_s (farad/m):

$$C = (A/d)\varepsilon_s \qquad (3.11)$$

If we apply a sinusoidal ac voltage to the plates as shown in Fig. 3.1, the ac current flowing as electrons in the capacitor metal wires will not have its maximum current value at the same time as the voltage maximum occurs. The current is *phase shifted* in time. We measure the phase shift in degrees, defining one complete cycle of the ac sinusoidal voltage as 360 degrees (°) or 2π radians. An ideal, steady-state sinusoidal capacitive current is 90° *ahead* of the sinusoidal voltage. It is possible with suitable electronic circuitry to measure separately the *in-phase* (conductive) current (or voltage) component, and the 90° out of phase (*quadrature*) (capacitive) current (or voltage) component. It is not sufficient to characterise an ac current with only one number, the amplitude. The current must be characterised by *two* numbers, with the additional number giving information about the phase. The sum of two equal-amplitude sine waves is double the amplitude if they are in phase, but constant and equal to zero if they are 180° out of phase.

The electric flux density **D** introduced in Section 3.1.2 is partially due to a displacement of *bound* charges in the bulk of the dielectric. At the capacitor plates, D is also the charge per plate area: q/A, but there it is related to the *free* (electron) charge flow corresponding to the measured current in the capacitor leads. The status in the volume of the dielectric is not directly observable from the capacitor plates. For the capacitor, the effect of the whole dielectric volume can be reduced to the two layers of bound charges at the surface of the dielectric, each (almost) in contact with the metal plates. These charged layers are "the ends" (this must not be taken too literally, e.g. the charge separation at ordinary field strengths with electronic polarisation is very, very small and less than the electron diameter) of the surface dipoles in the dielectric, induced by the external field, but bound in the dielectric.

The E-field strength in the dielectric is reduced by the bound charges at the surface of the dielectric. The higher the permittivity, the lower the internal E-field. The E-field is discontinuous at the dielectric surface. The electric flux density **D** in the dielectric will be the same irrespective of the permittivity of the dielectric (eq. 3.4): the magnitude of the **E**-vector is reduced and the **P**-vector increased. The **D**-field is continuous at the dielectric surface as long as there are no free charges there.

3.3. Complex Permittivity

Suppose we have a capacitor with a non-ideal dielectric between the two plates, the plate area $= A$, the plate distance $= d$ (Fig. 3.1). The *admittance* **Y** of the capacitor is (sinusoidal ac, and ideal conditions with no edge effects)

$$\mathbf{Y} = G + j\omega C = (A/d)(\sigma' + j\omega\varepsilon') \qquad (3.12)$$

Different uses of the symbol ε are found in the literature. Often it is not clear whether it represents the relative permittivity ε_r (without dimension, often called the *dielectric constant*) or (as in this book) the complete expression $\varepsilon_r \varepsilon_0$. In addition, it is often not clear whether ε is scalar or complex. If σ or ε are complex they are printed in bold type in this book: $\boldsymbol{\varepsilon}$ or $\boldsymbol{\sigma}$. If the parameters are not printed in bold, there may be an ambiguity: Y, Z, ε and σ may mean $|\mathbf{Y}|$, $|\mathbf{Z}|$, $|\boldsymbol{\varepsilon}|$ and $|\boldsymbol{\sigma}|$, or the real values Y', Z', ε' and σ'. The literature is not consistent: sometimes the meaning is that of $\sigma = |\boldsymbol{\sigma}|$, but sometimes that of $\sigma = \sigma'$.

Equation (3.12) shows that the basic dielectric capacitor model is a parallel model, an admittance model. The *impedance* of a capacitor as shown in Fig. 3.1 is not a parameter of choice, because the conductive parts are physically in parallel with the capacitance (even if the dielectric loss part can hardly be defined as being in series or in parallel).

As seen from Table 3.2, the permittivity of water is reduced by the addition of electrolytes. This *dielectric decrement* $\Delta\varepsilon_r$ for 1 mol/L concentration is -17 for H^+, -8 for Na^+ and K^+, -3 for Cl^- and -13 for OH^-. It is directly related to the hydration of the ion, because these water molecules are more tightly bound and therefore not so easily polarised as free water molecules. The number of hydrated water molecules around a monovalent ion is of the order of 6 (cf. Table 2.4).

Complex permittivity and conductivity
It may be useful to treat σ and ε as complex quantities in the time domain, e.g. in order to incorporate dielectric losses. We then define

$$\boldsymbol{\sigma} \equiv \sigma' + j\sigma'' \qquad (3.13)$$

$$\boldsymbol{\varepsilon} \equiv \varepsilon' - j\varepsilon'' \equiv (\varepsilon'_r - j\varepsilon''_r)\varepsilon_0 \qquad (3.14)$$

Accordingly, eq. (3.12) is written: $\mathbf{Y} = (A/d)(\sigma' + j\omega\varepsilon')$. Tissue data is often given with these ε' and σ' parameters, but may be given with either *only* complex conductivity or complex permittivity. All three forms are in common use.

Table 3.2. Relative permittivity at different frequencies

	10^2 (Hz)	10^6 (Hz)	10^{10} (Hz)
Gases			
Air (100 kPa)	1.00054		
Liquids			
Carbon tetrachloride	2.2	2.2	2.2
Benzene	2.28	2.28	2.28
Olive oil	3.1		
Chloroform	4.8		
Ethanol	25.7		
Water (0°C)	87.7		
Water (20°C)	80.1		
Water (25°C)	78.5	78.5	65
Water (25°C, KCl 0.5 mol/L)	75	75	
Water (37°C)	74.3		
Hydrogen cyanide	116		
Solids			
Teflon	2.1	2.1	2.1
Polystyrene	2.6	2.6	2.6
Plexiglas	3.1	2.76	
Ice		4.15	3.2
Glass (Pyrex)	5	4.8	4.8
NaCl		5.9	5.9
KCl	5		
AgCl	11		

Complex permittivity

Complex permittivity is used when it is wished to consider the material as a dielectric (an insulator) with losses. The capacitor is characterised with a complex capacitance or a complex permittivity:

$$\mathbf{Y} = (A/d)(\sigma' + j\omega\varepsilon') = j\omega\mathbf{C} = j\omega(A/d)\mathbf{\varepsilon} = j\omega(A/d)(\varepsilon' - j\varepsilon'') = (A/d)(\omega\varepsilon'' + j\omega\varepsilon')$$

(3.15)

Therefore

$$\varepsilon'' = \sigma'/\omega$$

(3.16)

Notice that ε'' and σ' have positive values, but due to the minus sign in eq. (3.14) the complex conjugate of $\mathbf{\varepsilon}$ is often used in the Wessel diagram. σ' is proportional to energy loss per *second* (power loss), ε'' to energy loss per *cycle* (*period*). Frequency-independent dc conductivity σ_{dc} implies constant energy loss per second and therefore an energy loss per cycle inversely proportional to frequency (eq. 3.16). σ' is related to the slow temperature rise in a material over many cycles (cf. Section 3.4.4). ε'' is related to processes which occur within one cycle.

Complex conductivity

Complex conductivity is used when it is wished to consider the material as a conductor with some capacitive properties. The medium is characterised with a complex conductivity:

$$\mathbf{Y} = (A/d)(\sigma' + j\omega\varepsilon') = \mathbf{G} = (A/d)\sigma = (A/d)(\sigma' + j\sigma'') \qquad (3.17)$$

Usually G is a real number in the equation $\mathbf{Y} = G + jB$, but now G is considered complex $= \mathbf{G} = \mathbf{Y}$, thus taking care also of B. Then $\sigma'' = \omega\varepsilon'$, diverging as $f \to \infty$ (in contrast to ε').

Complex resistivity

Resistivity ρ is the inverse of conductivity and the unit is ohm-meter (Ω m). For capacitive biomaterials the complex resistivity is

$$\rho = \rho' - j\rho'' = 1/\sigma = (\sigma' - j\sigma'')/|\sigma|^2, \ \rho' = \sigma'/|\sigma|^2 \ \text{and} \ \rho'' = \sigma''/|\sigma|^2 \quad (3.18)$$

Here also it must be realised that resistivity is linked with impedance and the series equivalent model, whilst conductivity is linked with admittance and the parallel equivalent model. Pay attention to the fact that, for example, $\rho' \neq 1/\sigma'$, but $\rho = 1/\sigma$ in the sense $|\rho| = 1/|\sigma|$!

Losses

$\varepsilon_r'' = \sigma'/(\varepsilon_0\omega)$ is sometimes called the loss *factor*. The loss *angle* δ of a capacitor is defined so that the ideal capacitor with zero losses also has zero loss angle. This means that $\delta = 90° - \varphi$:

$$\varphi = \arctan(\varepsilon'/\varepsilon'')$$
$$\delta = \arctan(\varepsilon''/\varepsilon') = \text{arccot}(\varepsilon'/\varepsilon'') \qquad (3.19)$$

Tan δ is also called the loss *tangent* or *dissipation factor*. Tan δ is energy lost per cycle divided by energy stored per cycle (rms or peak values).

Care must therefore be taken with regard to the term "phase angle". In ordinary immittance texts it is always the tangent of the out-of-phase component divided by the in-phase component. However, in many classical presentations the loss angle is used instead (Schwan (1963), symbol δ; Fricke (1932), symbol Ψ) for characterising the polarisation properties of an electrode. Also, in the classical Warburg (1899) paper, the angle Ψ is the loss angle. Sometimes the loss angle is called the loss angle, sometimes it is called the phase angle.

Modulus function

It may be useful to have a parameter for the inverse of the permittivity. This is the *modulus function* **M**:

$$\mathbf{M} = 1/\varepsilon = M' + jM'' \qquad (3.20)$$

Relationship between conductivity, resistivity and permittivity
The quantities σ and ε are scalars. σ', σ'', ε' and ε'' are components of the complex quantities $\boldsymbol{\sigma}$ and $\boldsymbol{\varepsilon}$. Notice the ambiguity that σ may mean $|\boldsymbol{\sigma}|$ or σ', which are not the same. Their somewhat bewildering relationships are

$$
\begin{aligned}
&\boldsymbol{\sigma} = \mathrm{j}\omega\boldsymbol{\varepsilon} \\
&\sigma' = \omega\varepsilon'' = G(d/A) && \sigma'' = \omega\varepsilon' = \omega C(d/A) = B(d/A) \\
&\varepsilon' = \sigma''/\omega = Cd/A = (B/\omega)(d/A) && \varepsilon'' = \sigma'/\omega = (G/\omega)(d/A) \\
&\rho' = \sigma'/|\boldsymbol{\sigma}|^2 && \rho'' = \sigma''/|\boldsymbol{\sigma}|^2
\end{aligned} \tag{3.21}
$$

If the capacitance of the empty cell is C_e (no dielectric, pure capacitance), it is possible to express some interesting relationships:

$$
\mathbf{Y} = \mathrm{j}\omega C_e \varepsilon_r, \qquad \mathbf{M} = \mathrm{j}\omega C_e \mathbf{Z}, \qquad \mathbf{M} = \mathrm{j}\omega C_e / \mathbf{Y} \tag{3.22}
$$

Surface immittivity
Usually a material is characterised with variables for volume properties. However, surface properties of a membrane or a double layer may also be of interest (w = sample width, Fig. 3.2):

$$
\begin{aligned}
&\textit{Volume} \text{ conductivity} && \sigma = GL/A && \text{(S/m)} \\
&\textit{Volume} \text{ resistivity} && \rho = RA/L && \text{(}\Omega\text{m)} \\
&\textit{Surface} \text{ (lateral) conductivity} && \sigma_{\mathrm{sul}} = GL/w && \text{(S)} \\
&\textit{Surface} \text{ (lateral) resistivity} && \rho_{\mathrm{sul}} = Rw/L && \text{(}\Omega\text{)} \\
&\textit{Surface} \text{ perpend. conductivity} && \sigma_{\mathrm{sup}} = G/Lw && \text{(S/m}^2\text{)} \\
&\textit{Surface} \text{ perpend. resistivity} && \rho_{\mathrm{sup}} = RLw && \text{(}\Omega\text{m}^2\text{)}
\end{aligned} \tag{3.23}
$$

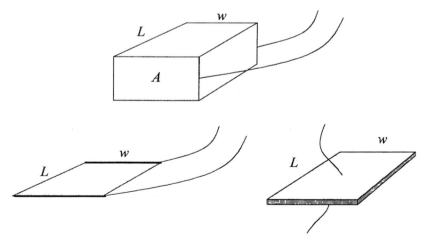

Figure 3.2. Top: volume immittivity. Bottom left: surface lateral immittivity, right: surface perpendicular immittivity.

Instead of conductivity, the definitions may be extended to *immittivity* or *specific immittance* in the form of *impedivity* (specific impedance) and *admittivity* (specific admittance), e.g.:

Volume admittivity	$\kappa = (G + j\omega C)\, L/A$	(S/m)
Surface (lateral) admittivity	$\kappa_{su} = (G + j\omega C)\, L/w$	(S)

$$(3.24)$$

Note that the surface lateral parameters are given in siemens or ohm, as L/w is dimensionless. The surface parameters are pseudomaterial constants because they characterise a given surface or membrane presupposing a constant given thickness (e.g. cell membranes always being 7 nm thick), and with negligible influence e.g. from a substrate underneath. If these conditions are not fulfilled, surface parameters should be used with care.

3.4. AC Polarisation and Relaxation in a Uniform Dielectric

3.4.1. Relaxation and dispersion

In the section about polarisation we considered polarisation as a static state of a material, not dependent on time but only on the externally applied field. In an electric field of time varying strength or direction, the charge positions were considered to be in time phase with the instantaneous values of the applied field. However, polarisation and the displacement of charges in a material do not occur instantaneously. If the measuring frequency is low enough so that all charges are allowed the necessary time to change their position, polarisation is maximal. But with increasing frequency the polarisation and permittivity will decrease.

This time dependence may be characterised by introducing the concept of *relaxation*. It was first used by Maxwell in connection with the elastic forces in gases. Later Debye used it referring to the time required for dipolar molecules to orient themselves. Instead of applying a sinusoidal ac measuring signal and measuring, for example, admittance and phase shift, the concept of relaxation is linked with a *step function* excitation signal. After such a step has disturbed the system under investigation, the system is allowed to *relax* to the new equilibrium, this is the relaxation process. Relaxation occurs in the *time* domain, after a step increase or decrease in the *E*-field strength. It is described by the parameter of *relaxation time*, different from the parameter of immittance and sine wave excitation.

It is often stated that all electrical properties of biological materials are due to relaxation phenomena. However, the concept of relaxation is linked with a delayed response and is not so meaningful with frequency-independent dc conductance: the conductance is constant with time and does not relax.

Relaxation theory often does not include resonance phenomena, as these are usually not found in macroscopic tissue samples in the frequency range from μHz to MHz.

Relaxation time is dependent on the polarisation mechanism. Electronic polarisation is the fastest mechanism, with relaxation in the higher MHz and GHz region.

Large organic molecules like proteins may have a particular large permanent dipole moment because of the large distance L between the charges. Because they are so large and with a complex bonding, the rotation and twist can be relatively slow. However, interfacial relaxation may be the most important process, with the longest mean relaxation times of the order of seconds.

Dispersion (frequency dependence according to the laws of relaxation) is the corresponding frequency domain concept of relaxation: permittivity as a function of frequency. Even if the concept of relaxation is linked with step functions, it can of course also be studied with sine waves. An ideal step function contains all frequencies, and dispersion can be analysed with a step function followed by a frequency (Fourier) analysis of the response signal, or with a sinusoidal signal of varying frequency.

As we shall see in the next subsection, in a simple case of a *single dispersion* with a single relaxation time constant there will be one permittivity level at low frequencies (time for complete relaxation), and a second *lower* level at higher frequencies (not sufficient time for the relaxation process studied). It will be a transition zone characterised by a frequency window with a characteristic centre frequency. There-fore, dispersion in relaxation theory often has a somewhat more precise meaning than just frequency dependence. *Simple dispersions are characterised by a permit-tivity with two different frequency-independent levels and a transition zone around the characteristic relaxation frequency.* In biomaterials these levels may be more or less pronounced.

3.4.2. Debye relaxation model

Let us assume a dielectric with only bound charges placed between two capacitor plates, and that a step increase of dc voltage is applied at $t = 0$. Let us assume that the material has only one relaxation process with a single characteristic time constant, and that the polarisation increases according to an exponential curve as a function of time. This is called *a Debye single dispersion*. As a result of the polarisation in the dielectric, the surface charge density $D(t)$ ($= q_s$) at the capacitor plates will increase from one value (D_∞) to another (D_0) as shown in Fig. 3.3 and according to the equation

$$D(t) = D_\infty + (D_0 - D_\infty)(1 - e^{-t/\tau}) \qquad (3.25)$$

The subscripts refer to frequency, a sine wave parameter. D_∞ is the surface charge density at $t = 0+$, that is after the step but so early that only apparently instantaneous polarisation mechanisms have come to effect (high frequency, e.g. electronic polarisation). The capacitor charging current value at $t = 0$ is infinite. D_0 is the charge density after so long a time that the new equilibrium has been obtained and the charging current has become zero. With a single Debye dispersion this low frequency value is called the *static* value (cf. Section 6.2.1). τ is the time constant of the relaxation process.

By Laplace transforming eq. (3.25) it is possible to find the response in the

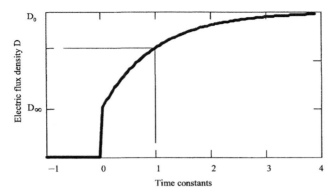

Figure 3.3. Electric flux density in the capacitor dielectric with an applied voltage step.

frequency domain. With $\varepsilon' = D/E$, and $C = \varepsilon' A/d$, it is possible to show that (*Debye single dispersion equation*)

$$\varepsilon(\omega) = \varepsilon'_\infty + \Delta\varepsilon' / (1 + j\omega\tau), \quad \text{where} \quad \Delta\varepsilon' = \varepsilon'_s - \varepsilon'_\infty \qquad (3.26)$$

$$\mathbf{C}(\omega) = C_\infty + \Delta C / (1 + j\omega\tau) \qquad (3.27)$$

The subscript s means "static", since ε_0 is reserved for the permittivity of vacuum. Here ε and $\mathbf{C}$ are vectors in the time domain. We have already seen the ambiguity that at ε may mean $|\varepsilon|$ or ε'. In dispersion theory this is less a problem, because the parameters used are at frequency extremes, where the values are real: $\varepsilon_s = \varepsilon'_s$ and $\varepsilon_\infty = \varepsilon'_\infty$.

At low frequencies we measure a frequency-independent capacitance $C_s = C_\infty + \Delta C$. The frequency must be low enough to guarantee that the polarisation process can follow. In a sufficiently higher frequency range we measure another frequency-independent capacitance C_∞, lower than C_s. The frequency must be sufficiently high that the polarisation process in question can not follow.

The best equivalent circuit for such behaviour is shown in Fig. 3.4. Figure 3.4 is the circuit of Fig. 7.17 in Sections 7.2.4 and Fig. 10.6 (eqs 10.26 to 10.30). With $\tau = R\Delta C$, the complex admittance and capacitance of this equivalent circuit are

$$\begin{aligned}\mathbf{Y} &= j\omega\mathbf{C} = j\omega C_\infty + j\omega\,\Delta C/(1 + j\omega\tau) \\ \mathbf{C} &= C_\infty + \Delta C / (1 + j\omega\tau)\end{aligned} \qquad (3.28)$$

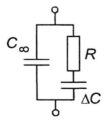

Figure 3.4. Equivalent circuit for a Debye single dispersion. Ideal components.

The *quadrature* components of eq. (3.28) are

$$Y'' = \omega C_\infty + \omega \,\Delta C/(1 + \omega^2 \tau^2)$$
$$C' = C_\infty + \Delta C/(1 + \omega^2 \tau^2) \tag{3.29}$$
$$\varepsilon' = \varepsilon'_\infty + \Delta \varepsilon' \,/\, (1 + \omega^2 \tau^2)$$

The *in-phase* loss components are

$$Y' = \Delta C \tau \,\omega^2/(1 + \omega^2 \tau^2)$$
$$C'' = \Delta C \tau \,\omega/(1 + \omega^2 \tau^2) \tag{3.30}$$
$$\varepsilon'' = \Delta \varepsilon' \tau \,\omega/(1 + \omega^2 \tau^2)$$

The quadrature component values dependence is decreasing with increasing frequency; the in-phase component goes through a maximum. C_∞ and ΔC are ideal capacitors. At very high frequencies C_∞ dominates the total admittance and capacitance. At very low frequencies R is negligible with respect to the impedance of ΔC, and the two capacitors C_∞ and ΔC are effectively in parallel. So both at high and low frequencies the circuit is purely capacitive. The *characteristic relaxation frequency* ω_c is the frequency corresponding to $\omega \tau = 1$ and the maximum value of ε''. The characteristic relaxation time constant is $\tau = R \Delta C$, and therefore $\omega_c = 2\pi/\tau$. The value of ε' and maximum ε'' are found from eqs (3.29) and (3.30):

$$\varepsilon_{\omega\tau = 1}{}' = \varepsilon'_\infty + \Delta \varepsilon'/2$$
$$\varepsilon_{\omega\tau = 1}{}'' = \Delta \varepsilon'/2 \tag{3.31}$$

The maximum value of ε'' is therefore not related to the resistor R, but R determines at what characteristic frequency the maximum value will occur.

The power loss W_L in the circuit is (constant amplitude sinusoidal *voltage* = v) (cf. eq. 10.1)

$$W_L = v^2 Y' = v^2 \,\Delta C \omega^2 \tau \,/(1 + \omega^2 \tau^2) \tag{3.32}$$

The power loss W_L (constant amplitude sinusoidal *current* = i) is

$$W_L = i^2 Z' = i^2 \,\Delta C \omega \tau/[(C_\infty + \Delta C)^2 + C_\infty^2 \omega^2 \tau^2] \tag{3.33}$$

The power loss and heat dissipation only occur in the resistor, the two capacitors are ideal components. The frequency dependence of the power loss is dependent on how the circuit is driven. With constant amplitude voltage, the power loss goes from zero level at very low frequencies to a defined level v^2/R at high frequencies (like σ', see below). With constant amplitude current, the power loss goes from zero level at very low frequencies, through a maximum at the frequency determined by the time constant $\tau = R \,\Delta C$, and back to zero (like ε'', see below).

Calculating the complex *conductivity* σ, the difference between high- and low-frequency conductance $\Delta \sigma' \, A/d$ must be $1/R$. $Y = (A/d)(\sigma' + j\sigma'')$, and this must be compared with Y' (eq.3.30), and Y'' (eq.3.29):

$$\sigma' = \Delta \sigma' \,\omega^2 \tau^2 \,/(1 + \omega^2 \tau^2) \qquad \text{in-phase (lossy) component}$$
$$\sigma'' = \omega C_\infty + \Delta \sigma' \,\omega \tau/(1 + \omega^2 \tau^2) \qquad \text{quadrature (capacitive) component} \tag{3.34}$$

The in-phase component σ' dependence with increasing frequency is from zero to a finite level, in contrast to ε'', which returns to zero at high frequencies. The last term of the quadrature component equation goes through a maximum, but the first term is proportional to frequency and diverges, just like a capacitive susceptance $B = \omega C$. It is not very logical to characterise a capacitive material with conductivity variables, and some training is necessary for interpreting quadrature conductivity data for dielectrics. $\Delta\sigma'$ is a real number associated with the ideal conductor R. $\Delta\sigma'$ is the parameter used in eq. (3.34), because at very high and low frequencies σ' is real.

Control equation
There is an interesting and useful link between three of the variables of eqs (3.29), (3.30) and (3.34). If $\tau = R\Delta C = R\Delta\varepsilon'\, A/d$ and $\Delta\sigma'\, A/d = 1/R$, the scalar relationship between $\Delta\varepsilon'$, τ and $\Delta\sigma'$ is very simple:

$$\Delta\varepsilon' = \tau\,\Delta\sigma' = \frac{\Delta\sigma'}{\omega_c} \qquad (3.35)$$

where ω_c is characteristic relaxation frequency. This equation is valid only if there is no dc conductance in parallel with the equivalent circuit and C_∞. If there is, that part must first be subtracted. σ' is limited to the ac lossy part of the dielectric.

Equation (3.35) represents an efficient tool for control of measurement results, e.g. of the data presented in Figs 3.5 to 3.8. It is a special case of the more general Kramers–Kronig transform, interconnecting, e.g., permittivity and losses as demonstrated in Fig. 3.7 and described in Section 6.1.6.

If there is an ionic dc conductance path in parallel with C_∞, it is also common practice to add this conductance to some of the equations, even if this does not represent a relaxation mechanism. In the in-phase eq. (3.30) a factor $\sigma_{dc}/\omega\varepsilon_0$ (diverging as $f\to 0$!) is added to ε''. The in-phase component of eq. (3.34) then has the form:

$$\sigma' = \sigma_{dc} + \Delta\sigma'\, \omega^2\tau^2/(1 + \omega^2\tau^2) \qquad (3.36)$$

Example
Figure 3.5 shows data based on the circuit of Fig. 3.4 and eqs (3.29) and (3.30) with chosen values $\varepsilon_{r\infty} = 1000$, $\Delta\varepsilon_r = 9000$, $\tau_c = 1.592$ ms ($f_c = 100$ Hz). With a ratio $A/d = 20$, this corresponds to $C_\infty = 0.18\ \mu F$, $\Delta C = 1.6\ \mu F$ and $R = 1004$ ohm.

Almost 80% of the ε'-dispersion takes place within one frequency decade, more than 98% in two decades. The ε''-dispersion is broader. It is easy to see that the data is purely capacitive at high and low frequencies.

In the Wessel diagram, the locus of the permittivity is a complete semi-circle (Fig. 3.6). It is also easy to see that the data are purely capacitive at high and low frequencies.

The complex conductivity (same component values) has been plotted as a function of $\log f$ in Fig. 3.7. In-phase conductivity σ' increases to the plateau corresponding to $\Delta\sigma = d/RA$.

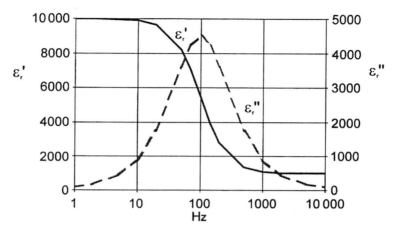

Figure 3.5. Relative permittivity, Debye single dispersion relaxation. Notice that ε_r'' maximum is half the value of $\Delta\varepsilon_r'$. Component values found in the text.

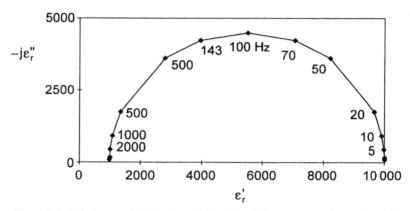

Figure 3.6. Relative permittivity plotted in the Wessel diagram, same data as Fig. 3.5.

Figure 3.8 shows the conductivity σ locus in the Wessel diagram. Here the characteristic frequency is > 100 Hz, and in contrast to the permittivity plot with a complete semicircle locus, there is a strong deviation with the σ'' that diverges proportional to frequency.

In conclusion, the complex conductivity semicircle is disturbed by the parallel capacitor C_∞. In order to obtain a circle this capacitor must be omitted in the *conductivity* Wessel diagram. Correspondingly, a *permittivity* semicircle would have been disturbed by a parallel conductance G. Tissue usually has a large dc conductance from the ions in the tissue liquids. For dc conductance there is no relaxation process. The relaxation equivalent circuit of Fig. 3.4 has no dc conductance, the in-phase component is the lossy part of the relaxation mechanism.

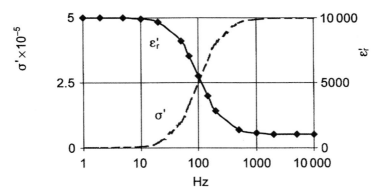

Figure 3.7. Combined permittivity and conductivity plot, same data as for Fig. 3.5. According to eq. (3.35), $\Delta\varepsilon = \tau\,\Delta\sigma$, which is here easily checked to be correct.

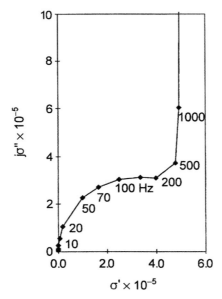

Figure 3.8. Complex conductivity, same data as for Fig. 3.5. σ'' diverges at higher frequencies because of C_∞.

The time constant and characteristic relaxation frequency can be defined in more than one way:

1. From the transient response (eq. 3.25)
2. From the midpoint of the $\Delta\varepsilon$ or $\Delta\sigma$ plotted with linear scales as a function of $\log f$ (Fig. 3.5)
3. From the apex of the circular locus of ε or σ in the Wessel diagram (Fig. 3.6)

Relaxation of permanent dipoles (MHz–GHz region)

The permanent dipoles in a polar dielectric experience a torque due to an applied electric field (cf. eq. 3.5). The actual position of each dipole is determined by this externally created torque and the thermal motion of the dipoles, and the polarisation **P** is found to be (cf. eq. 3.7)

$$\mathbf{P} = Np^2\mathbf{E}_L/3kT \tag{3.37}$$

where $\mathbf{E}_L$ is the local E-field at each dipole.

Viscosity effects in the dielectric are considered to hinder the rotational movement, and Debye has given the relaxation time of this process based on Stoke's law (eq. 2.6):

$$\tau = 4\pi\eta a^3/kT \tag{3.38}$$

The relaxation time for proteins in water is typically in the micro- to picosecond range (MHz–GHz), and the dielectric decrement of the order of ε_0 per gram per litre.

Relaxation and resonance

In the Debye relaxation theory presented, there are no absorption peaks, *no resonance phenomena*. Many techniques in analytical chemistry are based upon resonance phenomena. The prerequisite for resonance is two energy-coupled reservoirs with a periodic energy transfer between them, e.g. the inductance and capacitance of a *LC* circuit. The relaxation process in tissue is due to only one kind of energy reservoir: the capacitance of double layers and membranes. Resonances with only capacitive reservoirs are in principle also possible, but the coupling in tissue usually is not of such a character. At the higher end of the electromagnetic spectrum (optical IR, visible, UV and higher ranges) there are resonance phenomena due to, e.g., vibration of the atoms of a molecule with respect to each other. In the much lower RF range and below, such resonances are rarely found with E-field excitation, and in the relaxation theory presented here energy is simply stored in capacitors or lost in the relaxation process. However, with *magnetic* excitation of tissue in a coil, there is a *nuclear* magnetic resonance and absorption of energy at a characteristic frequency given by the Larmor equation:

$$\omega = \gamma B \tag{3.39}$$

where B is the magnetic field and γ is the gyromagnetic ratio. For protons the resonance frequency is about 42.6 MHz in a field of 1 tesla.

3.4.3. Displacement current density and dielectric loss

With a sinusoidal ac voltage v across a capacitor, the electric field strength in the capacitor's dielectric is $E = v/d$. Since $\mathbf{i} = v\mathbf{Y}$ and $\mathbf{J} = \mathbf{i}/A$, the current density in the dielectric according to eq. (3.17) is

$$\mathbf{J} = E(\sigma' + j\omega\varepsilon') \tag{3.40}$$

J, *A* and *E* are actually *space* vectors but here are treated as scalars so that **J** is the *time* vector (sine waveforms). The *quadrature* component of **J** is not a current density from charges crossing the dielectric. It is a *displacement* current, consisting of a vacuum component and a component due to small displacements of *bound charges* in the dielectric. With $E = 1$ V/m, σ' can be regarded as the current density and ε' the surface charge density on the capacitor plate surface (cf. eq. (3.10): $Q = VC$).

The *in-phase* component of **J** may be a current density from charges crossing the dielectric. But the local displacement of charges in the dielectric is also linked with viscous energy losses, called *dielectric losses*. Electrically this results in a conductance (in-phase current). Although it is an in-phase current it has nothing to do with dc conductance: it is a displacement current. The total ac conductivity σ'_{ac} of a material is therefore composed partly of the dielectric losses associated with the displacement of the *bound charges*, partly of the migration of the *free charges*. However, the dc conductivity σ'_{dc} is due only to the migration of free charges:

$$\sigma'_{ac} = \sigma'_{bound} + \sigma'_{free} \tag{3.41}$$

$$\sigma'_{dc} = \sigma'_{free} \tag{3.42}$$

Ac conductivity due to dielectric losses of bound charges is usually very frequency dependent. σ'_{free} is often considered to be frequency independent.

The current density **J** from eq. (3.40) can, according to eq. (3.41), be written

$$\mathbf{J} = E(\sigma'_{bound} + \sigma'_{free} + j\omega\varepsilon') \tag{3.43}$$

σ' takes care of all the in-phase current density, ε' all the quadrature current density. The equation may also be written

$$\mathbf{J} = \sigma'_{free}E + j\omega(\varepsilon_0 E + P) \tag{3.44}$$

If $\sigma'_{free} = 0$, there is no electron exchange with the plates, but still an ac current density in the dielectric. The last term takes care of all displacement current densities, also the in-phase component of the time vector $j\omega P$. The displacement current density J_d is proportional to frequency, and thus zero at dc. J_d may also be expressed more generally as a time differential equation, valid not just for sine waves but for any waveform:

$$J_d = \frac{\partial D}{\partial t} = \varepsilon_0 \frac{\partial E}{\partial t} + \frac{\partial P}{\partial t} \tag{3.45}$$

The first term is the displacement current in vacuum, not linked to matter at all, and therefore somewhat enigmatic in its nature. The second term is related to matter and is the polarisation current density that results from the displacement of bound charges in the dielectric.

3.4.4. Joule effect and temperature rise

With current flow, an ideal resistor *dissipates* heat energy; the electrical energy is lost. However, an ideal capacitor *stores* electrical energy. In a non-ideal capacitor there are dielectric losses and perhaps losses from dc current. The stored energy may be partially lost, and may be completely lost with time (relaxation). As long as a device, a black box or real tissue, has the ability to store energy, it contains some form of capacitors (or inductors).

Energy is the ability for doing work. Energy is measured in *watt-second* or *joule*, which are the same, or in *electronvolt (eV)*. Power is the rate of doing work or transferring energy, and is measured in *watt*. Current flow through a copper wire is a flow of electric charge. If the flow occurs with negligible voltage drop, the charge undergoes no energy change. There is no voltage difference, $\Delta V = 0$. A potential difference must be defined in order to define the energy of a charge. Power is defined as the product of electric current and voltage difference.

As seen from the outside of the dielectric in the external leads, the instantaneous power W_i delivered from a sinusoidal ac supply to the parallel combination of a capacitor (susceptance B) and a conductor (G), is the instantaneous ac supply voltage multiplied by the instantaneous current in the copper leads:

$$W_i = vi = V_{peak} \sin \omega t \, V_{peak} Y \sin(\omega t + \varphi) \tag{3.46}$$

The instantaneous power W_i is linked with a real electronic current in the copper leads. However, some of this sine wave current is in phase with the applied voltage, and some of it is $90°$ out of phase. The instantaneous in-phase power W_h is the instantaneous voltage multiplied by the instantaneous in-phase current component. The instantaneous out-of-phase power W_r is the instantaneous voltage multiplied with the instantaneous $90°$ out-of-phase (*quadrature*) current. To obtain the average power (rate of energy transfer) of the quadrature current ($\varphi = 90°$) (called *reactive power*) over the period T (one cycle), we integrate the instantaneous power over the period (to give average energy), and divide by the period (to give average power):

$$W_r = \left(\frac{1}{T}\right) \int [V_{peak} \sin \omega t (V_{peak} B) \cos \omega t] \, dt = \left(\frac{1}{T}\right) (V_{peak}^2 B/2\omega)[\sin^2 \omega t]_0^T = 0 \tag{3.47}$$

The *reactive* power W_r is pumped to and from the capacitor each quarter period, but the net supplied energy is zero. The capacitive, quadrature current merely pumps electrons, *charging* the plates. The quadrature current causes no heating of the dielectric, but the current in the leads is real and causes heat losses if the *wires* are non-ideal.

In the same way the power loss W_L is

$$W_L = \left(\frac{1}{T}\right) \int [V_{peak} \sin \omega t \, (V_{peak} G) \sin \omega t] \, dt = \left(\frac{1}{T}\right) (V_{peak}^2 G) \left[\frac{t}{2} - (1/4\omega) \sin 2\omega t\right]_0^T$$

$$= \frac{1}{2} V_{peak}^2 G \tag{3.48}$$

To find a *root-mean-square (rms)* value of a function of time, the function is first squared, then the mean value is taken, and then the root of the mean. The rms value must be used when power and heating effects are of interest ($W = v_{rms}^2/R$). From eq.

(3.48) it is clear that the rms value of a sine wave is $1/\sqrt{2}$ of the peak value. Dealing with other waveforms, the relationships between peak, mean and rms values will depend on the waveform. Many ac voltmeters *display* rms voltage, but actually *measure* the mean value. Such a practice introduces errors for non-sine waveforms.

In the dielectric there is ionic or electronic conduction. In a metallic conductor the free, migrating electrons collide with the lattice of the bound ionised metal atoms, and the electrons transfer their excess energy to the lattice. With electrolytes the charge carriers are ions, and ordinary migration or local displacement is hindered by viscosity-based friction. In both cases the dielectric is heated up and energy is dissipated; this is the *Joule* effect.

The energy $\hat{E}$ (watt-seconds, W s) stored in a capacitor is given by

$$\hat{E} = \tfrac{1}{2}CV^2 \tag{3.49}$$

In an ideal dielectric, all of this energy is stored, and there is no Joule heating. In a non-ideal ac-polarised dielectric, there is an in-phase current causing Joule heating. The power loss density (W_v, watt/m^3) in a homogeneous and isotropic dielectric is

$$W_v = EJ = \sigma'E^2 \tag{3.50}$$

This is a dc equation. With ac, **E** and **J** are time vectors and the *dissipative* power density and dielectric temperature rise are dependent only on the local in-phase components. In vacuum the current density $J = \varepsilon_0 \partial\mathbf{E}/\partial t$ is a pure displacement current. The power density in vacuum is not zero, but the dissipated power density is zero.

The corresponding adiabatic temperature rise per second, $\Delta T/t$, in the dielectric is

$$\Delta T/t = J^2/\sigma'cd \tag{3.51}$$

where d is mass density (for water: 1000 kg/m^3), and c the heat capacity (for water: 4.2 kJ kg^{-1} K^{-1}). In tissue, the temperature rise will be smaller than predicted from eq. (3.51). The conditions are not adiabatic because of the thermal contact with the surroundings, and blood perfusion will also cool the site considerably. In electro-surgery (Section 8.11) or with electromagnetic fields imposed on tissue (Section 8.1.1), only the local *in-phase* components contribute to the local diathermy effect.

3.5. Interfacial Polarisation

Biomaterials are inhomogeneous dielectrics. In general, the smaller the particles, the larger the surface-to-volume ratio and the larger the interfacial effects. In contrast to the theory of molecules forming homogeneous materials presented up to now, the relaxation mechanisms to be presented in this section are linked to the interfaces: the classical Maxwell–Wagner effects and the counterion polarisation theories. With non-conductive dielectrics the interface contains a surface charge density of only *bound* charges. With conductive dielectrics the interface contains also free charges.

3.5.1. Maxwell–Wagner effects, MHz region

Maxwell–Wagner effects deal with processes at the interface of different dielectrics. There may be free or bound surface charges at the interface (Fricke 1953). The potential Φ at the interface must be continuous, or else the E-field strength had to be infinite there. If there is no dc conductance in either of the two dielectrics, the interface can not be charged by free charges. It can be shown that the normal **D**-field and the tangential **E**-field components are unchanged across such a boundary. If there is a dc conductance in one or both dielectrics, the interface generally will be charged by free charges, and the normal **D**-component will not be continuous. Let us analyse Maxwell–Wagner relaxation in two simple models.

Capacitor with two dielectric layers
Consider two materials, each homogeneous, as two dielectric slabs between the capacitor plates. The models with equivalent circuits are shown in Fig. 3.9. Let us consider ideal components and therefore the resistors with frequency-independent values from dc.

Slabs in parallel
Figure 3.9b shows the two slabs in parallel, and the result is simply a parallel summation of $G_1 + G_2$ and $C_1 + C_2$. There is no Maxwell–Wagner effect. The *Maxwell–Wagner effect is dependent on the geometrical orientation of the interface with respect to the applied field direction.*

Slabs in series
The case becomes quite different and much more complicated for the series case (cf. Section 10.2.3). The total immittance, and also the voltage at the interface between the two dielectrics, will be determined by the resistors at low frequencies, but by the capacitors at high frequencies. The analysis of Fig. 3.9a will be very different depending on whether a series (impedance) or parallel (admittance) model is used.

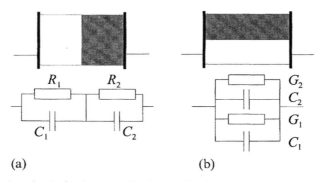

(a)   (b)

Figure 3.9. Equivalent circuits for the Maxwell–Wagner effect in a simple dielectric model. (a) The slabs in series. The resistors cause the interface to be charged. (b) The slabs in parallel.

The externally seen *series* capacitance C_{sext} is the series coupling of the two capacitances C_1 and C_2 at high frequencies, but at low frequency the value of C_{sext} diverges and becomes infinite.

If the resultant *parallel* capacitance C_{pext} is calculated (Section 10.2.3), we get

$$\begin{aligned}
C_{\text{pext}} &= (C_1 R_1^2 + C_2 R_2^2)/(R_1 + R_2)^2 & f \to 0 \\
C_{\text{pext}} &= C_1 C_2/(C_1 + C_2) & f \to \infty
\end{aligned} \tag{3.52}$$

Both capacitance values converge, at high frequencies with values smaller than at low frequencies. Thus with the parallel model of the two slabs in series, we have a classical Debye dispersion, with a capacitive decrement ΔC or $\Delta \varepsilon'$. *This is without postulating anything about dipole relaxation in the dielectric.* Debye dispersion appears and is modelled by two capacitors and two resistors, or even with two capacitors and *one* resistor (one layer without conductivity) as shown in Section 10.2. If the components are ideal (frequency independent), the dispersion will be characterised by one single relaxation time constant.

In the simple example of Fig. 3.9, Maxwell–Wagner dispersion is due to a conductance in parallel with a capacitance for each dielectric, so that the interface can be charged by the conductivity. With zero conductivity in both dielectrics, there is no charging of the interface from free charge carriers. If side 1 of the dielectric is without conductivity ($\sigma_1 = 0$ and $R_1 = \infty$), then C_{pext} at very low frequencies becomes equal to C_1. At very high frequencies the conductivities are without influence.

In an interface without *free* charges, the dielectric displacement **D** is continuous across the interface according to Poisson's equation. Since $\mathbf{D} = \varepsilon \mathbf{E}$, this indicates that the E-field strength will be smaller on the high-permittivity side. Then the ratio of the current densities on sides 1 and 2 is equal to 1

$$\frac{J_1}{J_2} = \frac{\sigma_1 E_1}{\sigma_2 E_2} = \frac{\sigma_1 \varepsilon_2}{\sigma_2 \varepsilon_1} \tag{3.53}$$

In this special case when $\sigma_1 \varepsilon_2 = \sigma_2 \varepsilon_1$, the interface has zero density of *free* charges. On the other hand, if $\sigma_1 \varepsilon_2 \neq \sigma_2 \varepsilon_1$, the difference in current densities implies that the interface is actually charged. If $\sigma_1 = 0$ and $\sigma_2 > 0$, the interface will also be charged.

The conductive path in parallel with the dielectric capacitance causes the Maxwell–Wagner surface charge. *This interface single layer surface charge must not be confused with the double layer charge formed at a wet interphase.* With liquid interphases, such as with particles or cells in aqueous media, the double layer counterion effects are additive to the Maxwell–Wagner effects.

Suspension of spherical particles

This is a model for blood or cell suspensions, for example, and the electrolytic solution of interest may then have a considerable ionic conductivity. An analytical solution (Maxwell 1873) is relatively simple for a *dilute* suspension of spherical particles, and with dc real conductivity as parameters (σ is for the total suspension,

σ_a for the external medium and σ_i for the particles) the relation is (Maxwell's spherical particles mixture equation) (Foster and Schwan 1989)

$$(\sigma - \sigma_a)/(\sigma + 2\sigma_a) = p(\sigma_i - \sigma_a)/(\sigma_i + 2\sigma_a) \tag{3.54}$$

where p is the particle volume fraction. Wagner (1914) extended this to ac cases and the use of complex parameters. Fricke (1924, 1925) extended it for the cases of oblate or prolate spheroids (Maxwell–Fricke equation):

$$(\sigma - \sigma_a)/(\sigma + \gamma\sigma_a) = p(\sigma_i - \sigma_a)/(\sigma_i + \gamma\sigma_a) \tag{3.55}$$

where γ is a shape factor and is equal to 2 for spheres and 1 for cylinders normal to the field. From such equations $\Delta\sigma' = \sigma_s - \sigma_\infty$ or $\Delta\varepsilon' = \varepsilon_s - \varepsilon_\infty$ can be found (subscript s for static values), but the permittivity decrement $\Delta\varepsilon'$ is often rather small, of the order of a few ε_0. If the particles have a dc conductivity, Maxwell–Wagner effects cause ε' as a function of frequency to have an additional slope downwards.

As for the relaxation time, Debye derived a simple expression for a viscosity-determined relaxation time of a sphere of radius a in liquid (eq. 3.38): $\tau = 4\pi a^3 \eta / kT$. The relaxation time is therefore proportional to the volume of the sphere and the viscosity of the liquid.

Maxwell's equation (3.54) is rigorous only for dilute concentrations, and Hanai (1960) extended the theory for high volume fractions:

$$\frac{\varepsilon - \varepsilon_i}{\varepsilon_a - \varepsilon_i} \sqrt[3]{\frac{\varepsilon_a}{\varepsilon}} = 1 - p \tag{3.56}$$

Hanai's equation gives rise to dispersion curves that are broader than the Maxwell–Wagner equation (Takashima 1989).

Dilute suspension of membrane-covered spheres
Equation (3.55) can be extended to include spheres within spheres, or sheath-covered spheres (Fricke 1955). If sheath thickness d is much less than large sphere radius a, the complex conductivity of one sphere inside another sphere is

$$\sigma = \frac{\sigma_i - (2d/a)(\sigma_i - \sigma_{sh})}{(1 + d/a)(\sigma_i - \sigma_{sh})/\sigma_{sh}} \tag{3.57}$$

where the subscript i is for sphere material, and sh is for the sheath membrane. Equation (3.57) is valid for a sphere without external medium. In order to have the complete description, the conductivity value from eq. (3.57) is therefore inserted into eq. (3.55) for the complex conductivity of the sphere.

For a *cell suspension* with cell membranes dominated by a membrane capacitance C_m, the equations can be simplified using certain approximations (Schwan 1957; Foster and Schwan 1989):

$$\Delta\varepsilon' = 9paC_m/4\varepsilon_0 \tag{3.58}$$

$$\sigma_s = \sigma_a(1 - 3p/2) \tag{3.59}$$

$$\tau = aC_m(1/2\sigma_a + 1/\sigma_i) \tag{3.60}$$

The full equations (Pauly–Schwan equations) are found in Pauly and Schwan (1959). They correspond to a large dispersion due to membrane charging effects and a small dispersion at higher frequencies due to the different conductivity of the cytoplasm and extracellular liquids. Schwan and Morowitz (1962) extended the theory to small vesicles of radii 100 nm. With even smaller particles, the two dispersion regions overlap in frequency (Schwan *et al.* 1970).

3.5.2. Adsorbed counterions and ion diffusion (mHz–kHz region)

In Section 2.4.3 we analysed the double layer charge in the solution as a function of the perpendicular distance from the solid surface. No double layer formations are considered in the Maxwell–Wagner theory (Section 3.5.1). However, in *wet* systems and in particular with a high volume fraction of very *small* particles, the surface effects from counterions and double layers usually dominate. This was shown by Schwan *et al.* (1962). By dielectric spectroscopy they determined the dispersion for a suspension of polystyrene particles (Fig. 3.10). Classical theories based upon polar media and interfacial Maxwell–Wagner theory could not explain such results: the

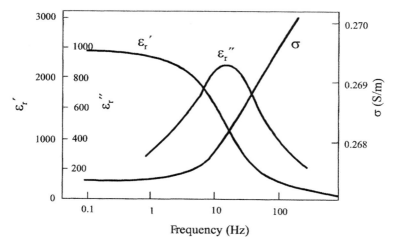

Figure 3.10. Permittivity and conductivity for a suspension of polystyrene particles. Note the expanded conductivity scale. Redrawn from Schwan *et al.* (1962), by permission.

measured permittivity decrement was too large. The authors proposed that the results could be explained in terms of surface (*lateral*) admittance.

Counterion diffusion, Schwartz's theory

Schwartz (1962, 1967, 1972) proposed a model with a tightly bound layer of adsorbed counterions on the sphere surface to explain the high permittivity increment found (hundreds of ε_0) in a suspension of colloidal particles. Diffusion, and not migration may govern ionic motion in a double layer. Diffusion processes are not necessarily exponential, but as an approximation the time constants of such effects are according to the term L^2/D, where L is the diffusion length and D the diffusion coefficient. This is in contrast to Maxwell–Wagner relaxation, where the RC product, that is resistance and capacitance, determines the time constant. The corresponding relaxation frequencies may be low, in the hertz and kilohertz range. Schwartz developed a theory based upon counterion relaxation where

$$\tau = a^2 e/2\mu kT \tag{3.61}$$

$$\Delta\varepsilon = \frac{9p}{(2+p)^2} \frac{e^2 a q_{sn}}{kT} \tag{3.62}$$

where a = particle radius, μ = counterion mobility, and q_{sn} = surface (unit: m^{-2}) density of charged groups on the particle surface.

In Schwartz's theory, the ions were bound to the double layer and restricted to lateral motion within it. A theory is also possible allowing ions to enter or leave the layer from the bulk.

Counterion diffusion in pores

Takashima (1989) deduces the expression for the dielectric increment of a suspension of long cylinders by transforming the relations for ellipsoids from rectangular to cylindrical coordinates:

$$\Delta\varepsilon = \frac{e^2 q_s a^2}{bkT} \frac{9\pi p}{2(1+p)^2} \frac{1}{1+j\omega t} \qquad \text{if } a \gg b \tag{3.63}$$

where a is cylinder length, b is cylinder radius and p is volume fraction of cylinders. This equation has been found to be applicable also to pores (Martinsen et al. 1998) and will give a rough estimate of the rate of dielectric dispersion to be expected in sweat ducts and other pore systems.

The electrical properties of microporous membranes in 1 mM KCl solution was investigated in the frequency range 1 mHz to 1 kHz by Martinsen et al. (1998), using a four-electrode measuring cell. An alpha-dispersion centred around 0.1 Hz was detected and this was assumed to be caused by counterion relaxation effects in the pores of the membrane. The membranes used were 6 μm thick Nuclepore polycarbonate membranes. These membranes have 3×10^8 pores/cm^2 with a pore diameter of 100 nm. The pores are cylindrical with a length equal to the membrane thickness. The surface density of charged groups of the membranes were calculated using streaming potential measurements (see Section 2.4.6), which enabled the use of

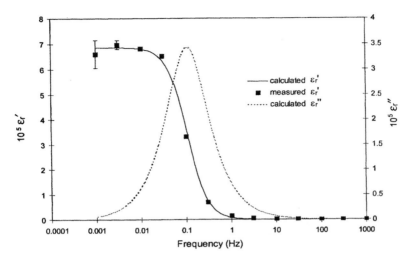

Figure 3.11. Dielectric dispersion of a polycarbonate membrane with pore diameters 100 nm; 1 mM KCl solution; measured and calculated values; dc conductivity omitted. From Martinsen *et al.* (1998), by permission.

eq. (3.63) for calculation of the dielectric increment. Figure 3.11 shows the measured dispersion together with the calculated real and imaginary parts from eq. (3.63). Other mechanisms which may contribute to the relaxation, e.g. the influence from diffusion of ions in the bulk electrolyte were not considered in the study.

3.6. The Basic Membrane Experiment

Setup
Let us again consider the electrolyte container with a NaCl 0.9% aqueous solution, but this time divided in two compartments by a membrane (Fig. 3.12). A sinusoidal ac current is applied to the large current-carrying electrodes. Two additional voltage pick-up electrodes (no current flow in their leads) are placed on each side of the membrane, and a phase sensitive voltmeter or a 2-channel oscilloscope measures the voltage. This four-electrode system eliminates any influence from electrode polarisation impedance: the potential drop at the current-carrying electrodes does not intervene, and the recording electrodes do not carry any current, so the potential drop in their polarisation impedances is zero.

Without membrane
Without a membrane an ac voltage Δv is measured. This voltage is proportional to the current through the system, it is in phase with the current, and the amplitude is independent of frequency. The solution behaves "ohmically", obeying Ohm's law (eq. 2.2) like a perfect resistor. The voltage is due to the ohmic voltage drop caused by the current density in the bulk volume between the recording electrodes.

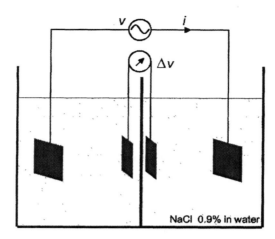

Figure 3.12. The basic membrane experiment.

Thin metal membrane

The measured voltage increases, the phase of the recorded voltage lags behind the current, and the amplitude is frequency dependent.

The conductivity of the metal is more than 1000 times that of the electrolyte, and the metal is thin. The voltage drop in the metal itself must be very small, and the increased measured voltage must be caused by the two double layers formed on each side of the metal plate. These double layers clearly have capacitative properties, and they are in series with the metal.

Let us suppose that there is a thin pore in the membrane. The walls of the pore must also be covered by a double layer. Will counterions in that double layer, and ordinary ions in the pore volume, migrate with the ac *E*-field? No: because of the high conductivity of the metal the local *E*-field strength along the axis of the pore will be very small.

Thin membrane of insulating material (no pores)

The measured voltage difference is very large at low frequencies. At higher frequencies (e.g. >1 kHz), measured voltage is smaller and almost 90° behind the applied voltage.

With an insulating material in the membrane, no dc current can flow through the cell. An ac current will flow through the three series coupled capacitors: two double layer capacitors at each side of the wetted membrane, and the capacitor with the membrane itself as dielectric. The measured voltage difference will consist of the voltage drop in the solution between the measuring electrodes, plus the voltage drop across the three capacitors. Most of the reactance will stem from the membrane, because its capacitance will be far lower than that of the double layers. At sufficiently high frequencies, the reactance of the membrane capacitors will be so small that the effect of the membrane will disappear, and the phase angle will approach 0°.

Membrane of insulating material with pores
This time the voltage is not so high at low frequencies (e.g. < 10 Hz). There is a very small phase shift, less than 5°. At higher frequencies the voltage difference is falling with frequency and with a 90° phase shift.

The pores form a dc current path. The sum of their conductances is so large that the capacitative effect of the membrane as a dielectric is small at low frequencies. If there are not too many pores, most of the potential difference is over the membrane pores and the current constrictional zones in the solution near the pore orifices. The *E*-field strength in the pores will be high. The counterions of the double layer on the pore walls will migrate synchronously with the *E*-field, and the solution inside the pore will be pumped back and forth by electro-osmosis. At higher frequencies the membrane susceptance will shunt the pores, and voltage across the pore will be reduced.

Discussion
The most interesting membranes in the field of bioimpedance are of course the membranes in the living body, the cell membranes: the excitable ones in the muscles and the nerve system, and the less excitable ones. In addition, there are membranes both *inside* and *outside* the cells. Inside the cells there are membranes around some of the organelles. Outside there are thick, solid macro-membranes around all organs such as the heart, the lungs, the brain, the intestines (mediastinum), etc.

In tissue the cell membranes are very small and a part of a living system, quite different from the basic *in vitro* model just shown. However, our simple model is well suited for artificial membranes. With a special technique it is useful for artificial bilayer lipid membranes (BLM) very much like the cell membranes. The membrane pore is then closed with a lipid droplet. The droplet gradually becomes thinner, until it changes from multi-molecular thickness into a single bilayer covering the orifice of the pore. Such a bilayer lipid membrane has important similarities with real cell membranes.

A better model for tissue is the suspension-measuring setup to be presented in the next section.

3.7.   The Basic Suspension Experiment

Setup
Let us consider our electrolyte container again, but let us fill it with a suspension of particles as shown in Fig. 3.13, instead of the separating membrane. The electrolyte is still NaCl 0.9% in water. The particles used are small insulating glass or plastic beads. If the particles are sufficiently small (< 0.5 μm) the solution is optically clear, and the particles will be evenly but randomly distributed in the volume by Brownian motion according to the Boltzmann factor kT. A *suspension* is usually defined as a

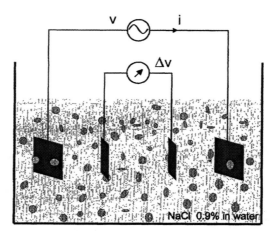

Figure 3.13. The basic particle suspension experiment.

liquid with *larger* particles. Then gravitational energy will dominate the kT-factor, and a sedimentation process will go on so that the particles will sink to the bottom. The system is unstable, and the concentration in bulk will fall with time. These definitions are rather broad, and several subgroups may be formed. Many molecules are, for instance, known to *associate* (molecules of the *same* kind forming more complex structures, e.g. water molecules) in solution, and particles may form *aggregates* if their charges are small enough. Macromolecules may form *colloids* with properties dominated by surface properties.

Findings
A voltage is measured almost like that found for the basic setup without a membrane (Section 3.6). However, a small phase lag is measured, and the voltage decreases somewhat with frequency. The voltage increases with increased particle concentration.

With a high voltage applied, and at a very low frequency (< 1 Hz), it is possible with a microscope to see the particles moving back and forth synchronously with the ac.

Discussion
In general, the double layer on each particle surface will add capacitance to the system, and therefore the measured voltage will decrease with frequency and lag the current. Note that some of the current passes perpendicular to the double layer of the particles, but some passes parallel to the double layers. Smaller particles have increased surface-to-volume ratios, and therefore result in higher capacitance.

If the particles are *highly conductive* with respect to the solution (metal), this metal is not directly accessible to the current and the double layer must be passed twice. If the concentration of metal particles (fill factor) is increased, the measured voltage will not necessarily decrease since it depends on metal/solution impedance and therefore also on the frequency. Above a certain particle concentration there will be a sharply increased probability of a direct contact *chain* of particles throughout the measured volume segment and the segment will be short-circuited.

If the particles are of low conductivity (e.g. glass), the resistance of the suspension at low frequencies will be higher than without particles. Measured voltage is increased, and in addition it will lag the current by a certain phase angle. If the frequency is increased, the susceptance of the sphere volume capacitors will be higher and will ultimately be determined by the permittivity of the spheres with respect to the solution. With cells, the cell membrane surface properties will complicate the case further.

With a sufficiently low frequency the particles can follow, and the particles of the suspension are pumped back and forth by *electrophoresis* as an indication of their net electric charge.

3.8. Dispersion and Dielectric Spectroscopy

Schwan emphasised the concept of *dispersion* in the field of dielectric spectroscopic analysis of biomaterials. Dispersion has already been introduced in Section 3.4.1: dispersion means frequency dependence according to relaxation theory. Biological materials rarely show simple Debye response as described in Section 3.4.2. Often they are also not in accordance with a pure Cole model (cf. Section 7.2.6) but exhibit a certain distribution of relaxation times (DRT). All such models are covered by the dispersion concept.

Dispersion is therefore a broad concept, and many types of DRTs are possible. The Cole brothers proposed a certain DRT corresponding to the apparently simple Cole–Cole and Cole equations (see Section 7.2). These equations presuppose a constant phase element (CPE). However, other distributions than the Cole types are also found to be in agreement with measured tissue data. The Cole models are, however, attractive because the mathematical expressions are so simple. Dispersion models as described below therefore pertain to many types of dispersion mechanisms, among those being Cole and Cole–Cole systems. However, electrolytic conductance is considered frequency independent below 1 MHz, and therefore without relaxation and dispersion. Even so, below 1 MHz it dominates immittance measurement results for most tissue. In order to study the challenging dispersion properties, high-resolution techniques must be employed, and the contribution from dc conduction must be subtracted.

Dispersion data are based upon the electrical examination of a biomaterial as a function of frequency, that is, dielectric spectroscopy. Schwan (1957) divided the dielectric relaxation mechanisms into three groups, each related, for example, to cell membranes, organelles inside cells, double-layer counterion relaxation, electrokinetic effects, etc. He termed them α-, β-, and γ-dispersions. The permittivity of muscle tissue decreases in three major steps corresponding to these dispersions, roughly indicated by Schwan (1957) to correspond to 100 Hz, 1 MHz and 10 GHz.

In the data of Gabriel *et al.* (1996) they are situated at lower frequencies: 1 Hz, 1 kHz and 1 MHz, (Figs 4.16 to 4.19). Later a fourth δ-dispersion was added in the lower GHz range. There are two possible interpretations of these dispersions:

1. The dispersions are linked with defined frequency ranges.
2. The dispersions are linked with defined relaxation mechanisms.

Sometimes dispersions are related to defined frequency ranges, and to a lesser extent to possible relaxation mechanisms. Newer findings have shown that the α-dispersion must be extended down to the mHz frequency range (see Table 3.3).

Figure 3.14 shows how the permittivity in biological materials typically diminishes with increasing frequency as, little by little, the charges (dipoles) cease to be quick enough to follow the changes in the *E*-field. The dispersion regions are here shown as originating from clearly separated Cole–Cole-like systems. Such clearly separated single dispersions can be found with cell suspensions. In tissue the dispersion regions are much broader and overlap, sometimes as a continuous fall almost without plateaux and over many decades of increasing frequency. Compare Section 4.3 on tissue properties and Section 7.2.12 on multiple Cole systems.

The permittivity ε_r' for tissue may attain values larger than 10^6 at low frequencies. This does not imply that the capacitive properties of living tissue are dominant at low frequencies (cf. Table 4.3). The ε_r'' values shown in Fig. 3.14 are not solely due to the dielectric losses of the relaxation mechanisms. The contribution of the dc conductance of the extracellular liquids found in living tissue has been subtracted. The admittance of living tissue is actually dominated by the conductive properties of such liquids, and the out-of-phase current is very much smaller than the in-phase current. These effects could not be measured without the introduction of high-resolution measuring bridges, brought into use by Schwan and pioneers before him (cf. Section 6.3.1). Dispersion is the most common model for explaining the electric behaviour of biomaterials, although electrolytic theory with dc conductivity and electrode phenomena are important factors.

The strength of the dispersion concept is that it is a very broad phenomenological concept that does not require us to consider detailed mechanisms. It is very much to Schwan's merit that this field has been opened in a broad way. But this generality is

Table 3.3. Dielectric dispersions

Type	Characteristic frequency	Mechanism
α	mHz–kHz	Counterion effects (perpendicular or lateral) near the membrane surfaces, active cell membrane effects and gated channels, intracellular structures (e.g. sarcotubular system), ionic diffusion, dielectric losses (at lower frequencies the lower the conductivity).
β	0.1–100 MHz	Maxwell–Wagner effects, passive cell membrane capacitance, intracellular organelle membranes, protein molecule response.
γ	0.1–100 GHz	Dipolar mechanisms in polar media such as water, salts and proteins.

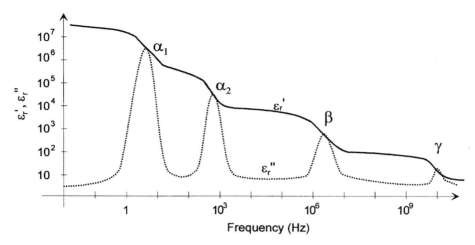

Figure 3.14. Dispersion regions for tissue, idealised and modified from Schwan (1957). Dc conductance contribution has been subtracted from the ε_r'' values shown. With this conductance included, the ε_r'' values would have increased regularly with *lower* frequencies (eq. 3.16). If the dc conductance had been given by σ, this variable would have increased regularly with *higher* frequency.

also a problem: dispersion and relaxation concepts can easily be taken as explanations of the machinery behind the phenomena, and they are not. Dispersion relates to the electrical and physical *behaviour* of molecules, as opposed to analytical methods such as NMR or light spectroscopy that rather examine molecular *structure*. This is partly due to the techniques in analytical chemistry, which are based upon resonance phenomena, while dielectric spectroscopy below about 1 GHz is concerned with relaxation, usually presented as a non-resonance phenomenon. However, in the lower GHz region sharp resonance phenomena are observed for DNA molecules (Takashima 1989, p.214). With ultrafast polarisation processes the capacitance and permittivity may be considered to be frequency independent and without dispersion. However, the reactance or susceptance ($B = \omega C$) is frequency dependent. Dispersion implies frequency dependent permittivity, and the frequency dependence of immittance will be even more complex. Permittivity is therefore often preferred for dispersion studies.

Relaxation is conventionally considered to be due to processes at a molecular level, and implies two levels of a frequency dependent variable with a transition frequency zone in between. We have seen from immittance calculations that Maxwell–Wagner dispersion may give rise to the same behaviour without postulating any molecular process (cf. eq. 3.52). The concept of dispersion is therefore linked both with permittivity and immitivity frequency dependence. The passive properties of biomaterials have also been discussed by e.g. Pethig and Kell (1989).

CHAPTER 4

Electrical Properties of Tissue

4.1. Basic Biomaterials

Hydrogen (63% of the human body's atoms by number), oxygen (25%), carbon (9%) and nitrogen (1.4%) are the four most abundant atoms of the human body. They are all able to form covalent bonds (e.g. water), based upon the sharing of electron pairs by two atoms with unpaired electrons in their outer shells (cf. Section 2.1). Most biomolecules are compounds of carbon, because of the bonding versatility of this element. Nearly all the solid matter of cells is in the form of proteins, carbohydrates and lipids.

4.1.1. Water and body fluids

To recapitulate from Chapter 3: water is a polar liquid with a static relative permittivity of about 80 (at 20°C), falling to 73 at 37°C (Fig. 4.1). The addition of electrolytes such as NaCl or KCl lowers the permittivity proportionally to concentration, e.g. with a dielectric decrement $\Delta\varepsilon_r$ of about 4 for 250 mmol/L concentration of KCl (cf. Section 3.3).

The high permittivity is one reason for the dissociative power of water. Ionic bonds are split up so that ions exist in aqueous solutions in a free, but hydrated form. Because of the strong dipolar electric field, water molecules are attracted to

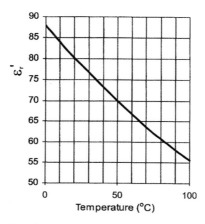

Figure 4.1. Relative permittivity of pure water, temperature dependence. According to CRC (1998).

ions and local charges, forming a hydrated layer (sheath) that tends to neutralise the charge and increase the effective dimensions of the charged particle. The binding of water to protein molecules may be so strong that the water is better characterised as *bound* water. Bound water has different properties from liquid water and must be considered to be a part of the solid and not the liquid phase. A range of bonding energies may be present corresponding to a scale from free water molecules as liquid, via more loosely bound molecules to a very tight bonding when it is very difficult to extract the last water molecules.

Hydrogen ions in the form of protons or oxonium ions contribute to the dc conductivity of aqueous solutions both by migration and by hopping (Fig. 2.4).

Pure water exhibits a single Debye dispersion with a characteristic frequency of about 17 GHz (Fig. 4.2). Ice has a static relative permittivity of about 92 at 0°C (at low frequencies, but falling to < 10 at 20 kHz), slightly anisotropic ($\sim 14\%$), and increasing with decreasing temperature. The characteristic frequency of the relaxation is much lower than for water, around 3 kHz, and with disturbing conductive effects < 500 Hz (Hasted 1973). The dielectric properties of aqueous electrolytes have been studied by e.g. Kirkwood (1939), Mandel and Jenard (1963), Grant (1965) and Jonscher (1974, 1977).

The *living* cell must contain and be surrounded by aqueous electrolytes. In human blood the most important *cations* are H^+, Na^+, K^+, Ca^{2+}, Mg^{2+} and the *anions* are HCO_3^-, Cl^-, protein$^-$, HPO_4^{2-}, SO_4^{2-}. Note that protein in the blood is considered as a negative (macro)-ion.

The electrolytes, both intra- and extra-cellular, are listed in Table 2.6. They produce an electrolytic conductivity of the order of 1 S/m. Up to at least 10 MHz it is considered to be frequency independent.

Most of the cells of the body undergo mitosis only if they are attached to a surface. Cancer cells are not dependent on such attachment. Blood containing erythrocytes is the best example of a natural cell suspension.

4.1.2. Proteins

Proteins are the most abundant macromolecules in living cells, and 65% of the protein mass of the human body is intracellular. Proteins are the molecular

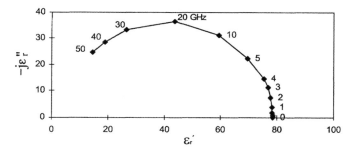

Figure 4.2. Complex relative permittivity of pure water, 25°C. According to CRC (1998).

instruments through which genetic information is expressed. All proteins are constructed from 20 *amino acids*, joined by covalent bonds. All 20 amino acids have the following group in common:

$$\begin{array}{c} COO^- \\ | \\ H_3^+N-C-H \\ | \\ R \end{array}$$

This common group is shown in ionised form as would occur at pH 7: the amino group has acquired a proton (NH_3^+), and the carboxyl group has lost one (COO^-). The charges are separated and represent a permanent dipole with zero net charge: such substances are *polar*. At pH 7 all amino acids are more or less polar. Even so, the common group is also paradoxically called a "dipolar ion" or a *zwitterion*, because at other pH values it attains a net charge: at low pH values, the NH_3^+ group dominates and the acid is a cation with a net positive charge. At high pH values, the COO^- group dominates and the "acid" is an anion with a net negative charge. Clearly, the term amino acid may be misleading, because in water solution these compounds actually can be an acid (proton donor) or a base (proton acceptor). Although these conditions are mostly non-physiological, the acid–base behaviour *in vitro* is valuable for the examination and mapping of amino acids.

The R in the structure above symbolises a side chain that determines the properties of each amino acid. The electrical properties of the protein are also strongly dependent on this R-group, but only if it is polar or charged. The amino acids are classified according to their R-groups, but the properties of an R-group and therefore of the amino acid are very dependent on pH and molecular configuration, so the classification differs in the literature. The following classification for the 20 amino acids may be used at a physiological pH around 7.4:

- Eight of the amino acids are *hydrophobic* and therefore the R-groups are grouped as *non-polar*: they are without net electric charge and have a negligible dipole moment. The net dipole moment of all eight acids is equal to that of the common group.
- Seven are *hydrophilic* and the R-groups are therefore grouped as *polar*. These polar R-groups have an expectedly large influence on dielectric permittivity. Some of them tend to dissociate H^+ ions so that the amino acid also becomes charged and ionic.
- Two have negatively charged R-groups and therefore have a net negative charge.
- Three have positively charged R-groups and therefore have a net positive charge.

Those amino acids having a net electric charge migrate in a homogeneous electric field and thus have an electrophoretic mobility. As explained, at pH $\neq$ 7 all amino acids may be charged. The pH of zero net charge is the *isoelectric point*. For example, glycine has no net charge in the pH range 4–8, so this is its isoelectric *range*.

Attempts to interpret the measurements with aqueous solutions of amino acids and peptides with the Clausius–Mossotti equation fail (Section 3.1.2). Instead, it is usual practice to characterise the polarisability of biomolecules by a *dielectric increment δ*. At low concentration there is a proportionality between the concentration (*c*) and the permittivity increase for many biomolecules:

$$\varepsilon_r = \varepsilon_{rps} + \delta c \tag{4.1}$$

where ε_{rps} is the relative permittivity of the pure solvent. The unit for the dielectric increment δ is [concentration]$^{-1}$. The dielectric increment for many amino acids is in the range 20–200, with dipole moments in the range 15–50 Debye units (D) (Pethig 1979).

Electrical properties of isolated amino acids. In dry form the polar amino acids may crystallise as a hard material like NaCl. With non-polar types the dry crystal lattice is softer. The electrical properties of dry and moistened amino acids are dramatically different: Rosen (1963), Takashima and Schwan (1965), Smith and Foster (1985). Glycine is the simplest amino acid with only one hydrogen atom as the R side chain. The R-chain is uncharged, but polar, so glycine is hydrophilic. When dry it forms hard crystals, and Figs 4.3 and 4.4 show the dielectric dispersion of glycine powder in dry form and with different water content. The permittivity of dry powder is very low and is frequency independent (at least down to 20 Hz). With water, a clear dispersion is apparent with a characteristic frequency around 1 kHz. The size of the crystals had no effect on high- and low-frequency permittivity levels, but the larger the particle size, the lower the characteristic frequency. Protons are believed to play an important role in the water-dependent conductivity of relative dry materials (South and Grant 1973), due to the hopping conductance mechanism, (Figs 2.4 and 4.27).

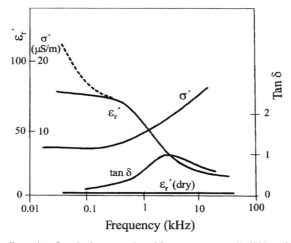

Figure 4.3. Dielectric dispersion for glycine crystals, with water content (0.67% and 0%) as parameter. Two-electrode method, dashed line without correction for electrode polarisation. From Takashima and Schwan (1965), by permission.

Peptides are small groups of amino acids, and *polymers* are even more complicated forms of peptides forming *proteins*. A polymer is a general term covering any material built up from a series of smaller units (monomers). Proteins differ from each other because each has its own amino acid *sequence*. Like the amino acids, many proteins can be isolated in crystalline and dry forms. Cross bindings and foldings are very important for the geometrical form, size and rigidity, and also for the function and the electrical properties. A protein may have redox centres capable of single electron exchange. If such centres are closer than about 0.15 nm, direct tunnelling through energy walls occurs (Page *et al.* 1999). A change in water solvent concentration may change the electrostatic forces because they are permittivity dependent (cf. Section 3.3) and the protein will swell or shrink accordingly.

A protein may have a very regular form such as an α-helix, or a chaotic morphology as a result of *denaturation* at high temperatures or extreme pH values. The denaturation of a protein reduces its solubility in water, and heat coagulation of tissue is therefore accompanied by liberation of water. The denatured protein nearly always loses its characteristic biological activities, and the electric properties are often completely changed. This shows the importance of the higher orders of the geometrical (secondary) structure of a protein. A simple peptide may, for instance, have the form:

$$R^1 \!-\! \overset{\displaystyle \overset{H}{|}}{\underset{\displaystyle \underset{(+)}{\overset{||}{NH_3}}}{C}} \!-\! C \!-\! \overset{\displaystyle \overset{H}{|}}{N} \!-\! \overset{\displaystyle \overset{H}{|}}{\underset{\displaystyle \underset{}{\overset{|}{R^2}}}{C}} \!-\! COO(-)$$

As the forms become more complex, very different types of charge distributions and bonds are possible. The rigidity of the bonds will be important for electrical relaxation phenomena. In comparison with the displacement of electrons and nuclei during atomic polarisation (10^{-15} m), the distance between the charges in a macromolecule can be very large (10^{-8} m). Therefore the dipole moment of proteins can be very large. But the symmetry of the electric charges in protein molecules is also surprisingly high. Even with a large number of ionised groups, the dipole moment often corresponds to only a few unit charges multiplied by the length of the molecule. Some proteins actually have a negligible dipole moment and must be regarded as non-polar.

A protein with a large number of ionised groups is a *polyelectrolyte* if it has a net electric charge. If it is free, it will migrate in an electric field (Bull and Breese 1969). Usually it is not free, and may therefore only undergo local polarisation. Because a polyelectrolyte has charges distributed all over the molecule, each charge is sufficiently isolated to attract ions of an opposite charge. A local ionic atmosphere is formed, just like that formed around ions in strong electrolytes; in a better analogy they are *counterions*, just like the counterions forming the electric double layer. The counterions decrease the dc mobility, and counterion polarisation is believed to be very important in many proteins in water solution.

Electrical properties of isolated amino acids, peptides and proteins.   Figure 4.4 shows Cole–Cole plots for glycine (amino acid), glycylglycine (peptide) and ovalbumin (protein). As a biomolecule becomes more and more complicated and larger (amino acid→peptide→protein), the frequency exponent 1-α of the Cole–Cole equation becomes higher (α lower), indicating a broader distribution of time constants. Protein properties have been studied by e.g. Takashima (1962, 1967), Takashima and Schwan (1965), Lumry and Yue (1965), South and Grant (1973), Wada and Nakamura (1981) and Barlow and Thornton (1983).

A *DNA molecule* consists of two polynucleotide chains (helices). The two chains intertwine with a fixed pitch of 3.4 nm. Two types of base pairs bridge the two helices at a fixed distance of 0.34 nm. The phosphate groups in the nucleotide chains carry negative electric charges in water solution. Because of the bridges, the double helix demonstrates considerable rigidity, and may be considered as a charged rigid rod. When a DNA solution is heated to higher than about 80°C, the two strands unwind and the DNA is changed from a double helix to random coils. The

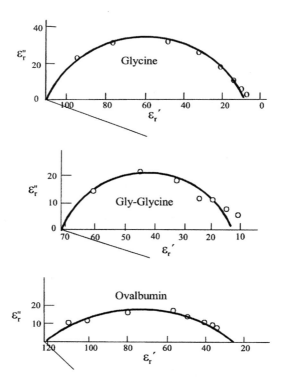

Figure 4.4. Cole–Cole plots for an amino acid (glycine with 1.5% water), a peptide (glycylglycine with 1% water) and a protein (albumin with 11% water). From Takashima and Schwan (1965), by permission.

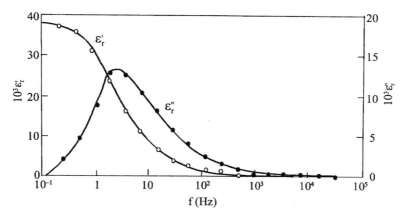

Figure 4.5. Dielectric dispersion of DNA molecules in dilute suspension. From Hayakawa *et al.* (1975), by permission.

denatured form may in some cases revert to the natural form through a slow cooling process. The dielectric dispersion is shown in Fig. 4.5, and is very different in the natural and denatured forms. The polarisation found may be due to either the movement of charges along the rod or to the turning of the whole molecule. However, the polarisation has been shown to be more due to counterion polarisation than the orientation of the permanent dipole. DNA was also studied by Mandel (1977) and Maleev *et al.* (1987). A possible resonance phenomenon was discussed by Foster *et al.* (1987) and Takashima (1989).

4.1.3. Carbohydrates (saccharides)

These are the fuels for the cell metabolism. They also form important extracellular structural elements, like cellulose (plant cells) and have other specialised functions. Bacterial cell membranes are protected by cell walls of covalently bonded polysaccharide chains. Human cell membranes are coated with other saccharides. Some saccharides form a jelly-like substance filling the space between cells. Some polysaccharides may have a negative charge at pH 7, but many carbohydrates are not believed to contribute predominantly to the admittivity of tissue.

4.1.4. Lipids and the passive cell membrane

Lipids are water insoluble, oily or greasy organic substances. They are the most important storage forms of chemical energy in the body. In our context, a group of lipids is of particular importance: the *polar* lipids. They are the major component of the passive cell membranes, and an important basis for the capacitive nature of cells and tissue. Polar lipids form micelles (aggregates of molecules, e.g. formed by surface-active agents), monolayers and bilayers. In aqueous systems the polar lipids spontaneously form micelles, whereby the hydrophobic hydrocarbon tails are

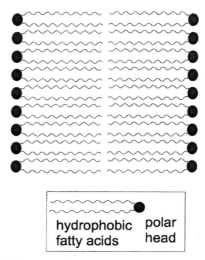

Figure 4.6. Bilayer lipid membrane (BLM), the main component of the cell membrane.

hidden from the water. On a water surface they form monolayers with the hydrophobic tails pointing out towards the air.

Phospholipids form bilayer lipid membranes (BLM) about 7 nm thick (Fig. 4.6). Each of the monolayers has its hydrophobic surface oriented inwards and its hydrophilic surface outwards towards either the intra- or extra-cellular fluids. The inside of such a bilayer is hydrophobic and lipophilic. A BLM is a very low electric conductivity membrane and is accordingly itself closed for ions. It allows lipids to pass, and also water molecules even if the membrane interior is hydrophobic. The intrinsic conductance is of the order of 10^{-6} S/m, and a possible lipophilic ionic conductivity contribution can not be excluded.

An artificial BLM may spontaneously take the form of a sphere, enclosing a solution inside. Such a sphere is called a *liposome*, and has important similarities with a cell. If agitated by ultrasound during formation they may have a diameter of 10–100 nm, and are thus smaller than most living cells. Agitated by hand they may be in the micrometre range.

4.2. The Cell

The dynamic cell membrane

The cell membrane is an absolute prerequisite for life, because with it the cell can control its interior by controlling the membrane permeability. The membrane is also the site of some basic life processes. It is a layer that separates two solutions, and forms two sharp boundaries towards them. A dead membrane has a static separation effect, but the living membrane is a highly dynamic system. It is selectively permeable, and the selectivity can change abruptly. Of particular interest to us is a membrane's selective permeability to ions, i.e. its properties as an *electrochemical membrane*. A considerable part of membrane phenomena are linked with electric currents and associated phenomena.

The passive part of the cell membrane is the bilayer lipid membrane (BLM) as shown in Fig. 4.6. In parallel with the BLM there are embedded proteins, transport organelles, and ionic channels for the electrogenic pumps. These structures are a part of a very dynamic system, and they can change positions and functions very rapidly (Fig. 4.7). Electrically, they represent shunt pathways in parallel with the BLM.

Even if the conductivity of the BLM itself is very low, the membrane is so thin that the capacitance is very high and the breakdown potential is low. The electric field strength with 70 mV potential difference and thickness 7 nm is 10 kV/mm. This represents a large dielectric strength, but not larger than, for example, Teflon would withstand. And as we shall see, this is not the potential across the bilayer itself, but the potential of the bilayer + the potential difference of the two electric double layers formed on each surface of the membrane. A coating of special carbohydrates also covers the external cell membrane surface, the *glycocalyx*, strongly modifying surface properties. Many of the glycocalyx carbohydrates are negatively charged, so that living cells repel each other. Some of the carbohydrates are receptor substances—binding hormones, for example—and some are important for the immunological properties of a cell.

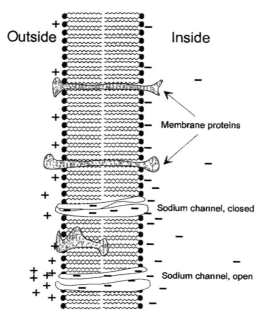

Figure 4.7. Bilayer lipid membrane with embedded proteins and sodium channels. The real structure is dynamic and very complicated; for example, the outer surface of the cell membrane is covered with negatively charged glycocalyx carbohydrates.

An electric double layer covers the wetted outer cell membrane surface. The total cell has a net charge because it has electrophoretic mobility. The cell membrane thickness is about 7 nm, and the capacitance is of the order of 20 $\mu F/cm^2$. The frequency dependence of the membrane capacitance has been a subject of dispute (Cole 1972), but often the BLM as such is considered to have a frequency-independent membrane capacitance.

When the potential difference is increased > 150 mV, the *membrane* breaks down (cf. Section 8.12.1 on electroporation). This must not be confused with the excitation process, when the ionic *channels* of the membrane suddenly open.

Some of the organelles inside the cell have their own membranes, presumably built the same way as BLMs.

The ion pumps embedded in the cell membrane

The net cell membrane conductance is very complicated and is governed by the active ionic channels. The most important channels are those of the sodium/potassium pumps, an energy-consuming mechanism that polarises the cell so that the interior has a negative potential with respect to the extracellular liquid. For excitable cells this voltage is about -70 mV, for non-excitable cells only -10 to -20 mV. Such an energy consuming pump is a chemically driven molecular device, embedded in the cell membrane, capable of generating an electric potential difference across the membrane, it is an *electrogenic* pump. It may be characterised by an emv in series with an electrical resistance (*chord* resistance) (cf. Fig. 4.8). The electrogenic current density in excitable animal cells is usually of the order of a few $\mu A/cm^2$. The emv is of the order of 250 mV, but because of the large chord resistance the net membrane potential is still around -70 mV. The passive properties of the excitable cell membrane were studied by South and Grant (1973) and Takashima and Schwan (1974), the voltage dependence of the membrane capacitance by e.g. Cole (1972) and Takashima and Yantorno (1977). The sodium/potassium pump is an ATP-driven transmembrane pump that translocates a greater electric charge in one direction than in the other. Per cycle it exports three sodium ions Na^+, and imports two potassium ions K^+. Many such pumps operate in parallel in the cell membrane. Plant cells have a similar proton pump. Figure 4.8 illustrates a simple model with the ionic permeabilities given as variable conductances.

The excitable cell membrane

When a muscle or nerve cell is excited, within 1 ms the membrane conductance for certain ions increases abruptly and the ionic channels open up for some hundreds of microseconds (Fig. 4.9). The cell is depolarised, and cannot be re-triggered (the *refractory* period) before the cell is repolarised after the ionic channels are once more closed. Figure 4.9 shows the most basic signal source in the human body, and also the fastest. The frequency content of the signal is up to a few kHz, but this can only be recorded very near the cell. At larger distance the geometry acts as a low-pass filter (cf. Section 5.3.2). The recorded potentials are called *action potentials*.

The intracellular dc potential means that the membrane capacitances are charged. With an intracellular microelectrode, the -70 mV of the polarised cell can be

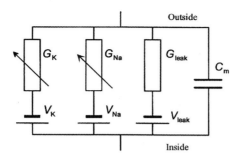

Figure 4.8. A simple electrical model of an excitable cell membrane. According to Hodgkin and Huxley (1952), by permission.

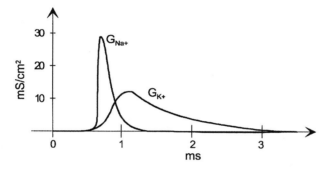

Figure 4.9. Membrane conductance during cell excitation. According to Hodgkin and Huxley (1952), by permission.

measured (Fig. 4.10). That potential includes the potential of the two charged double layers on the internal and external membrane surfaces. Presumably the two potentials have roughly the same values with opposite polarity, and as they are in series they probably more or less cancel. With respect to capacitance, the double layer has a thickness of the order of 0.1–10 nm, and the cell membrane 7 nm. The permittivity of the membrane dielectric is presumably less than that of the water in the double layer. Depending on the actual values, the measured membrane capacitance is therefore influenced by double layer capacitance to a greater or lesser degree.

Measured voltage is the potential *difference* between intra- and extracellular fluids, and does not define the net potential or charge of the cell in the extracellular liquid. We know that the cell has a net charge, because it moves in an extracellular electric field (electrophoresis).

The ionic pumps in the membrane create local current densities in the fluids near the pump channels of the membrane. If the channels suddenly open (Fig. 4.11) so that the intracellular potential changes abruptly, the ions must also supply a transient discharge current of the membrane capacitance. The current is not with

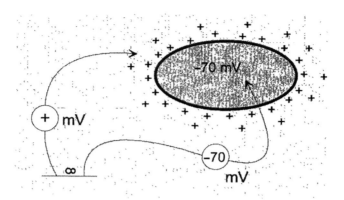

Figure 4.10. The polarised excitable cell, static conditions. Intra- and extracellular potential; the negative charge from glycocalyx is not indicated. Non-excitable cells may have low intracellular resting potentials $|V_m| < 50$ mV.

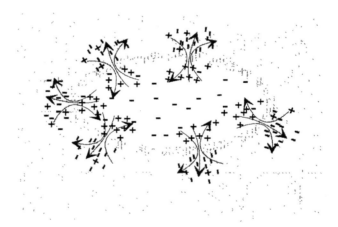

Figure 4.11. Opened ionic channels at the start of an action potential. Inflow of cations, outflow of anions. Intracellular potential increasing from -70 mV towards zero or positive values.

respect to a distant reference electrode but concentrated in an interstitial fluid zone near the cell. The current flow can be modelled with local current dipoles, and is clearly measurable with unipolar or bipolar pick-up electrodes in the interstitial liquid. When the cell is depolarising, cations and anions flow in all directions through the open channels, but there must be a net flow of anions out of the cell. If the internal pressure is not to fall, there must also be a net flow into the cell.

In the actual situation there are many channels and they may be represented by a surface layer of distributed current dipoles. This layer can be simplified at a distance to one net current dipole representing one or more cells.

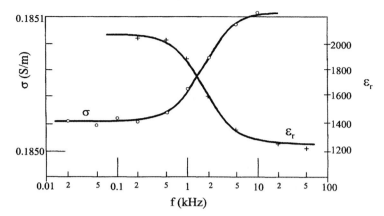

Figure 4.12. Dielectric constant and conductivity of a suspension of lysed erythrocytes. Characteristic frequency f_c is approx. 1.5 kHz. Note conductivity scale resolution. From Schwan (1957), by permission.

Electrical properties of cell suspensions

Figure 4.12 shows data for a cell suspension. It shows a single dispersion with a characteristic frequency of about 1.7 kHz.

The use of dielectric spectroscopy for characterisation of living cells and the possible derivation of cellular parameters such as living cell volume concentration (Fig. 4.13), complex permittivity of extracellular and intracellular media and morphological factors is discussed by Gheorghiu (1996). Another possible application is the electrical measurement of erythrocyte deformability (Amoussou-Guenou et al. 1995). Low frequency dielectric dispersion was studied by e.g. Einolf and Carstensen (1973).

4.3. Tissue and Organs

Tissue is a very inhomogeneous material, and interfacial processes are very important. The cells are of uneven size and with very different functions. There are large differences in tissue conductivity: from the liquid tissue flowing through the blood vessels to the myelin sheaths as insulators surrounding the axons of the nerve cells, from connective tissue specialised to endure mechanical stress to bones and teeth, muscle masses, the dead parts of the skin, gas in lung tissue and so on. From an electrical point of view, tissue can not be regarded as a homogeneous material. A useful presumption is also that *anisotropy* must be more the rule than the exception.

Let us consider a simple case of a volume containing many cells in an interstitial fluid (Fig. 4.14). The cell membranes are considered to have a high capacitance and a low but complicated pattern of conductance. At dc and low frequencies current must pass around the cells (Fig. 4.14). Lateral conductance in the double layers is possible. The cell interior does contribute to a smaller degree to current flow. At higher frequencies the membrane capacitance lets ac current pass. The membrane

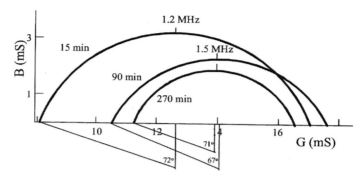

Figure 4.13. Admittance of erythrocytes in natural condition during sedimentation. From Gougerot and Fourchet (1972), by permission.

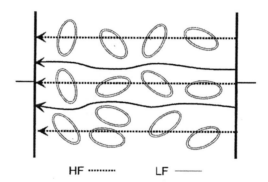

Figure 4.14. Low- and high-frequency current paths in tissue.

effect disappears, and the current flows everywhere according to local ionic conductivity.

All interfaces give rise to Maxwell–Wagner and counterion polarisation effects as described in Section 3.5.

In general, tissue is an anisotropic medium because of the orientation of cells, macromembranes and organs (see Figs 4.33 and 4.34). Such anisotropy is a low-frequency phenomenon if it is due to membranes, but not if it is due to air, for example. Organs are very often compartmentalised, with macromembranes as compartment walls. The lung, the heart, the brain and the stomach all have multi-layer membranes, but the largest is the abdominal membrane, the *peritoneum*. These membranes may have large influence on current flow, and also on endogenic currents between the organs. The gradients of ionic concentration develop emv's that influences current flow (Nordenstrøm 1983).

Tissue electrical data are tabulated in Table 4.3. The maximum magnitudes for the different tissue dispersion regions (cf. Section 3.8) are given in Table 4.1. Note the agreement with eq. 3.35.

Table 4.1. Maximum dispersion magnitudes according to Schwan (1963)

	α	β	γ
$\Delta\varepsilon_r$	5×10^6	10^5	75
$\Delta\sigma$	10^{-2}	1	80

Extracellular liquids usually dominate the in-phase conductivity. The low-frequency *temperature coefficient* is therefore that of electrolytes, around $+2\%/°C$. The quadrature ε' or Y'' components have a smaller temperature coefficient, around $-0.5\%/°C$ but dependent on the measuring frequency in relation to the characteristic frequencies of the dispersions.

4.3.1. Muscle tissue

Muscle tissue exhibits a large α-dispersion, and as shown in Fig. 4.15 it may be strongly anisotropic with a low-frequency conductance ratio of about 1:8 between transversal and longitudinal directions. In the longitudinal direction the high conductance is not very frequency dependent, indicating that direct free liquid channels dominate the current path. It has been proposed that the α-dispersion arises from interfacial counterion polarisation, or from sarcotubular membrane systems in the interior of the muscle fibre. The transversal properties are presumably more dominated by interfacial β-dispersion of the Maxwell–Wagner type.

Figure 4.16 shows another data set presented as permittivity and conductance. In the transversal situation 3–4 dispersions are clearly seen. The longitudinal data illustrate how distinct dispersion regions may be almost non-existent. Dispersion may take the character of a single straight line through many decades of frequency.

4.3.2. Nerve tissue

Linear properties
The brain is the only tissue with a large volume of nerve tissue. Nerve tissues elsewhere may be regarded as cables, and electrical cable theory is often applied. The insulating properties of the myelin sheath increase the impedance for myelinated nerves.

Macroscopical data for brain tissue are shown in Fig. 4.17. Measurements *in vivo* through the low-conductivity scalp will show quite different data. Brain tissue was also studied by Foster *et al.* (1979).

Non-linear properties
Muscles and nerves are excitable tissue. If the intracellular potential in such cells is reduced towards zero by chemical, mechanical or electrical (excitatory potential) stimuli, the cell membrane may fire: gated ionic channels open, and the *action*

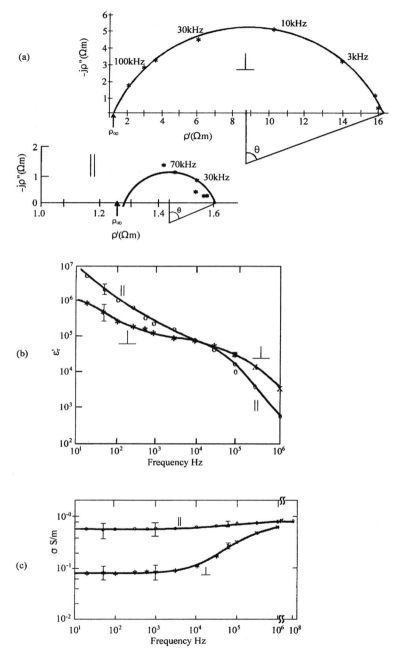

Figure 4.15. Dog excised skeletal muscle (37°C), frequency dependence and anisotropy. Current path: ⊥ transversal, ‖ longitudinal to muscle fibres. (a) Complex resistivity in the Wessel plane. (b) Relative permittivity. (c) Conductivity. Redrawn from Epstein and Foster (1983), by permission.

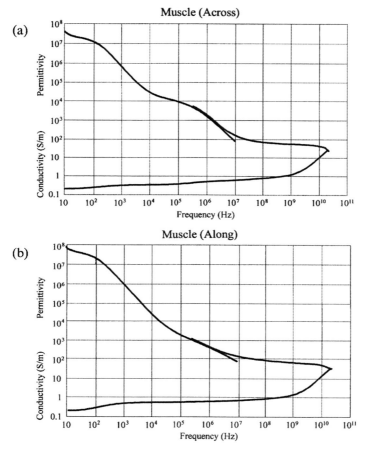

Figure 4.16. Dielectric dispersion of muscle tissue: (a) transversal; (b) longitudinal. Based on Gabriel *et al.* (1996), by permission.

potential is created. This is illustrated in Figs 4.8 and 4.9. The action potential of such cells is the most important endogenic signal source in the body. The propagation of the action potential signal through tissue occurs with an accompanying low-pass filtering effect (cf. Section 5.3.2). ECG and other uses of the action potentials are described in Chapter 8. Electrotonus and rheobase are explained in Section 8.7.

4.3.3. Adipose and bone tissue

Figure 4.18 shows the dispersion of adipose tissue. Adipose tissue and in particular bone and bone marrow have a wide spread of permittivities and conductivities, very dependent on the degree of perfusion with blood and other liquids.

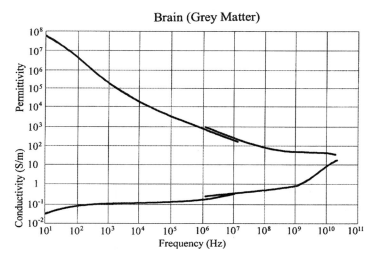

Figure 4.17. Dielectric dispersion of brain tissue. Based on Gabriel *et al.* (1996), by permission.

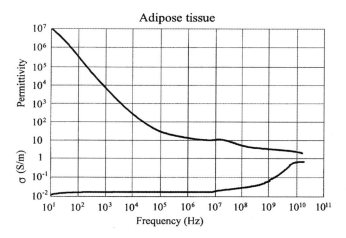

Figure 4.18. Dielectric dispersion of adipose tissue. Based on Gabriel *et al.* (1996), by permission.

4.3.4. Blood

Whole blood consists of erythrocytes (containing the haemoglobin) and other cells in plasma. Blood may be studied as whole blood *in vivo*, as erythrocytes in suspension, as lysed erythrocytes in suspension, as plasma, etc. The erythrocytes are shaped like doughnuts with an outer diameter of about 10 μm. Plasma forms the liquid part containing electrolytes and large organic electrically charged molecules. Lysed erythrocytes are disrupted cells with their intracellular material (haemoglobin) discharged into the liquid. The electrical properties of whole blood and lysed blood are naturally very different. From Maxwell–Wagner theory it is possible to find an expression for the conductivity of a suspension of membrane-covered spheres (Section 3.5.1).

Whole blood exhibits β-, γ- and δ-dispersion, but curiously no α-dispersion (Foster *et al.* 1989). The β-dispersion has a dielectric increment of about 2000 centred around 3 MHz (haematocrit (the volume ratio of blood cells to plasma) 40%). Erythrocytes in suspension have a frequency-independent membrane capacitance with very low losses (Schwan 1957). The impedance of lysed erythrocytes in suspension shows two clearly separated single relaxation frequencies (Debye dispersions). The α-dispersion is in the lower kHz range and the β-dispersion in the lower MHz range (Schwan 1957; Pauly and Schwan 1966).

Figure 4.12 presented an example of the dielectric constant and conductivity of a suspension of *lysed* erythrocytes; the characteristic frequency is low, ~ 1.5 kHz. Figure 4.13 showed the admittance of erythrocytes in natural condition during sedimentation.

4.3.5. Human skin and keratinised tissue

Epithelia are cells organised as layers: skin is an example. Cells in epithelia form *gap junctions*. In particularly tight membranes these junctions are special *tight junctions*. The transmembrane admittance is dependent both on the type of cell junction and to what extent the epithelium is shunted by channels or specialised organs (e.g. sweat ducts in the skin).

The impedance of the skin is dominated by the stratum corneum at low frequencies. It has generally been stated that skin impedance is determined mainly by the stratum corneum at frequencies below 10 kHz and by the viable skin at higher frequencies (Ackmann and Seitz 1984). This will of course be dependent on factors such as skin hydration, electrode size and geometry, but may nevertheless serve as a rough guideline. A finite element simulation of a concentric two-electrode system used by Yamamoto *et al.* (1986) showed that the stratum corneum accounted for about 50% of the measured skin impedance at 10 kHz, but only about 10% at 100 kHz (Martinsen *et al.* 1999).

The stratum corneum (SC) may have a thickness of from about 10 μm (0.01 mm) to as much as 1 mm or more, e.g. on the sole of the foot. It is a solid-state substance, not necessarily containing liquid water, but with a moisture content dependent on the surrounding air humidity. SC is not soluble in water, but the surface will be charged and a double layer will be formed in the water side of the interphase. SC can absorb large amounts of water, for example doubling its weight. It may be considered as a solid-state electrolyte, perhaps with few ions free to move and contribute to dc conductance. The SC contains such organic substances as proteins and lipids, which may be highly charged but bound, and therefore contribute only to ac admittance.

An open question is whether the conductance in SC, in addition to the ionic component, also has an electronic component, e.g. as a semiconductor.

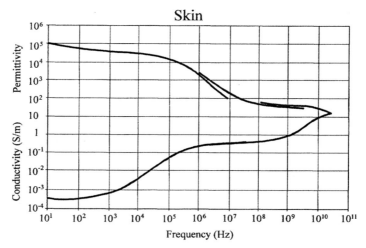

Figure 4.19. Dielectric dispersion of skin tissue. Based on Gabriel *et al.* (1996), by permission.

The stratum corneum has been shown to display a very broad α-dispersion (Figs 4.19 and 4.20) that is presumably largely due to counterions. Viable skin has electrical properties that resemble those of other living tissue, and hence displays separate α- and β-dispersions. The interface between the SC and the viable skin will also give rise to a Maxwell–Wagner type of dispersion in the β-range. While the impedance of the SC is much higher at low frequencies than the impedance of the living skin, the differences in dispersion mechanisms make the electrical properties converge as the frequency is increased. This is the main reason why increased frequency in general leads to measurements of deeper layers in the skin.

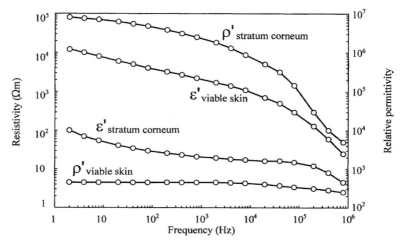

Figure 4.20. Average resistivities and relative permittivities in stratum corneum and viable skin. Redrawn from Yamamoto and Yamamoto (1976), by permission.

Yamamoto and Yamamoto (1976) measured skin impedance on the ventral side of the forearm with a two-electrode system and an ac bridge. They used Beckman Ag/AgCl electrodes filled with gel and measured 30 min after the electrodes had been applied. The skin was stripped with cellulose tape 15 times, after which the entire SC was believed to have been removed. Impedance measurements were also carried out between each stripping so that the impedance of the removed layers could be calculated. The thickness of the SC was found to be 40 μm, which is more than common average values found elsewhere in the literature. Therkildsen *et al.* (1998), for example, found a mean thickness of 13.3 μm (min. 8 μm/max. 22 μm) when analysing 57 samples from non-friction skin sites on caucasian volunteers. However, the increase in moisture caused by electrode occlusion and electrode gel would certainly have increased the stratum corneum thickness significantly.

Knowing the stratum corneum thickness enabled Yamamoto *et al.* to calculate the parallel resistivity and relative permittivity of the removed stratum corneum. Furthermore, the resistivity and relative permittivity of the viable skin were calculated by assuming homogeneous electrical properties and using the formula for the constrictional resistance (cf. Fig. 5.2). The resistance of a disk surface electrode is (eq. 5.9): $R = \rho/4a$. Since $RC = \rho \varepsilon_r \varepsilon_0$, the relative permittivity of the viable skin can be calculated from the measured capacitance using the formula: $C = 4a\varepsilon_r\varepsilon_0$. The calculated data from Yamamoto and Yamamoto (1976) are presented in Fig. 4.20. These data can also be presented as conductance and susceptance versus frequency as in Fig. 4.21. The very broad nature of the dispersion can easily be seen in this figure. The conductance levels out at low frequencies, indicating the dc conductance level of the skin. The susceptance seems to reach a maximum at approximately 1 MHz, which should then correspond to the characteristic frequency of the dispersion. This frequency response is difficult to interpret and the apparent broad dispersion is most probably a composite of several dispersion

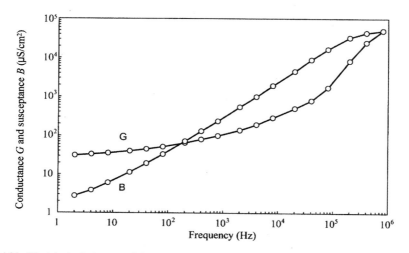

Figure 4.21. Electrical admittance of the stratum corneum as calculated from data presented in Fig. 4.20.

mechanisms. The highly inhomogeneous nature of the stratum corneum with a significant hydration gradient in brick-like layers of dead, keratinised cells should produce significant dispersion mechanisms in both the alpha and beta ranges.

The frequency response shown in Fig. 4.21 can be compared with the admittance data from a 180 μm thick sample of palmar stratum corneum *in vitro* shown in Fig. 4.22 (Martinsen *et al.* 1997a). These measurements were performed with a two-electrode system and hydrogel electrodes at 50% relative humidity. The dc level of this stratum corneum sample is much lower than that shown in Fig. 4.21, even after adjusting for the 4.5 times greater thickness of the *in vitro* sample. This is easily explained from the difference in hydration for the two samples. Stratum corneum *in situ* is hydrated by the underlying viable skin and, in this case, also by the electrode gel. The *in vitro* skin is in balance with the ambient relative humidity and the hydrogel electrodes do not increase the hydration (Jossinet and McAdams 1991).

An important finding by Yamamoto and Yamamoto (1976) was that the impedance of the removed stratum corneum layers did not produce a circular arc in the complex impedance plane. This is obvious from Fig. 4.23, where the admittance data from Fig. 4.21 have been transformed to impedance values and plotted in the complex plane. Hence, if one plots the data from multi-frequency measurements on skin *in vivo* in the complex plane and uses a circular regression in order to derive, e.g., the Cole parameters, one must be aware that stratum corneum alone does not necessarily produce a circular arc, and as described earlier in this chapter, the measured volume or skin layer is highly dependent on frequency. The derived parameters therefore represent a mixture of different skin layers and different dispersion mechanisms and are thus totally ambiguous when used for characterising conditions of specific skin layers.

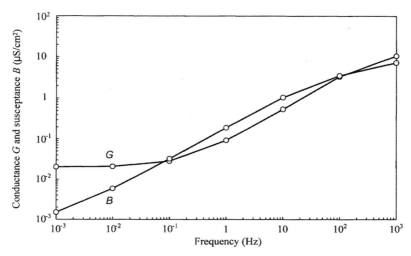

Figure 4.22. Electrical admittance of palmar stratum corneum *in vitro*. Martinsen *et al.* (1997a), by permission.

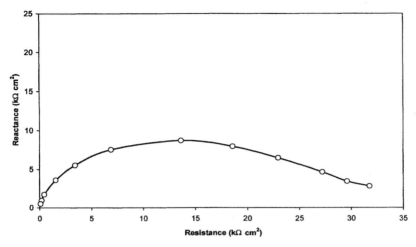

Figure 4.23. The stratum corneum data from Fig. 4.21 plotted in the complex impedance plane.

Skin admittance varies greatly both between persons and between different skin sites on the same person. Changes in, for example, sweat gland activity and ambient relative humidity during the day or year season are also reflected in large variations in skin admittance, mainly because of changes in skin hydration.

Table 4.2 shows the results from impedance measurements at 10 Hz on different skin sites as measured with a 12 cm^2 dry metal ECG plate electrode positioned directly on the skin site after a short breath had been applied to the skin surface (Grimnes 1983a) in order to create a surface film of water molecules for improved

Table 4.2. Site dependence of skin impedance (kΩ cm^2) at 10 Hz, dry electrode plate. Initial values and values obtained after two intervals of 2 hours. Each measurement is compared with the value (kΩ) obtained from a control pre-gelled 12 cm^2 ECG electrode on the ventral forearm. Reproduced from Grimnes (1983a)

	kΩ cm^2/kΩ		
	Start	2 hours	4 hours
Hand: dorsal side	720/80	210/17	300/33
Forearm: ventral–distal	250/80	240/17	190/35
Forearm: ventral–middle	840/80	230/17	360/36
Forearm: ventral–proximal	560/80	180/17	260/36
Upper arm: dorsal	840/75	260/16	660/36
Upper arm: ventral	1000/70	300/16	780/34
Forehead	60/70	36/16	48/35
Calf	325/45	375/17	325/36
Thorax	130/37	110/16	130/35
Palm	200/80	150/17	200/33
Heel	120/60	180/15	120/35

skin–metal electrolytic contact. The first column presents measurements immediately after the electrode had been applied, and the two subsequent columns gives the values after 2 and 4 hours respectively. The dry measuring plate electrode was removed immediately after each measurement. The first column shows a large variation in the control impedance, which was interpreted as unstable filling of sweat ducts during the measurement. The two other columns show stable results at two different levels of control. The values in Table 4.2 clearly demonstrate the large variability in skin impedance on different skin sites, and how this changes during a period of skin occlusion.

The sweat ducts of the skin introduce electrical shunt paths for dc current. Although lateral counterion relaxation effects have been demonstrated in pores, this effect is presumably negligible in sweat ducts, and sweat ducts are hence predominantly conductive (Martinsen *et al.* 1998). The dc conductance measured on human skin is not only due to the sweat ducts, however. Measurements on isolated stratum corneum as well as nail and hair reveal conductance values comparable to those found on skin *in vivo* (Martinsen *et al.* 1997a,b).

Since sweat duct polarisation is insignificant, the polarisation admittance of the skin is linked to the stratum corneum alone. This implies that measurements of capacitance or ac conductance at low frequencies reflect only the properties of the stratum corneum.

The series resistance, i.e. the limiting impedance value at very high frequencies, is very small for the stratum corneum. In a practical experimental setup, the impedance of the viable skin will in fact overrule this component. The value of this effective series resistance is typically in the range 100–500 Ω.

Skin penetrated by external electrolytes

In low-frequency applications, < 100 Hz, the skin impedance is very high compared to the polarisation impedance of wet electrodes and deeper tissue impedance. The stratum corneum consists of dead and dry tissue, and its admittance is very dependent on the state of the superficial layers and the water content (humidity) of the surrounding air in contact with the skin prior to electrode application. In addition, the sweat ducts shunt the stratum corneum with a very variable dc conductance. Sweat both fills the ducts and moisturises the surrounding stratum corneum. The state of the skin and measured skin admittance at the time of electrode application is therefore very variable. With low sweat activity and dry surroundings, the skin admittance may easily attain values < 1 μS/cm^2 at 1 Hz.

From the time of application of a *dry* metal plate, the water from the deeper, living layers of the skin will slowly build up both a water contact with the initially dry plate and water content in the stratum corneum. A similar process will take place in the skin with hydrogel as contact medium (but here the metal/gel interphase is already established). The processes may take a quarter of an hour or more. The water vapour pressure of the hydrogel may be such as to supply or deplete the stratum corneum of water, depending on the initial skin conditions.

To avoid a long period of poor contact, a skin drilling technique may remove the stratum corneum. Even mild rubbing with sandpaper may reduce the initial impedance considerably. To shorten the long period of poor contact, an electrolytic solution or wet gel is often applied to the skin. The electrolyte concentration of the contact medium is very important. With high salt concentration, the water osmotic pressure in the deeper layers will strongly increase the water transport up through the skin to the high-concentration zone. This may be admissible for short-time use (e.g. $< \frac{1}{2}$ hour). For prolonged use, skin irritation may be intolerable. For long-term use, the concentration must be reduced to the range of a physiological saline solution (around 1% of electrolytes by weight).

A surface electrode with contact electrolyte covering a part of the skin may influence measured skin immittance by four different mechanisms:

1. Changing the water partial pressure gradient in the stratum corneum (SC).
2. Osmotic transport of water to or from the contact electrolyte.
3. Penetration of substances from the electrode gel into the SC.
4. Changing the sweat duct filling.

With a dry electrode plate, the moisture buildup and increase in admittance in the SC start at the moment of electrode application. With a hydrogel, admittance may increase or decrease. With wet gel or a liquid the initial admittance is high, and with strong contact electrolytes the admittance will further increase for many hours and days (Fig. 4.24). As the outer layers of stratum corneum may be wet or dry according to the humidity of the ambient air, it will not be possible to find a general contact medium that just stabilises the water content in the state it was in before electrode application, and the onset of the electrode will generally influence the parameters measured.

With dry skin the admittance may be $< 1 \ \mu S/cm^2$ at 1 Hz. A typical admittance of $> 100 \ \mu S/cm^2$ is possible when the stratum corneum is saturated by electrolytes and water. The conductivity is very dependent on water content and is believed to be

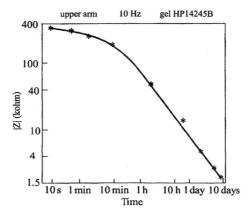

Figure 4.24. Skin impedance as a function of time with an ECG electrode of commercial, wet gel, strong electrolyte type. From Grimnes (1983a), by permission.

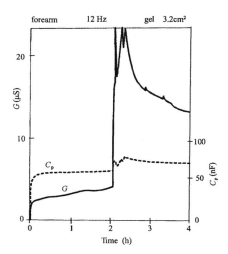

Figure 4.25. *In vivo* skin conductance and parallel capacitance during abrupt sweat duct filling (sudden physical exercise) and emptying. From Grimnes (1984), by permission.

caused, for example, by ions, protons (H^+) and charged, bound proteins that contribute only to ac-admittance.

Figure 4.25 shows the dominating effect of sweat duct filling on skin admittance, and shows clearly how skin capacitance is in parallel and therefore unaffected by the parallel conductance change.

Hair and nail

Hair is composed of compactly cemented keratinised cells, and grows in *hair follicles* that essentially are invaginations of the epidermis into the dermis. The *sebaceous glands* are located on the sides of the hair follicles. They secrete *sebum* to the skin surface, but the purpose of this secretion is unclear, apart from the fact that it gives a scent that probably is unique to each human being. The problem of acne vulgaris, well known to most young people, is connected with these glands.

The frequency response of hair is not unlike the properties of skin, but typical measured admittance values are of course very small, which makes these measurements complicated to perform. Figure 4.26 shows measurements on 100 hair fibres in parallel at different ambient relative humidities (Martinsen *et al.* 1997b). Hair length was approximately 2 cm. At 86% RH the conductance is almost frequency independent (dominated by dc properties) up to 1 kHz, and at 75% and 62% only a small increase at the highest frequencies can be detected. The susceptance increases linearly at the highest frequencies, but levels off at the lowest frequencies. The frequency at which this flattening starts increases with RH.

Adsorption of water into hair is a very slow process. Robbins (1979) found that the hydration of hair fibres stabilised after a period of 18–24 hours after being introduced to an increase in ambient RH. Martinsen *et al.* (1997b) found that the

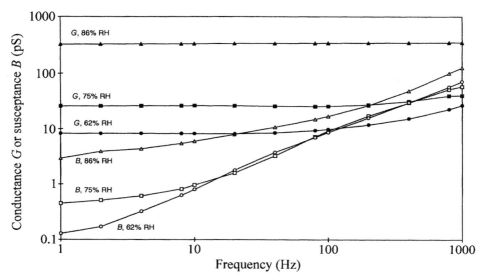

Figure 4.26. Electrical admittance of 100 hair fibres in parallel as function of relative humidity (RH) (Martinsen 1997b).

conductance continued to increase for several days after such a step in RH and concluded that a possible cause of this is that adsorbed water molecules regroup in a way that increases their contribution to the conductivity. They furthermore found desorption to be a very rapid process where the main change in conductance appeared during the first minutes after reducing ambient RH, but also with minor changes in conductance over the succeeding few hours.

The electrical admittance of keratinised tissue is typically logarithmically dependent on water content or ambient relative humidity. An example from human hair is shown in Fig. 4.27, where the 1 Hz conductance of 50 fibres in parallel is plotted against ambient relative humidity (Martinsen *et al.* 1997b).

Nail is also keratinised tissue but is harder than stratum corneum. This is partly due to the "hard" α-keratin in nail as opposed to the more "soft" β-keratin in stratum corneum (Baden 1970; Forslind 1970). Although nail is readily available and easy to perform electrical measurements on, the electrical admittance of human nail has not been extensively investigated. Figure 4.28 shows the conductance and susceptance of a 450 μm-thick nail at 38% RH, measured with a two-electrode hydrogel system (Martinsen *et al.* 1997a).

The electrical properties of nail resemble those of stratum corneum and hair. Note, however, that the low-frequency susceptance plateau in Fig. 4.28 represents a deviation from a simple model with a distribution of relaxation times for a single dispersion mechanism (cf. Section 7.2), and must be due to another dispersion mechanism such as electrode polarisation, skin layers, etc. The admittance of nail is also logarithmically dependent on water content as shown in Fig. 4.29 (Martinsen *et al.* 1997c).

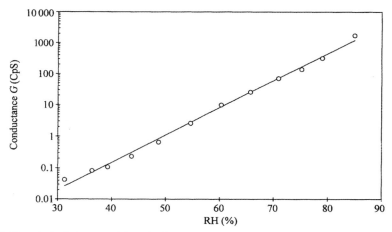

Figure 4.27. Electrical admittance of 50 hair fibres as a function of RH at 1 Hz. Circles are measured values and the line is logarithmic regression (Martinsen 1997b).

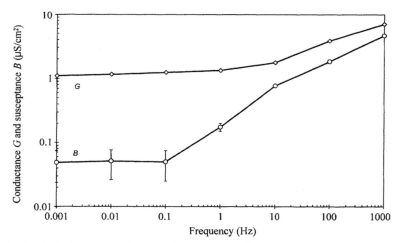

Figure 4.28. Electrical admittance of human nail *in situ*. Error-bars show uncertainty in low-frequency susceptance data (Martinsen *et al.* 1997a).

4.3.6. Whole body

Resistance

The value of the impedance between two skin surface electrodes is usually dominated by the contribution of the skin. However, the skin impedance may be negligible if:

1. the voltage is high enough for skin electrical breakdown;
2. the skin is thoroughly wetted;
3. the skin is perforated;

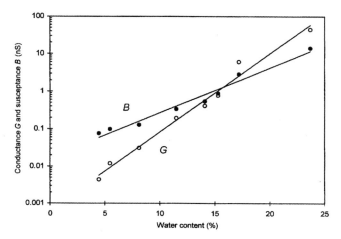

Figure 4.29. Electrical admittance for nail *in vitro* at 80 Hz as function of absolute water content (weight%). Area is 3.14 mm^2 and thickness is 0.34 mm. Circles are measured values and lines are logarithmic regression (Martinsen *et al.* 1997c).

4. effective skin contact area is very large;
5. the signal frequency is sufficiently high.

Without the skin and with the living body tissue considered purely resistive, the resistance R of a body segment is determined by the mean resistivity ρ, mean length L and mean cross-sectional area A according to $R = \rho L/A$. The resistance can be measured by a four-electrode technique (Freiberger 1933; Grimnes 1983a).

Notice in Fig. 4.30 the importance of the cross-sectional area. The resistance is dominated by the contribution from a finger, underarm and leg, while the influence from the chest is negligible. In addition, the current flow may be constricted near the electrodes if the electrodes are small, and an additional *constrictional resistance* must be accounted for (cf. Fig. 5.2). Higher resistance values may also be due to the relatively small well-conducting tissue cross-sectional areas in such joints as the wrist, elbow and knee.

When skin's electrical protection has been broken down during an electrical accident, the constrictional and segmental resistances are the current-limiting factor (cf. Section 8.14.2 and Fig. 5.2).

Body immittance data may be used for the estimation of body composition (see Section 8.8).

4.3.7. Post-excision changes, the death process

Tissue metabolism decreases after the tissue has been excised. Often the temperature falls, and measured variations must account for such possible variables. If the tissue is supported by temperature maintenance and perfusion systems, the tissue may be stabilised for a limited period of time in a living state *in vitro* (*ex vivo*).

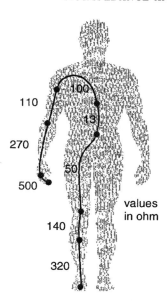

Figure 4.30. Body segment resistance distribution (no skin contribution or current constriction). Values as found with four-electrode technique. 500 ohm is a one finger contribution. Linear values according to eq. (2.2), not very dependent on current density levels.

If the tissue is not supported, irreversible changes will occur, followed by cell and tissue death. If the blood flow is interrupted, the metabolism goes further but in an anaerobic way. Osmosis will cause cell swelling and tissue damage and a consequence of this is the narrowing of extracellular pathways which typically gives an increase in the low frequency impedance (< 10 kHz). A steep increase in low frequency impedance is also found in tissue exhibiting gap junctions when these gap junctions are closed after a certain duration of ischemia. Gap junctions can also produce an additional dispersion at very low frequencies, e.g. in porcine liver at about 7 Hz, which vanishes completely when the gap junctions close. Since metabolic products are no longer removed, one will sometimes find a decrease in impedance at high frequencies (e.g. 10 MHz) (Gersing 1998). In brain tissue, irreversible changes may occur within 5 minutes at 37°C and in cardiac tissue after 20 minutes. Impedance spectroscopy has been shown to be of great value for the monitoring of these processes and for the assessment of organ state (Gersing 1998; Casas *et al.* 1999). In keratinised tissue such as the outer layers of the human skin, hair, nail and horn, the change and ultimate death of the cells are genetically programmed to occur during a period of about 30 days. The dead keratinised cells need no intra- or extracellular fluids, and may be in a dry state with low admittance values.

Bozler and Cole (1935) measured the electrical impedance of frog sartorius muscle from 1.1 kHz to 1.1 MHz. Measurements were first done approximately 2 hours after dissection. The tissue was then stimulated in order to induce contraction, and the tissue was measured again, approximately 3 hours after dissection. They found a minor arc of a circle when the data were plotted in the complex impedance plane. Between the relaxed and contracted states, R_0 was found to increase by 75%, while R_∞ increased only 2%. R_0 and R_∞ denote as usual the resistances measured at very low and very high frequency, respectively. The significant increase in R_0 was interpreted as due to a reduced ionic conduction through the cell membranes.

Schäfer *et al.* (1998) measured skeletal muscle of rabbits and dogs from 100 Hz to 10 MHz and found a β-dispersion whose characteristic frequency in the impedance plane typically moved from 20 to 10 kHz and then back to 20 kHz during ischemia. They furthermore found the low-frequency resistance to increase during the first 300 min of ischemia, subsequently decreasing until 850 min and then increasing again. The initial increase in resistance is explained as being due to increasing oedema because of osmotically induced water shifts, which reduce the extracellular volume. This effect continues after 300 min, but is surpassed by a reduced membrane resistance caused by the opening of ion channels, leading to a net reduction of the low-frequency resistance. In their model, the membrane resistance reaches a constant level after about 850 min while the extracellular resistance continues to increase, resulting in a net increase of resistance after 850 min.

Martinsen *et al.* (2000) presented measurements of the electrical properties of haddock muscle from 1 Hz to 100 kHz as a function of time after the fish was sacrificed (Fig. 4.31). Clear α- and β-dispersions were found. Most of the α-dispersion disappeared after a few hours. The low-frequency resistance of the β-dispersion increased during the first 5 hours as the fish went into rigor, and then decreased as cell destruction developed.

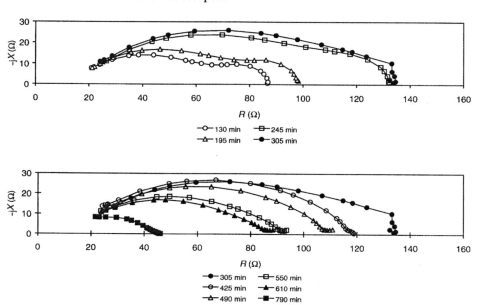

Figure 4.31. Post-mortem Z-plot for fish muscle. (top) First 5 hours after death. (bottom) Next 8 hours. Martinsen *et al.* (2000), by permission.

4.3.8. Tabulated data for measured conductivity and phase angle, mammalian tissue

In the frequency range < 100 MHz, the admittivity of living mammalian tissue is usually dominated by conductivity. This conductivity is due to the ions of the body liquids inside and outside the living cells.

Let us recall that tissue can be characterised by permittivity or admittivity. In a unity ($A/d = 1$ m) measuring cell, the permittivity is $\varepsilon' = C_p$ and $\varepsilon'' = G/\omega$. The conductivity is $\sigma' = G$ and $\sigma'' = \omega C_p$, and the problem with the double-primed parameters is their frequency dependence. In the case of a large dc conductivity from the interstitial electrolytes, ε'' will diverge as frequency is lowered, and thus completely cover the characteristic maximum and reversal towards zero that is so characteristic for the ε'' of a relaxation dispersion (cf. Fig. 3.5). It is of course possible to give ε' and ε'' as purely dispersion parameters, omitting or subtracting the dc component. A more common approach is to characterise tissue with ε' and σ', σ' being directly proportional to G. From such data the phase angle or loss angle is not directly seen, which is of interest for assessing the necessary resolution of the measuring setup.

The poorly conducting membranes of the body and the interfacial polarisation there cause permittivity corresponding to α-dispersion in the mHz to kHz range (cf. Section 3.8). At frequencies < 1 Hz, the capacitive current component vanishes, and the admittivity is purely conductive, determined by the liquids outside the outermost membranes. In a frequency range 0.1–10 MHz, the phase angle is maximum (except for human skin where it occurs at much lower frequencies). This is also the Maxwell–Wagner β-dispersion range for the dielectric interfaces, where the susceptance of the cell membranes becomes large, and their influence more and more negligible. If there is an anisotropy at lower frequencies, it has disappeared in this frequency range. The dipolar dispersion of the proteins also appears in this frequency range, and it is still active above 100 MHz. This is the γ-dispersion range, extending all the way up to the single Debye characteristic frequency of water, centred on 18 GHz. The susceptance dominates at these high frequencies, and the phase angle may be $> 80°$.

In conclusion, at low frequencies living tissue admittance is usually dominated by the conductivity of tissue electrolytes, at high frequencies by the dielectric constant.

Electrical tissue data have been tabulated by many authors in many review articles and books: Schwan and Kay (1957), Geddes and Baker (1967), Schwan (1963), Stoy *et al.* (1982), Foster and Schwan (1986), Duck (1990), Stuchly and Stuchly (1990), Gabriel *et al.* (1996). The difference between data from human and mammalian animal tissue is considered small. The tissue may be excised material from freshly killed animals, human autopsy material obtained a day or two after death, or tissue *in vivo*. Tissue data changes rapidly after death (cf. Section 4.3.7).

Table 4.3 gives some typical tissue values.

4.3.9. Plant tissue

In general, plant tissue immittance data are not in accordance with Cole system models. Non-symmetrical models of the Davidson–Cole and Havriliak–Negami

Table 4.3. Tissue conductivity, LF and 1 MHz

Tissue	σ (S/m) 1 Hz–10 kHz	σ (S/m) ca. 1 MHz	φ_{max} at <10 MHz	Anisotropy
Human skin, dry	10^{-7}	10^{-4}	80°	?
Human skin, wet	10^{-5}	10^{-4}	30°	?
Bone	0.01		20°	?
Fat	0.02–0.05	0.02–0.05	3°	Small
Lung	0.05–0.4	0.1–0.6	15°	Local
Brain	0.1	0.15	15°	?
Liver	0.2	0.3	5°	?
Muscle	0.05–0.4	0.6	30°	Strong
Whole blood	0.7	0.7	20°	Flow dependent
Urine	0.5–2.6	0.5–2.6	0°	0
Saline, 0.9%, 20°C	1.3	1.3	0°	0
Saline, 0.9%, 37°C	2	2	0°	0

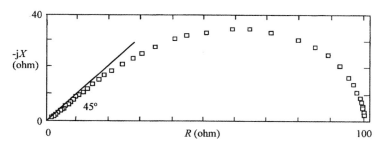

Figure 4.32. Impedance plot for Scots pine needles. From Zhang *et al.* (1995), by permission.

types have been used (Zhang *et al.* 1995). Figure 4.32 shows a Z-plot with data from Scots pine needles obtained with a two-electrode technique. Air spaces in the needles were believed to have an important influence both on the α of the Cole equation and the skewness of the ZARC. Strong temperature dependence has also been reported for plant tissue. Chilcott and Coster (1991) found that a 10°C increase resulted in more than a doubling of the conductance of Chara Corallina (an aquatic plant).

4.4. Special Electrical Properties

Electric current flow patterns are determined by tissue morphology, tissue composition and the nature of the electricity source. If it is an exogenic source, the electrode geometry is of course important. But the current distribution in the human body is not very well known. Is current density much higher in the blood because of its higher conductivity? Or are some of the epithelia so tight that at least dc and LF currents do not penetrate into the vessels? At high frequencies (>1 MHz) most epithelia are capacitatively shunted, and the current density should be determined mainly by the concentration of ions and tissue permittivity.

4.4.1. Tissue anisotropy

In tissue conductivity may be 10 times larger in one direction than another (cf. Fig. 4.15). At low amplitude levels the tissue is still linear, and the principle of superposition and the reciprocity theorem are still valid. However, Ohm's law for volume conductors $\mathbf{J} = \sigma\mathbf{E}$ is not valid even if it still is linear: the current density direction determined by the local $\mathbf{E}$-field direction (Figs 4.33 and 5.20) will not necessarily coincide with the macroscopical (Figs 4.34 and 5.20) $\mathbf{E}$-field direction. The possibility of selective measurement of defined tissue volumes is disturbed. This may lead to serious difficulties in impedance tomography, finite element (FEM) calculations and immittance plethysmography.

4.4.2. Tissue dc properties

Endogenic dc potentials/currents
The Nernst equation (2.24) defines potential differences as a function of ionic activities, and the Maxwell–Gauss equation (7.2) defines potential changes as a

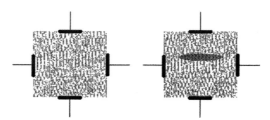

Figure 4.33. Measured anisotropy created by *macroscopic* inhomogeneity. Homogeneous medium (left). One object with different immittivity (right).

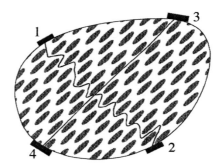

Figure 4.34. Measured anisotropy created by *microscopic* inhomogeneity. The anisotropy disappears at higher frequencies if caused by capacitive membranes. When caused by, e.g., air in the lungs, the anisotropy will persist over a broader frequency range.

function of charge densities. We may hypothesise that there are dc potential differences between some of the organs of the body, and that accordingly there may be large endogenic electric dc currents flowing between organs. Imagine, for example, the stomach with its large content of concentrated HCl (Nordenstrøm 1983).

The most basic dc potential difference is the polarisation potential of about -70 mV in the interior of excitable cells with reference to the extracellular fluids. Also the palmar and plantar skin sites are at a negative potential with respect to other skin sites, by a magnitude up to -30 to -40 mV dc. This is created by a difference in ionic activities across the skin barrier, with a possible addition of streaming potential waves from sweat propelled up the sweat ducts. The skin potential is shunted by skin sweat duct conductance. Increased sweat activity, as occurs during a *galvanic skin response* (GSR), therefore reduces the magnitude of the dc voltage (GSR-dc-wave). GSR is the dc part of the more general concept: *electrodermal response* (EDR) (cf. Section 8.13.3). On the surface of the body, dc recording is utilized in electro-oculography (EOG), to determine the position of an eye behind a closed eyelid. Together with GSR this is one of the few electrophysiological methods making use of dc potentials.

For dc measurement it is important that the two electrodes are reference electrodes. If the electrodes are made of different metals or surfaces, a large exogenic dc voltage (possibly > 1 volt) may be generated. Often the best choice is two AgCl reference electrodes. They can be coupled to the skin or tissue via electrode gel as shown in Fig. 8.10 or via a salt bridge to reduce the dc offset from liquid junction potentials. An *invasive* electrode as neutral electrode is the most stable dc reference with a unipolar skin potential recording system.

Remember that with an exogenic dc source it is not possible from a pure dc measurement to discern between changes in emv's and resistance/conductance (cf. Section 6.2.6). Measured dc voltage is not necessarily proportional to the dc resistance of the unknown if a dc constant current is passed through the unknown.

Closed electric circuits in the body
As explained, there are endoelectrogenic sources in the cell membranes, but it is quite likely that some macroscopic membranes around organs also are the sites of electricity sources. Nordenstrøm (1983) proposed that there are closed dc circuits in the body with the well conducting blood vessels serving as cables, e.g. a vascular-interstitial closed circuit. These dc currents can cause electro-osmotic transport through capillaries.

Dental galvanism is the production of electricity by metals in the teeth.

Dc conductivity
Living tissue cells are surrounded by extracellular liquids, and these liquids have a dc conductivity around 1 S/m (Table 4.3). However, all organs in the body are surrounded by epithelia, and epithelia with tight junctions between the cells have very low dc conductance. Measured dc conductivity is therefore very dependent on what sort of epithelia and lipid bilayers the current has to cross.

Injury potentials, bone fracture growth

It is well known that a zone of tissue injury is a source of dc potentials. It has often been postulated that bone growth is enhanced if a dc current is supplied to each side of the fracture. Nordenstrøm (1983) used dc in an electrochemical treatment of tumours.

4.4.3. Nerves and muscles excited

In the late 1930s Cole and Curtis extended their investigations to the non-linear effects of excitable membranes. In 1939 Hodgkin and Huxley, then aged 25 and 21, succeeded in measuring the voltage inside a nerve cell for the first time. They showed that during a nerve impulse the intracellular voltage reversed, e.g. from -70 mV to $+30$ mV. Therefore this could not be due to a simple short-circuiting effect caused by increased membrane conductance. In a series of five remarkable articles in 1952, they and Bernhard Katz gave the now classical description of the voltage-sensitive opening and closing of separate channels for Na^+ and K^+ in the nerve cell membrane (cf. Fig 4.9). Hodgkin and Huxley won the Nobel Prize in 1963.

4.4.4. Non-linear tissue parameters, breakdown

At sufficiently low volume power density, every biomaterial is linear. At sufficiently high volume power density, every biomaterial is non-linear. Many applications make use of the non-linear region, where the principle of superposition no longer is valid. The non-linearity may be a property of the biomaterial, or of the electrode/electrolytic systems used, cf. Section 2.5.4 on the non-linearity of electrolytics, and Section 6.1.4 on measuring principles for non-linear systems.

At the *atomic and molecular* level, some of the charge displacements will reach saturation at high E-field strengths (cf. Section 2.5.4). The alignment of dipoles in a polar dielectric will reach a maximum when the field energy is of the same order of magnitude as the Boltzmann factor kT.

At the *cellular* level the cell membranes of polarised cells are of the order of 10 kV/mm, and additional field strengths may easily bring the membrane into a non-linear region even without cell excitation. Cell excitation, the opening of membrane channels and the creation of an action potential are the results of non-linear processes. Electroporation and electrofusion of cells *in vitro* (Section 8.12.1) are also processes in the non-linear region.

In human *skin* Yamamoto and Yamamoto (1981) found the upper limit of linearity to be about 10 $\mu A/cm^2$ at 10 Hz and 100 $\mu A/cm^2$ at 100 Hz. They ascribe this phenomenon to the ionic conduction in the keratins of the stratum corneum. Ionic flow through human skin *in vivo* is probably constrictional through special zones of high conductance (Grimnes 1984). Thus non-linear phenomena may occur at such low average current densities. Electro-osmosis in the sweat ducts causes strong non-linear effects with only a few volts dc (Fig. 4.35) (Grimnes 1983b). A dielectric breakdown will occur at very high electrical fields in the stratum corneum. Grimnes (1983c) found that dry skin on the dorsal side of the hand did withstand

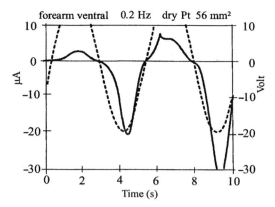

Figure 4.35. Monopolar voltage–current curves showing electro-osmosis and strong non-linearity in human skin *in vivo*. From Grimnes (1983b), by permission.

580 V dc for more than 3 s, but that breakdown was immediate at 935 V. This is an astonishingly high value remembering that the skin is of the order of 20 μm thickness. Pliquett and Weaver (1996) studied electroporation in human skin. Figure 4.36 shows that dry human stratum corneum *in vivo* could withstand a positive polarity dc voltage of 600 volts for 4 seconds. With negative polarity the breakdown is much more pronounced because of electro-osmotic transport of electrolyte solution from deeper living parts of the skin.

There are some important *clinical* applications involving non-linearity:

- A *defibrillator* current of > 50 amps over 60 cm^2 electrode area implies a current density around 1 A/cm^2 (Section 8.10). This is clearly in the non-linear region for both skin and living tissue, and the skin is actually reddened, particularly under the electrode edge after a shock has been given. Geddes *et al.* (1976) reported that the resistance found from measured peak voltage with a 20 A peak current pulse

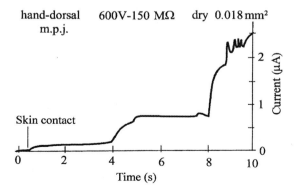

Figure 4.36. Thin stratum corneum dielectric strength, human skin *in vivo*. From Grimnes (1983c), by permission.

with two 60 cm^2 electrodes corresponded to the small signal impedance measured at 30 kHz. During successive defibrillator shocks, a decrease in transthoracic impedance has been reported (Geddes *et al.* 1975b). After ten 400 J shocks, the impedance fell to 80% of the initial value. These effects are probably attributable to cell membrane and myocardium damage. Because of the current constriction near the electrodes, the highest current densities and largest effects are presumably found there.

- In *electrosurgery* the current density near the electrode is much higher than during defibrillation, and current duration is much longer. In cutting mode the cell interior is brought to 100°C in just a few milliseconds, and the cells explode. Tissue destruction is the goal (Section 8.11).

- *Iontophoresis* for sweat sampling using dc current through the skin is presumably carried out in the non-linear region (Section 8.13.4 and 5).

- An *electrostatic discharge* with an electric arc as contact medium may indicate very high current densities on the affected area (cf. Section 8.14 on threshold of perception).

4.4.5. Piezoelectric and triboelectric effects

In Chapter 3 we discussed the fact that an applied electric field induces dipole moments in a dielectric. Such a displacement of charges generally generates a dimensional change in the material. This is called *electrostriction*.

Vice versa, mechanical stress changes the dimensions of the material. Usually this does not result in an electrical polarisation of the material, because most materials have a so-called centre of symmetry, cancelling opposite charge displacements. However, crystals lack such a centre of symmetry, and they generate an internal polarisation **P** when mechanically deformed. These materials are called *piezoelectric*, with a direct conversion from mechanical to electrical energy.

Piezoelectric properties have been found in human hair and other keratinised materials. This is also the case for bone and tendon. Bone remodelling has been attributed to piezoelectricity. The theory is that the mechanical stress on a bone generates bioelectricity that in turn influences bone growth. As most biomaterials exhibit piezoelectric properties, it is not strange that there is an abundance of theories postulating piezoelectric effects in tissue, e.g. that the transduction mechanisms in the inner ear, in the hair follicles and of touch and vibrational sensitivity are piezoelectric. Results in the literature are often obtained with dry samples, and the question remains of the importance of piezoelectricity in living, highly conductive tissue.

Triboelectricity (frictional charging) is the generation of charges when two materials suddenly are separated. All materials, solids and liquids, isolators and metals, display this phenomenon. *Triboelectric series* are often set up, but as the position of a material in the series is very dependent on surface properties, the series often differ. Humidity and cleanliness affect the series drastically. One such series is:

rabbit fur, bakelite, glass, *wool*, silk, cotton, wood, metals, polystyrene, Teflon

+ +◄——► — —

Presumably human hair and perhaps dry stratum corneum have positions not too far from that of wool.

During winter, triboelectricity may generate several kilovolts on an insulated person during walking, or by movement of blankets when a hospital patient's bed is made. If the patient is monitored, noise will appear on, for instance, the ECG recordings (Gordon 1975). The seriousness of the electrostatic problem is concerned not only with the generation mechanism, but also with the *discharge* mechanism. The conductivity, both volume and surface, of the materials is of utmost importance, because it determines the discharge time constant. If a person is charged by walking on a floor or removing some clothing, the person's capacitance to earth (e.g. 150 pF) together with the resistance to earth (e.g. 8 GΩ, $\tau = 1.2$ s), determines the time constant. If the time constant is several seconds, the charge will remain long enough to cause an electric arc discharge at, for instance, the touch of a metal object. If it is in the millisecond region the person is rapidly discharged, and no arc discharge will occur at the touch of a metal object. The conductance of many low-conductivity materials is very humidity dependent, so the problems are much larger at low ambient relative humidity. This occurs indoors in cold countries during the winter. Under these conditions, cotton is better (it has higher conductivity) than most synthetic materials.

Triboelectricity is of great practical interest because it is the basis of problems associated with *electrostatic* spark discharges. Such discharges represent a risk because they may excite nerves or ignite flammable vapours. The discharge may be perceptible (Section 8.14). Some of the electricity generation is due to the rubbing and separation of skin from textile clothing. The triboelectric properties of skin and hair are therefore of interest (e.g. when combing the hair). When the relative humidity in a room has been 10% for some time, a person can easily be charged to 30 kV with respect to earth just by walking or rising from a chair. During walking, the electricity generation is determined by the materials of the shoe soles and the floor.

CHAPTER 5

Geometrical Analysis

Important electrical characteristics of an electrode/tissue system are determined solely by the geometrical configuration. To clarify this important geometrical function, the systems to be treated in this chapter are simple models suited to basic analysis and mathematical treatment or, for example, finite-element analysis (Section 5.5). They are idealised to static dc systems without polarisation phenomena and frequency dependence, and therefore a potential difference between two points in the tissue is equal to the voltage difference found between two wires connected to those two points, $\Phi_{12} = V_{12}$. All models are actually volume models, but from symmetry the solutions are presented in two dimensions, often as electrodes on a surface, with a half-infinite homogeneous medium below.

A clear distinction must be made between *source* electrode systems (current-injecting, current-carrying, stimulating, driving) and *recording* (signal pick-up, receiving, registering, lead) electrode systems. The same electrode pair may serve as both (one-port, two-terminal network, cf. Section 6.1), or there may be separate 3- or 4-electrode systems (two-port, three- or four-terminal network). For the one-port system, *driving point* immittance is the only possible parameter; for the two-port systems, transfer parameters are also possible (Section 6.1). A two-port system also measures *transmittance*: e.g. current is injected in one port, and voltage is recorded at the other port. The electrode pairs of current injection and voltage recording may be interchanged: if the reciprocity theorem is valid, the transmittances found will be the same. The reciprocity theorem is not based on geometry but on network theory, and is therefore treated in Section 6.1. For the reciprocity theorem to be valid there are no constraints on geometry, only on system linearity as outlined in Section 6.1.2.

The source may be *endogenic* (arising from causes inside the organism, a bioelectric source) or *exogenic* (arising from causes outside the organism, e.g. externally applied electricity). If the conducting medium is linear and the reciprocity theorem is valid, the recorded voltages are determined solely by the input stimulation level, tissue electrical parameters, and the geometry of the injection–recording electrode system. An exogenic electrode current injecting system may be monopolar or dipolar; a bioelectric source is usually modelled with dipoles, which in their nature are bipolar.

The terms monopolar and dipolar are of Greek origin, and are the preferred usage for current-carrying systems in this book. The terms unipolar and monopolar are used in a confused way and the terminology in the literature is unclear (cf. Table 5.1).

Table 5.1. The number of electrodes and confusion from words with both Latin and Greek origins

Latin	Greek	Meaning
Unipolar	Monopolar	One electrode dominant in a non-symmetrical system of two electrodes
Bipolar	Dipolar	Two equal electrodes in a symmetrical system
Tripolar	Tripolar	Three-electrode system
Quadropolar	Tetrapolar	Four-electrode system

Current-injecting electrodes (source)

A *dipolar* current-injecting electrode pair has two electrodes that are as similar as possible, each electrode contributing in the same way (current dipole). Bioelectric sources may be modelled as discrete or distributed dipoles; important examples are the heart muscle or the cell membrane.

A *monopolar* current-injecting electrode system has one electrode as the *active* (*working*) electrode, with the other one as the *indifferent* (*silent, passive, dispersive, neutral*) electrode. An electrode may be more or less indifferent if it is made with a much larger electrode area than the active electrode.

The *ideal monopolar* system analysed in a polar or Cartesian coordinate system usually presupposes an enormous spherical reference electrode, infinitely distant at zero potential.

Recording electrodes (lead)

The term *lead* refers to the recording electrodes used, to their specified position, and to how they are coupled to display a signal channel. There are *unipolar* and *bipolar* leads. Recording signals from the body is done in its simplest form with an electrode pair, usually with negligible current flow. With unipolar leads the reference signal may come from a *reference* electrode, or from the sum of voltages from several electrodes. The other electrode is the unipolar measuring electrode.

Any recording of electric biopotentials is called an *electrogram*. If zero potential difference is measured, the tissue is *isoelectric*. A line designating tissue at the same potential is called an *equipotential* line; a cable connecting two conductors so that they have the same potential is called an equipotential cable.

A lead may be characterised by some *transmittance* parameter: a transfer function or transfer *sensitivity* factor describing the relationship between a source and the recorded signal. Transfer sensitivity can be studied with fixed recording electrodes and current sources at varying positions, or with a fixed current source and recording electrodes at varying positions (cf. Section 5.3.2).

A lead may also be characterised by a *volume sensitivity* factor S_V: the ability to differentiate between the contributions from different parts of a measured volume. This is an important parameter in impedance tomography and plethysmography.

5.1. The Forward and Inverse Problems

Calculation of the potential distribution caused by a known signal source in a defined medium is called the *forward* problem. Examples are given in Section 5.3. Usually there is a unique solution to a posed problem.

In electrophysiology a body surface potential is often recorded, with the purpose of characterising the unknown bioelectric signal source in position, magnitude and extension. Going from measured potentials in a known conductive medium and calculating back to the source properties is called the *inverse* problem. Usually there are *infinitely* many possible solutions to a posed problem (cf. the ambiguity of resistance or emv contributions in a black box, Section 6.2.6). The important classical patient examination methods such as ECG and EEG are aimed at characterising the properties or status of the source organ, and are thus based upon more or less empirical solutions to the inverse problem.

In impedance tomography, the problem is a *boundary value* problem. From recorded potentials caused by exogenic current injection, the distribution of conductivity is to be found.

These are geometrically based problems, but in all real cases the task is complicated by the fact that the body is a very inhomogeneous conductor. Precision is difficult to achieve, particularly the longer the distance between the electrodes and the tissue region studied. An interesting approach is therefore to measure the *magnetic field* without galvanic connection, because the non-magnetic tissue does not perturb magnetic fields. However, this does not solve all problems. The conductivity of tissue spreads the current out into the whole volume conductor, and these currents create their own magnetic fields.

5.2. Monopole and Dipole Sources

For the graphical illustration of a vector field, see Section 10.1.

5.2.1. Sphere monopoles

The conductance $G_{\frac{1}{2}}$ of a superconducting sphere of radius a, half ($\frac{1}{2}$) submerged in a half-infinite conducting medium of resistivity ρ, and with a half-spherical reference electrode infinitely far away (Fig. 5.1), is

$$G_{\frac{1}{2}} = 2\pi a\sigma, \qquad R_{\frac{1}{2}} = \rho/2\pi a \qquad (5.1)$$

With a metal sphere electrode in an electrolyte solution, the electrode polarisation impedance and the electrolytic resistance are physically in series. The electrolytic resistance is inversely proportional to *radius a* (or circumference $2\pi a$) of the electrode. However, the electrode *polarisation impedance* is inversely proportional to a^2 (or surface area $2\pi a^2/3$). Therefore, the influence of the electrode polarisation can be made as small as wanted by increasing the sphere radius. Alternatively, by reducing the electrode radius sufficiently, it will be possible to measure polarisation

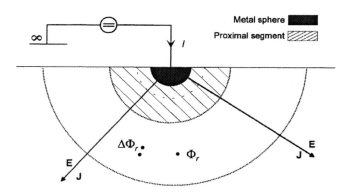

Figure 5.1. Fields and equipotential hemispheres caused by a monopolar, hemispherical source.

or electrode surface properties selectively, without influence from the electrolytic series resistance.

With an applied voltage V, the total current of the half-sphere electrode is $I_{\frac{1}{2}} = 2\pi\sigma aV$. The voltage Φ_r at distance r in the medium (a forward problem) is

$$\Phi_r = V\frac{a}{r} = \frac{I}{2\pi\sigma r} \qquad (r \geqslant a) \qquad (5.2)$$

The *E-field strength* in the medium outside the sphere at distance r from the centre is

$$E = V\frac{a}{r^2} \qquad (r \geqslant a) \qquad (5.3)$$

and the current density $\mathbf{J}$ is

$$\mathbf{J} = \sigma\mathbf{E} = \sigma V\frac{a\mathbf{r}}{r^3} \qquad \text{or} \qquad J = \sigma V\frac{a}{r^2} \qquad (r \geqslant a) \qquad (5.4)$$

Φ_r is proportional to r^{-1}, but E is proportional to r^{-2}. Note that E is independent of σ; it is only geometry- and V-dependent. Note also that as sphere radius $a \to 0$, then $G \to 0$, $E \to 0$ and $J \to 0$! The whole voltage drop Φ occurs at the electrode, and no current flows. This is not the result of a polarisation, it is a pure function of geometry.

Power density W_v (watt/m^3) is

$$W_v = \sigma V^2 \frac{a^2}{r^4} \qquad (5.5)$$

Power density falls off extremely rapidly with distance from the electrode, and the equation formulates the basis of unipolar electrosurgery (Section 8.11). It shows that diathermy effect is concentrated to a very narrow sheath around the electrode, with little tissue damage elsewhere in the tissue. The temperature rise ΔT in a tissue volume is

$$\Delta T = \frac{J^2}{\sigma} \frac{t}{cd} \qquad (5.6)$$

where c is specific heat capacity (e.g. for water 4.2 kJ kg^{-1} °C^{-1}), and d is the density. With a current density of $J = 200$ mA/cm^2, $\sigma = 1$ S/m and $d = 1000$ kg/m^3, the temperature rise in the tissue per second is 1°C. This is under adiabatic conditions, that is with no cooling effects from the surroundings (e.g. no blood flow). If the current density is not in phase with the electric field, the in-phase part of the current density is the only part contributing to joule heating.

The contribution of a small tissue volume to the total conductance $G_{\frac{1}{2}}$ is dependent on the volume's distance from the electrode centre. It will be shown later that immittance volume sensitivity is proportional to the local current density in the volume. According to eq. (5.4), current density J of a sphere electrode falls with the square of the distance r in the tissue. We introduce the relative *volume sensitivity* S_{VG} of an electrode as the ratio between the conductance contribution of a small defined volume at two different distances r_1 and r_2. It is easily shown that for the sphere conductance $G_{\frac{1}{2}}$, S_{VG} is

$$S_{VG} = 100(r_1/r_2)^2 \qquad \text{(in \%; monopolar spherical source)} \qquad (5.7)$$

Strong local E-field strength and current density cause a large influence from a local change in conductivity. It is often practical to use $r_1 = a$ as reference, so that S_{VG} is 100% at the electrode surface and falls with the square of the distance into the tissue. This is important, for example, when a sphere is used as a skin surface electrode. The geometry of the electrode itself causes a large influence from the tissue surface (skin), with far less influence from deeper layers, even with uniform tissue conductivity. The proximal zone with current constriction implies that such electrodes show a degree of volume selectivity.

The zone of high sensitivity *proximal* to a monopolar electrode is the *constrictional zone* of the current path (Fig. 5.2). The resistance in the proximal zone of a half-sphere varies according to eq. (5.1); in the segmental zone with uniform current density it goes according to $R = \rho L/A$. With a small sphere the total resistance will be dominated by the constrictional resistance in Fig. 5.2; this then is a monopolar system.

The proximal zone may be defined as the zone where, say, the sensitivity $S_{VG} > 10\%$. In the case of a spherical electrode, the constrictional zone is then the spherical volume enclosed by $r = 3.3a$. This zone contains 70% of the total resistance $R = 1/G$, because the part of R lying within a proximal sphere of radius r, $R_{prox}(r)$ is

$$R_{prox}(r) = R(1 - a/r) \qquad (r \geqslant a) \qquad (5.8)$$

If instead the constrictional zone is defined as the volume contributing to 90% of the total resistance R, then $r = 10a$. It is the volume outside which the power density has fallen to $< 1/10\,000$ of the value at the sphere surface.

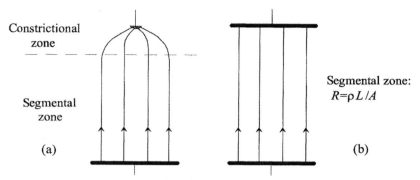

Figure 5.2. Current constriction. (a) Monopolar system where the segmental resistance of the bipolar system is increased by the current constrictional geometry. (b) Bipolar system.

5.2.2. Spheroid and disk monopoles

Disk electrodes

Analytical solutions are possible by letting a sphere degenerate to an oblate, symmetrical ellipsoid (spheroid). If the thickness of the spheroid → 0, we end up with a disk on the surface, the most common of all electrode geometries. We then discover the somewhat contraintuitive fact that current is not injected evenly under the surface of such a plate electrode.

Figure 5.3 is a two-dimensional cut through the centre of a surface disk electrode on a conducting homogeneous medium. It shows the potential field with a current-injecting electrode plate of radius $a = 5$ units. The equipotential lines are themselves spheroids: at a distance $> 2a$ they rapidly approach a circular shape. But near the electrode edge the equipotential spheroids are much closer than at the centre, indicating a higher current density at the edge.

In addition to the disk, each confocal spheroid in Fig. 5.3 may be regarded as a possible electrode, with unchanged equipotential lines outside the electrode. The only difference is that the conductance G of such a spheroid will increase with size. The conductance of the surface disk with radius a is found to be (one side)

$$G_{\frac{1}{2}} = 4a\sigma \tag{5.9}$$

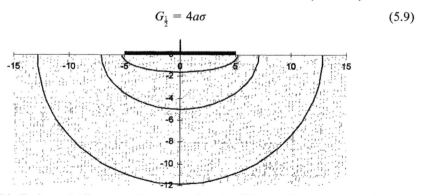

Figure 5.3. Equipotential lines caused by a current-injecting disk electrode at the tissue surface.

Comparing with eq. (5.1), a disk has therefore only $2/\pi$ times less conductance than a sphere of the same radius a, independently of a. As for the sphere, the conductance is proportional to the electrode circumference, not to the area.

Under a surface disk electrode, the current density as a function of the radius r (one disk surface) is (Fig. 5.4)

$$J_{\frac{1}{2}} = \frac{I}{2\pi a(a^2 - r^2)^{1/2}} \qquad r < a \qquad (5.10)$$

The current density at the disk centre is the same as on the surface of a sphere of the same radius a.

Electrode area
Equation (5.10) shows that the conductance of a disk is proportional not to disk area but to disk radius or circumference. However, the electrode polarisation immittance is related to electrode area. Equation (5.10) shows that current density is infinite at the edge of a disk electrode. The edge is therefore outside the zone of linearity. There are many misconceptions in the literature, with current density under a surface electrode considered uniform. This relates to the disk electrode as a *current-carrying* electrode.

As a *recording* electrode, the area is of importance in completely different ways: the larger the electrode area: (a) the larger the averaging effect and the lower the spatial resolution, and (b) the smaller the signal source impedance (less noise, better high-frequency response).

Sensitivity
The relative volume sensitivity S_{VG} of $G_{\frac{1}{2}}$ will be a function of the local current density. Conductance measured with a disk electrode will therefore be more sensitive to changes in the local conductivity of the tissue near the circumference of the electrode.

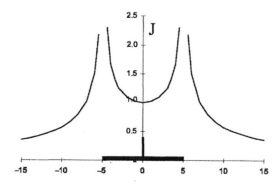

Figure 5.4. Current density in the tissue surface layer caused by a disk electrode. $J = 1$ corresponds to the current density of a spherical electrode with the same radius, half immersed in the tissue.

Measuring depth

The distance between surface disk or band electrodes will determine measuring depth. Figure 5.5 shows how the measuring depth depends on the interelectrode spacing. If the bipolar electrodes are driven from a constant-amplitude voltage source, the total current and in particular the current density will decrease with increasing distance. In addition, the deeper layers' relative contribution will increase. Even so, the sensitivity is proportional to the current density, so a given volume of tissue proximal to the electrodes is more important for the result than the *same volume* in the deeper layers. By varying the distance between the electrodes, it is therefore possible to control the measuring depth (Ollmar 1995).

Measuring depth is not so strongly dependent on electrode dimensions. This is illustrated in Fig. 5.6 with equal electrode centre distances.

By using multiple surface electrodes it is possible to reduce the measuring depth somewhat (Fig. 5.7). Such electrodes are often made in the form of metal strips rather than disks. Measuring depth is determined *both* by the strip width and by the distance between the strips.

If the homogeneous layer is covered by a second poorly conducting homogeneous layer, the current will go more or less perpendicularly right down to the well conducting layer (Fig. 5.8). The whole volume of the poorly conducting layer will dominate measuring results. Depending on the conductivity ratio and upper layer thickness/edge to edge distance, a certain part of the current will still go the shortest way through the poorly conducting layer.

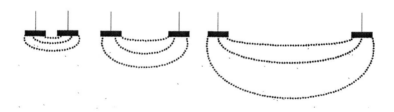

Figure 5.5. Measuring depth as a function of spacing distance between equal electrodes.

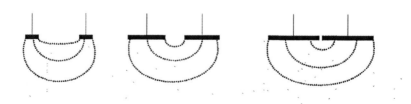

Figure 5.6. Measuring depth as a function of electrode dimensions, constant distance between electrode centres. With large electrodes and narrow space, total electrode current is not necessarily dominated by the small zone between the electrodes.

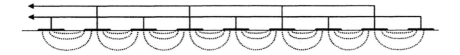

Figure 5.7. Multiple surface electrodes.

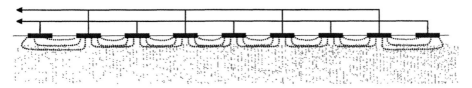

Figure 5.8. Surface electrodes on poorly conducting surface layer.

Spheroid needles
The conductance of an ellipsoidal (= spheroidal) needle in a half-infinite homo-geneous medium is

$$G_{\frac{1}{2}} = \frac{2\pi a\sigma}{\log(2a/b)} \qquad (5.11)$$

where a = rotational radius, and b = half long axis.

At the tip of such a needle electrode the current density will be very high. Figure 5.9 shows the potential field from a needle electrode in the form of a long, thin ellipsoid. The potential drop is significant near the needle tip.

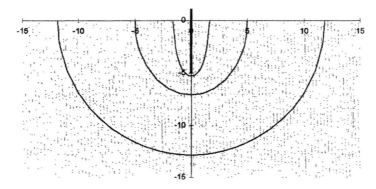

Figure 5.9. Equipotential lines of a current-injecting needle (elongated spheroid of maximum radius 0.2 units and length 5 units).

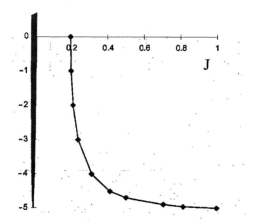

Figure 5.10. Electrode surface current density of a spheroid electrode (elongated spheroid of maximum radius 0.2 units and length 5 units).

Figure 5.10 shows the electrode surface current density as a function of depth. If such a non-insulated needle is used for nerve excitation, most of the current will be injected from the shaft. But the current density at the tip is determined by the length/radius ratio and may be 10 times as high, so at the threshold the excitation occurs near the tip.

5.2.3. Dipole

In *electrostatic* theory a dipole is two point charges separated by a distance **d**. The dipole moment **p** = q**d**. The dimension is C m, and there is no current flow, only static charges.

The *current dipole* **m** is defined with two current-carrying points separated by a distance **d** so that **m** = I**d**. The dimension is A m, and there is a flow of charges. The equations are very similar; the only difference is that for the current dipole, charge is replaced by charges per second.

The current dipole is the fundamental model for endogenic signal sources in biology. The living cell may be regarded as dipoles spread out on the cell surface area at the moment its polarised membrane is discharged. The function of the action potential of a nerve cell may be regarded as a dipole or dipoles under translocation. The sum of millions of muscle cells is considered equivalent to a single dipole (the heart activity vector). The dipole is an attractive model because it is simple. Mathematically it is also apparently simple.

The potential field of a current dipole embedded in a uniform infinite medium (a forward problem) is

$$\Phi_d = \frac{I}{4\pi\sigma}\left(\frac{1}{r_1} - \frac{1}{r_2}\right) \qquad \text{(exact)} \qquad (5.12)$$

where r_1 and r_2 are the distances from each pole to the position of Φ_d. Equation (5.12) is valid both near and far from the dipole. The equation does not refer to a coordinate system.

As already mentioned, the dipole moment of an *electric* dipole is $\mathbf{p} = q\mathbf{d}$, and of a *current* dipole $\mathbf{m} = I\mathbf{d}$. An *ideal* electric dipole is a dipole where $q \to \infty$ and $d \to 0$, keeping the dipole moment constant. An *ideal current* dipole is a dipole where $I \to \infty$ and $d \to 0$, keeping m constant. These are not realistic conditions for a current dipole, because the diameter a of the poles would approach zero, and the driving potential and the dissipated power in the constrictional zone would diverge.

At a distance large in comparison with the dipole length d, the potential is

$$\Phi_d = (I/4\pi\sigma)\mathbf{d} \cdot \mathbf{r}/r^3 = (1/4\pi\sigma)\mathbf{m} \cdot \mathbf{r}/r^3 \quad \text{(approximation)} \quad (5.13)$$

With an *ideal* current dipole this is an *exact* solution. The vector dot product indicates the directional sensitivity shown in Fig. 5.11. The potential falls as $1/r^2$ in all radial directions.

Figure 5.11 illustrates the potential distribution around a current dipole in a homogeneous conductive medium. The model is based upon two spherical current-carrying electrodes at a distance $\mathbf{d}$ apart. Each electrode may be regarded as a unipolar electrode in the proximity zone of the current path, where the equipotential lines are concentric circles. At a distance d or larger from the poles, the equipotential lines are clearly not circular. However, in cable theory, with the poles of Fig. 5.11 being long wires extending perpendicular out from the paper plane, all equipotential lines are circular but eccentric.

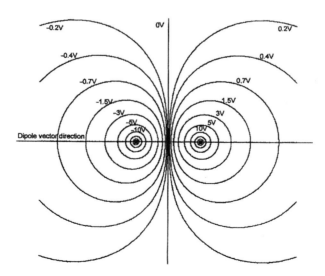

Figure 5.11. Equipotential lines of a current dipole. The high density of equipotential lines in the centre is due to the arbitrarily chosen potential difference between equipotential lines.

5.3. Recording Leads

The lead was defined in Section 5.1. The signals picked up by a recording lead may be of endogenic or exogenic origin. This Section is based upon endogenic sources modelled as single monopoles or dipoles. The forward problem with well-defined electrode geometries is treated. In Section 5.4 exogenic sources are treated, involving true three- or four-electrode systems.

5.3.1. Potentials recorded in the monopolar field

Unipolar recording electrode, spherical monopolar source
With a *unipolar* recording electrode, measured potential falls with the distance as $1/r$ (eq. 5.2: $\Phi_r = Va/r$, where a is the electrode radius). The sensitivity of the lead permits recording at relatively large distance from the source.

Bipolar recording electrodes, spherical monopolar source
With the bipolar orientation *tangential* to the equipotential circles of the monopole, $\Delta\Phi = 0$. With *radial* orientation:

$$\Delta\Phi = Va(r_1 - r_2)/r_1r_2 = Vad/r_1r_2 \qquad \text{(radial orientation)} \qquad (5.14)$$

where d is the distance between the bipolar recording electrodes. At a distance from the monopole when $r_1 \approx r_2$, $\Delta\Phi$ falls off as $1/r^2$. The sensitivity of the lead is therefore highly dependent both on electrode pair orientation and distance to source. The lead is more selective (discriminative) with respect to the position of the source.

5.3.2. Potentials recorded in a dipolar field

All endogenic sources are of a dipolar nature, e.g. the signals from nerves and muscles. The recording may be both bipolar and unipolar, from skin surface electrodes or from needles (ECG, EEG). Exogenic current sources may be monopolar or dipolar: pacemakers use both dipolar and monopolar stimulation and recording; defibrillator electrodes are usually dipolar; electrosurgery is both dipolar and monopolar.

The unipolar potential Φ_d from a current dipole is $\Phi_d = (I/4\pi\sigma)(1/r_1 - 1/r_2)$ exactly (eq. 5.12). When $r_1 = r_2$, $\Phi_d = 0$. At large distance between the recording electrode and the source, the potential falls off as $1/r^2$. We will now model the potential from a dipole source as the recording electrode(s) move with respect to the dipole, or vice versa. A practical example is the moving nerve signal in the vicinity of a stationary recording electrode.

Figures 5.12 and 5.13 show calculated potentials with a stationary current injecting dipole of length L in a homogeneous conducting medium. The dipole vector L is at the origin, and oriented in the x-direction. The *recording* electrode is moved in a straight line at a distance $5L$ and $10L$ from the dipole, parallel to the x- or y-axis. The absolute potential is arbitrary, but the relative values are correct and the potentials and distances on all the figures can be compared. The potential is given as a function of x or y. If the dipole and the recording electrode move at a constant velocity with respect to each other, the x-axis also is a time axis.

Unipolar recording electrode (dipole source)
Figure 5.12a shows how a recorded unipolar potential is *bi-phasic*, going through zero at the line of symmetry. At larger distance from the dipole, the signal amplitude is reduced, but also the sharpness of the waveform. This means reduced frequency content, and is a general finding in all potential recording in a volume conductor. *Distance to the signal source acts as a low pass filter*, as can be seen in Figs 5.12 and 5.13. A poorly conducting tissue layer between the source and the recording electrode has the same effect. This is why ordinary surface EEG has a low frequency content, the skull acts as a low pass filter. EMG is often obtained with the recording electrode near to a large muscle mass, and the signal has a high content of high frequency signal components.

Figure 5.12b shows that the signal obtained in the y-direction is *mono-phasic*, the amplitude is larger than in the x-direction.

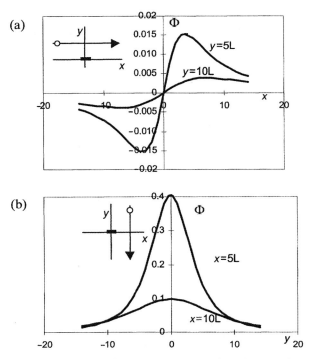

Figure 5.12. Recorded unipolar potentials from a current dipole of length L. (a) The recording electrode is moved parallel to the dipole at distance $5L$ and $10L$. The unit of the x-axis is L. (b) Recording electrode moves perpendicular to the dipole orientation as indicated. Notice the much larger signal voltage.

Bipolar recording electrode pair (dipole source)

With a *bipolar* recording electrode pair, the maximum reading is obtained with the bipolar axis in radial direction from the dipole source. Then the potential difference $\Delta\Phi_r$ at distance r and with distance L between the recording electrode pair is

$$\Delta\Phi_r = (IdL/2\pi\sigma)/r^3 \qquad (L \ll r, \text{ recording electrodes in radial direction}) \qquad (5.15)$$

The sensitivity falls off much more rapidly than the unipolar potential Φ_r, and the discriminative properties with respect to source position are large. However, the signal-to-noise ratio may be poor.

This is also confirmed by Fig. 5.13a, which shows a very sharp signal waveform, indicating large positional discriminative property. The signal is bi-phasic. Fig. 5.13b shows a tri-phasic waveform with a large high-frequency content, but the recorded voltage difference is small. The other two recorded bipolar potential curves also show the same higher discriminative power.

5.3.3. Lead sensitivity and the lead vector

We have already pointed out that owing to the falling current density with distance from a monopolar current source, tissue volumes far away from the source have less

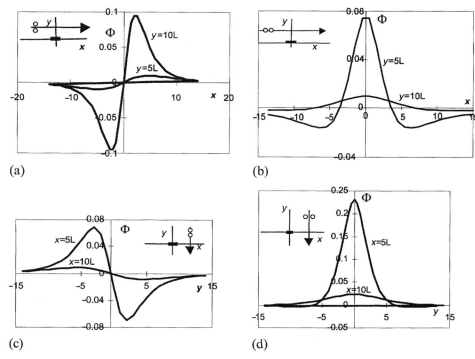

Figure 5.13. Recorded bipolar potentials in a dipole field. Orientations as indicated. Better spatial discrimination than with unipolar recordings.

influence on the total measured conductance. For a one-port/two-terminal system this is expressed as a volume sensitivity parameter of conductance S_{VG} (Section 5.2).

As we shall see in Section 6.1, a lead's recorded voltage can be described by a transmittance function and varies according to the distance (and orientation) of the recording electrode(s) with respect to the source.

It is possible to use the reciprocity theorem (Section 6.1) to describe a lead's sensitivity (to signals from a source) in a more general way by introducing the *lead vector*. Let us first recall the reciprocity theorem: if we consider a two-port system with separate current injecting (1) and recording (2) electrodes, the transmission impedance Z_{21} from port (1) to (2) is $Z_{21} = v_2/i_1$. According to the reciprocity theorem we may instead inject current at port (2) and record at port (1) and obtain the same ratio: $Z_{12} = v_1/i_2 = Z_{21} = v_2/i_1$. In a linear two-port network, a certain current injected in one port results in the same voltage difference at the other port, irrespective of which port is chosen for current injection.

Lead vector
The recorded *unipolar* potential V_d due to a current dipole source in an infinite homogeneous material is $V_d = \Phi_d = (I/4\pi\sigma)(1/r_1 - 1/r_2)$ (eq. 5.12). If the current dipole is ideal, then $\Phi_d = (1/4\pi\sigma)\mathbf{m} \cdot \mathbf{r}/r^3$, where $\mathbf{m}$ is the current dipole moment vector. Inserting this into a Cartesian coordinate system we may decompose the dipole moment in m_x-, m_y- and m_z-components if the superposition theorem is valid. If we record the unipolar potential contribution from each dipole component, e.g. in the x-direction, we can define and determine a transfer function $\mathbf{H}(\omega)$ (time domain vector in eq. 6.9, but here a space vector). For simplicity we consider H frequency independent, so that in the x-direction $\Phi_{dx} = H_x m_x$. This is also true if the tissue is inhomogeneous and the relationship is more complicated than given in eq. (5.12), as long as the system is linear. H has the dimension ohm/metre, and we consider the transfer coefficient H_x as a part of a general vector transfer function $\mathbf{H}$ (space vector) which we also call the *lead vector*. Accordingly,

$$\Phi_d = \mathbf{H} \cdot \mathbf{m} \qquad (5.16)$$

Equation (5.16) implies that it is possible to define a unique space vector $\mathbf{H}$ defining the potential recorded in a given lead from an ideal current dipole source $\mathbf{m}$ situated in a limited, non-homogeneous volume with linear electric properties. The lead vector is a transfer factor determined by the geometry and conductance distribution of the complete system, and is valid for just one setup with all electrode positions and distributions defined and constant. $\mathbf{m}$ must be a fixed vector, but both magnitude and direction may vary. Changing any electrode position but leaving all other factors constant results in another lead vector $\mathbf{H}$. One side in the Einthoven triangle (Section 8.2) is an example of a recorded voltage V_d (the ECG signal) modelled as the dot product of the *heart vector* ($\mathbf{m}$) and the *lead vector* ($\mathbf{H}$) of two limbs.

The image surface (fixed source)
We may assume a fixed dipole source, and determine the lead vector as a function of the unipolar or bipolar recording electrode position, e.g. along the surface of the

conductor. We may then let the loci of all the lead vectors define a new surface, the *image surface* of the leads. The recording electrodes may also be positioned inside the conductor; the magnitude of the image surface vectors will of course increase as the source is approached.

The lead vector field (fixed lead)

In the same way we may assume a fixed recording electrode pair; the lead vector or *sensitivity* may be determined as a function of the dipole source location. For each location, a unique transfer function vector **H** can be determined, and all the vectors form a vector field, the lead field. It can be shown that the lead vector field of **H** for a given recording electrode configuration is identical to the current density vector field $\mathbf{J}_{reci}$ obtained if that electrode configuration is used as an injection port for a unit current of 1 ampere. The injection of unit current in the recording lead is called a *reciprocal* excitation. The lead potential V_{LE} will be the sum of all potentials generated by the dipole sources in the volume v:

$$V_{LE} = \int_v (\nabla\Phi_{reci} \cdot \mathbf{m}/v)\,\mathrm{d}v = \int_v (\mathbf{J}_{reci} \cdot \mathbf{m}/\sigma v)\,\mathrm{d}v \qquad (5.17)$$

Here $\nabla\Phi_{reci}$ is the gradient of the potential field of the reciprocally excited recording electrodes, $\mathbf{J}_{reci}$ is the local current density as the result of reciprocal unit current excitation of the recording electrodes (NB! dimension: $1/m^2$), and $\mathbf{m}/v = I\mathbf{d}/v$ is the source current dipole volume density. For further details see Malmivuo and Plonsey (1995).

According to eq. (5.17), the *local volume sensitivity* as a function of source location can be found by the vector dot product of (1) the dipole source and (2) the current density vector produced by reciprocally excited recording electrodes. This is illustrated for a recording system in Fig. 5.14. From Fig 5.14 it is clear that more signal is recorded from a dipole source near the electrodes and their constrictional current zones of high reciprocal current density than from the same dipole further away. The lead field equal to the reciprocal unity current density field is an

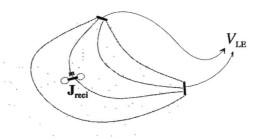

Figure 5.14. Sensitivity of a recording lead to a dipole source is found as the vector dot product of reciprocal current density $\mathbf{J}_{reci}$ and local dipole moment **m**. V_{LE} is the sum of all dot product contributions along a current density line according to eq. (5.17).

important tool when optimal lead geometry is to be determined, or for calculating the sensitivity of a given electrode lead for given signal sources in the tissue volume.

In Section 8.3 on plethysmography the sensitivity for detecting small local changes in tissue conductivity is discussed.

5.4. Three- and Four-Electrode Systems

We have examined the model geometry of the monopole and dipole current sources, and how recording electrodes may be the same electrodes or a separate pair. We now examine the three- and four-electrode systems further, with reference to surface electrodes or *in vitro* measuring setups.

5.4.1. Three-electrode system, monopolar recording

In a two-electrode monopolar system it may be difficult to estimate a possible contribution from the neutral electrode. Often only a zone of the current path proximal to the measuring electrode is of interest, e.g. when measuring skin admittance. In such cases the distal volume segment of the current path is a disturbing part. It may also be difficult to work with a sufficiently large neutral electrode in some situations. By adding a third electrode, it becomes easier to control the measured tissue zone. The setup is equivalent to the two-port, three-terminal network, with one electrode common for both current injecting and voltage recording. Figure 5.15 shows the principle (Grimnes 1983a). The measuring electrode M and the electrode C are the current carrying electrodes. With just these two electrodes the resultant current is dependent on the impedance of both electrodes plus the tissue in between. By measuring the voltage v on a potential recording electrode R (no current flow and no polarisation) with respect to M, the admittance Y of M is: $\mathbf{Y} = \mathbf{i}/v$. In a practical setup this is effectively done with an operational amplifier as shown. The circuit guarantees that the potential on R is equal to the excitation voltage, so it is a constant amplitude voltage, admittance reading circuit. What is then included in $\mathbf{Y}$?

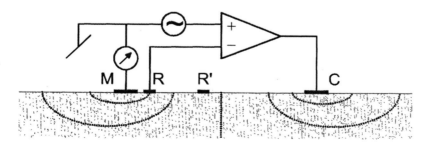

Figure 5.15. Three-electrode geometry and circuit, equipotential lines shown in tissue. The position of electrode R determines the measured volume segment proximal to M.

1. The polarisation impedance of electrode M.
2. A tissue volume zone proximal to M, delimited by the position of R.

One interesting feature with this circuit is that if there is a capacitative coupling of 50/60 Hz mains leakage current to the patient, the necessary ac current is supplied by the current-carrying counterelectrode. This current is supplied by the operational amplifier, does not pass the measuring electrode, and is therefore not recorded by the transresistance amplifier. The circuit of Fig. 5.15 is also the principle of a *potentiostat*, with the reference electrode controlling the voltage of the electrode M.

The measured proximal or constrictional tissue zone corresponds to the equipotential surface shown as a stippled line under the R electrode. By placing R closer to M, a smaller proximal zone of the current path will be included. The zone will be even smaller by placing the R electrode on the left side of M. In contrast, if position R' is used, almost half of the total current path and tissue zone will be included. However, a tissue volume in the measured zone does influence the results depending on position, and local sensitivity is proportional to local current density. From Fig. 5.2 we know that a sphere/disk electrode alone has a larger sensitivity in the constrictional zone of the electrode for purely geometrical reasons. Using a surface disk M electrode, tissue near the electrode edge also influences the result to a larger extent than tissue elsewhere (Section 5.2.2).

The polarisation of C and the impedance of R will not influence the results. The tissue volume distal to the measured zone may have some influence because tissue outside may also influence current density distribution and thus the form and position of the equipotential surface of R. Such a disturbing influence may add to or subtract from the original values.

The metal of the R electrode should be recessed (see Fig. 5.18), or the metal should be narrowed in the current flow direction, to impede current flow from being attracted away from the lower admittivity tissue surface layers.

Intracellular voltage clamp
Figure 5.16 shows a similar three-electrode voltage intracellular clamp circuit. A problem with the circuits is that two microelectrodes must be introduced through the cell membrane. A single invasive electrode may be used, with constant-current

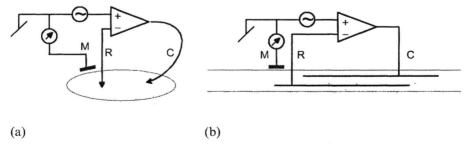

(a)   (b)

Figure 5.16. Intracellular three-electrode voltage clamp circuits. (a) Ellipsoid cell form. (b) Nerve cell axon.

injection simultaneously with the voltage recording by the same electrode. Electrode polarisation will be included in the voltage reading. To diminish this problem a sample-and-hold circuit can be used, so that a reference voltage is read before current is switched on. In current flow mode the sampled reference voltage is subtracted from the instantaneously recorded voltage, and the voltage clamp circuit drives a current necessary to obtain a clamp to the voltage difference.

5.4.2. Four-electrode (tetrapolar) system

If the distal volume segment of the current part is the zone of interest, and the effect of the zones proximal to the current carrying electrodes are to be eliminated, the four-electrode system is preferred. Such four-electrode systems correspond to a *two-port, four-terminal network* equivalent. *Transfer functions* may be set up to describe the signal path from current carrying to signal recording electrodes. This is very much the case in impedance plethysmography or impedance imaging. It is also the best system for measuring excised tissue samples *in vitro*. An *in vitro* version is shown in Fig. 5.17.

The measured segment is determined by the position of the two recording electrodes R and R', or more exactly by the position of their electrolyte/salt bridge connections to the measuring cell. The two stippled equipotential lines indicate this. Current is recorded with a current-reading operational amplifier. The recording electrodes are connected to one buffer amplifier each. No current is flowing in the electrode leads, and the electrodes can therefore not be externally polarised (but internal currents may polarise them, see below). Electrolyte/salt bridge connections are used in order to be able to increase the electrode metal area so as to increase electrode admittance and reduce noise.

Large electrodes directly inserted into the measuring cell will disturb the ionic current flow pattern, and polarisation will occur on the metal surface. The electrode metal should not be in direct contact with the electrolyte, but be recessed (Fig. 5.18).

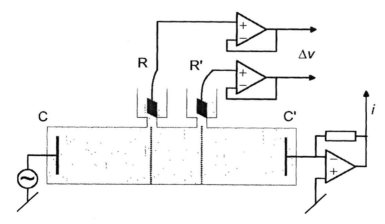

Figure 5.17. Four-electrode system, tubular *in vitro* version.

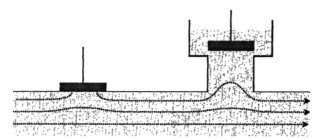

Figure 5.18. The non-recessed electrode to the left implies polarisation from the current entering and leaving the metal.

The electrolyte is contained in a tube with isolating walls, and if a part of this wall is substituted by electrode metal, the current will prefer the high conductivity path of the metal. The current lines will deviate from the path parallel to the tube walls, and in one part of the area the current will enter, in the other part it will leave. Thus the electrodes *are* polarised, but not by a current in the external leads. The polarisation may not be uniform over the electrode surface area, and the polarisation will occur according to local current direction and polarisation admittance. When the metal is recessed, the current will also deviate into the electrolyte of the bridge path, but the current will not pass any metal surface, and no polarisation will occur.

Figure 5.19 shows a four-electrode *in vivo* skin surface version. The two recording electrodes ensure that only the current path segment between these electrodes contributes to the result. The stippled equipotential lines indicate the segment that is measured. However, not all tissue volumes contribute equally. Sensitivity is proportional to local current density in the measured volume segment; the sensitivity is zero in the two segments proximal to the current-carrying electrodes as long as potential difference is taken from the R–R' electrodes. Of course it is possible to measure simultaneously all three voltages M–R, R–R' and R'–M'. Admittance/impedance of the two constrictional zones and the one distal volume

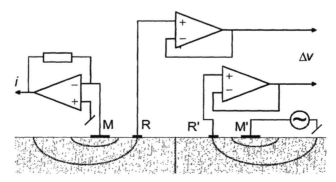

Figure 5.19. Four-electrode system, *in vivo* version. Equipotential lines shown in the tissue. The position of R–R' determines the tissue segment measured, and the size of the proximal zones measured by M and M'.

segment of the current path can then simultaneously be measured as a function of frequency and followed as a function of time.

To reduce the proximal zone contribution, and sometimes for anatomical reasons, it may be advantageous to place the R electrodes *outside* the M electrodes, instead of inside as shown in Fig. 5.19. This is almost the same as interchanging the electrode pairs, cf. the reciprocity theorem. Actually these three- and four-electrode systems measure transmittance. By applying current, for example, through port one (M and M'), the potential difference created is measured at port two (R and R'), and the transmittance, e.g. admittance, is calculated as the current in the M–M' divided by the voltage at R–R'. If the reciprocity theorem is valid, the same admittance should be found by injecting current in R–R', and record the voltage at M–M'.

5.5. Finite Element Method

Maxwell's equations are discussed in Section 7.2.1. Any problem in regard to the distribution of current or electric field in a homogeneous or composite material can be solved with these equations if the excitation and electrical properties of the materials are known. There are in general two different types of problems to be solved in differential equations: *initial value* problems and *boundary value* problems. Initial value problems arise when the values of the unknowns are given at a particular point, e.g. at a given time, and the values at future times are to be computed. Boundary value problems arise when the values on the boundary of a material are known, and the values of the interior are to be computed. The latter is a common situation in bioimpedance research where, for example, the current distribution in tissue is to be computed from a given excitation from surface electrodes.

Since the differential equations that describe the behaviour of our system, in this case the Maxwell equations, basically describe an infinite-dimensional object, we must use a finite-dimensional approximation to represent the solution. There are two main forms of such approximation, *finite differences* and *finite elements*. In the finite difference method, the differential equation describing the continuous change of the values can be replaced with an approximation describing the slope of change between a finite number of discrete points, called mesh points.

The finite element method involves dividing the modelled geometry into small subregions called elements. The unknown solution is then expressed as the weighted sum of basis functions, which are polynomials in each element. The differential equation is then used in each element to compute the weight used for the basis function of that particular element in the total sum. Each calculation introduces a small error but the errors will in the ideal case diminish as the number of points or elements is increased. There is no guarantee, however, for some ill-posed problems, that the sum of an increasing number of diminishing errors will not have a large effect. It is therefore highly recommended to verify any algorithms or computer software of the sort before use by first solving similar problems that have an analytical solution.

Earlier in this chapter we treated some simple electrode/tissue geometries with mathematical analytical solutions. Of course tissue morphology and composition is such that analytical solutions most often can not be found. It is therefore necessary to take a more engineering approach: to make a realistic geometrical model of the tissue and electrode system, immittance distribution included, and then let a computer calculate current density vectors and equipotential lines on the basis of a chosen mesh.

In principle the computer programs may calculate complex three-dimensional models with frequency dependent tissue parameters. However, the computing time may be long, and it will be correspondingly time-consuming to experiment with the model. It is important to define the problem in the simplest possible way, and in a way adapted to the software used. If the three-dimensional problem can be reduced by symmetry to two dimensions, and if a dc/purely resistive problem definition can be used, results are more easily obtained.

In Figure 5.20a we see a conducting rectangle with a circular area of higher conductivity placed inside the rectangle. The arrows inside the rectangle are current

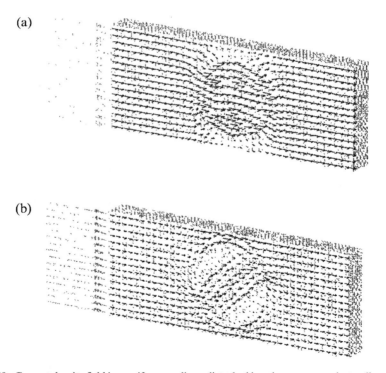

(a)

(b)

Figure 5.20. Current density field in a uniform medium, disturbed by a homogeneous but well-conducting cylinder (a) and an anisotropic cylinder (b). Both models bend the lines of the current density and the E-field, but in the homogeneous cylinder the E-field direction is the same as that of the externally applied field (Knudsen 1999).

densities calculated by means of the finite element method. The results show how current is attracted to the area of higher conductivity. In Fig. 5.20b, a diagonal slice of the circular area has the same conductivity as the rest of the rectangle, while the remaining part of the circular area has lower conductivity. It is now obvious how the current is attracted to the well conducting channel, and that the direction of the current in this channel, owing to the geometry of the medium, is different from the main current direction.

This simple example elucidates a general problem discussed by Plonsey and Barr (1982). In their study of the four-electrode technique utilised in cardiac muscle conductivity measurements, they point out that one does not directly measure conductivity along each axis in turn when measuring along the x-, y- and z-axes. Owing to anisotropy the current will not necessarily follow the direction of the measured axis and, hence, the measured electrical properties will be influenced by the conductivity in all directions. Plonsey and Barr (1982) use an anisotropic *bidomain* (considering both intracellular and interstitial conduction) model and show that if the ratios of intracellular to interstitial conductivity are equal along each of the three coordinate axes, then all six conductivities can be calculated from measurements along the three axes.

5.6. Tomography and Plethysmography

A mapping of the immittance distribution in a tissue layer (tomography) or a tissue volume is possible with a surface multiple electrode system. In electrical impedance tomography (EIT), current (about 1 mA) is typically injected in one electrode pair and the voltages between other electrodes are recorded (Rosell *et al.* 1988b). Current injection is then successively shifted so that all electrode pairs are used. The reciprocal theorem will be a sort of a control of system linearity. A frequency in the order of 50 kHz is commonly used, so a complete set of measurements with, e.g., around 50 electrodes can be performed in less than 0.1 s. The images obtained have a resolution of about 1 cm at 10 cm tissue depth.

One of the fundamental problems of anatomical imaging based on single-frequency impedance measurements is that the absolute impedance is difficult to determine accurately because of the uncertainty in the position of the electrodes. Impedance imaging has hence until recently mainly been concerned with impedance changes due to physiological or pathological processes. With the introduction of multifrequency EIT, the possibility of tissue characterisation based on impedance spectroscopy has been realised (Riu *et al.* 1992, 1995). If the frequency response varies significantly between different organs and tissues, it will be easier to achieve anatomical images because the images will be based on characteristic changes of impedance with frequency rather than the absolute value of the impedance. Indexes such as the relation between measured data at two frequencies may be used, but, because of the differences in characteristic frequencies for the dispersions of different tissues, it has proved difficult to choose two frequencies to match all types of tissue. Hence, typically 8–16 frequencies are used. Brown *et al.* (1994a) used seven frequencies from 9.6 to 614 kHz in a 16-electrode thorax measurement and computed from the measured data several parameters from a three-component electrical model. They also adapted the measured data to the Cole impedance equation (see Section 7.2.6). Indexes were furthermore derived by using the 9.6 kHz data as a reference, and they concluded that it may be possible to identify tissues on the basis of their impedance spectrum and the spectrum of the changes in impedance. Blad (1996) proposed another approach in a preliminary study on imaging based on the measurement of characteristic frequency.

Contactless data acquisition techniques are also utilised in EIT. Such systems may involve coils for inducing currents and electrodes for measuring voltage or they may be totally contactless when magnetic coils or capacitive coupling are used both for the current excitation and for measuring the voltage response (Scaife *et al.* 1994; Gencer and Ider 1994). Tozer *et al.* (1998) describe a system where currents at 10 kHz, 100 kHz and 1 MHz are passed axially in the body (between head and feet) and the resulting magnetic field in the body is measured by means of small search coils. The measured magnetic fields are related to current distribution by the Biot–Savart[15] law, which, among other things, states that the intensity of the magnetic field set up by a current flowing through a wire is inversely proportional to the distance from the wire. Korjenevsky and Cherepenin (1998) present an all-magnetic system that comprises coils both for inducing current and for measuring voltage. Their system works in the frequency range 10–20 MHz and measures phase shifts between excitation and response, which are found to be proportional to the integral of the conductivity of the medium. The high frequency needed for such all-magnetic systems may reduce their clinical utility, however.

In principle, the method of mapping tissue immittance distribution is not limited to a slice. By increasing the number of electrodes, volume acquisition is equally possible. One of the potential benefits from 3D EIT is that it could take into account the fact that the current spreads out of the imaging plane of 2D EIT. However, the complexity of both the EIT hardware and software will increase considerably on introduction of this third dimension. Some important achievements in the pursuit of 3D EIT were presented in *Nature* in 1996 (Metherall *et al.* 1996).

Typical EIT systems involve 16–32 electrodes in any one plane and operate at frequencies between 10 kHz and 1 MHz. The Sheffield Mark III system (Brown *et al.* 1994a,b) uses 16 electrodes and injects current and measures potential drop between interleaved neighbouring electrodes in order to reduce cross-talk (Fig. 5.21). They measure on frequencies between 9.6 kHz and 1.2 MHz and a current of 1 mA p-p. A total of 64 measurements are made at each frequency when driving odd-numbered electrodes and measuring at even-numbered electrodes.

As explained in Sections 5.3.3 and 8.3, the sensitivity of the measurement to a given local change in admittivity is given by the dot product of the current vectors resulting from injecting current through the two pairs of current and potential

[15] Jean-Baptiste Biot (1774–1862) and Félix Savart (1791–1841), French scientists. Biot was author of "Traité élémentaire d'astronomie physique" (1805) and accompanied J.-L. Gay-Lussac in 1804 on the first balloon flight undertaken for scientific purposes.

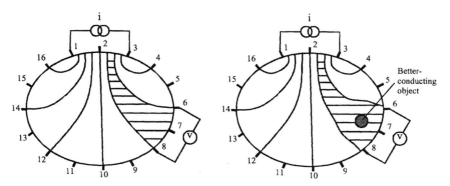

Figure 5.21. Applying current and measuring potential difference between interleaved electrodes on a medium; homogeneous (left) and non-homogeneous (right).

electrodes, respectively. The sensitivity is hence highest close to the current-injecting or potential pick-up electrodes and lowest towards the centre of the medium. The Sheffield group is now developing a modular system expandable to any multiple of 8 electrodes where adjacent electrodes are used for driving current and measuring potential.

Cross-talk is by no means the only instrumentation problem experienced in EIT. Rosell and Riu (1992) discussed the significant common-mode voltage sensed by the potential-measuring electrodes. They proposed a method called common-mode feedback and obtained an improvement of 40 dB in the measurements at frequencies up to 10 kHz. Bragos *et al.* (1994) designed wideband single and floating current sources for EIT based on current-mode components and with dc feedback, and the effect of electrode size on images of different objects in saline tanks was described by Newell *et al.* (1998).

The methods for constructing images from immittance measurements may be divided into *single-pass* and *iterative* processes. In single-pass methods, a single mathematical operation is performed on the measured data. Although the conceptual simplicity of the method makes it unsuitable for static imaging where the images are based on the absolute immittance values measured, it works well for dynamic imaging where the images are based on changes in the time or frequency domain. Single-pass methods can be further divided into *back-projection* and *sensitivity matrix* methods. Back-projection is a commonly used technique in x-ray computer tomography (CT) imaging where the values of a measured profile are distributed in pixels through the thickness of the object and the accumulated value in each pixel, after profiles have been measured in different angles, are used for imaging. A high-pass filtering technique is also normally used on the image to reduce artificial blurring of the contour of the organ produced by the back-projection. The sensitivity matrix is the matrix of values by which the conductivity values can be multiplied to give the electrode voltages. The matrix describes how different parts of the measured object influence the recorded voltages owing to the geometrical shape of the object. The sensitivity matrix has to be inverted in order to enable image reconstruction. This operation is not uncomplicated and many techniques have been suggested (cf. Morucci *et al.* 1994).

Iterative methods are mostly used for static imaging. An intelligent guess of the distribution of, e.g., conductivities in the tissue is made initially in the iterative algorithm. The forward problem (see Section 5.1) is then solved to calculate the theoretical boundary potentials that are compared to the actual measured potentials. The initial guess is then modified to reduce the difference between calculated and measured potentials and this process is repeated until the difference is acceptably small. The calculation for the boundary potentials may be performed by means of the finite element method (see Section 5.5).

Because of poor resolution, the range of clinical applications has so far been limited. Examples include imaging of gastric function (Mangnall *et al.* 1987; Smallwood *et al.* 1994), pulmonary ventilation (Harris *et al.* 1987), perfusion, brain haemorrhage (Murphy *et al.* 1987), hyperthermia (Griffiths and Ahmed 1987; Gersing *et al.* 1995), epilepsy and cortical spreading depression (Holder 1992, 1998), swallowing disorders and breast cancer (Jossinet 1996). A fundamental difficulty is the considerable anisotropy found in the electrical properties of, e.g., muscle tissue. No satisfactory solution of how to deal with anisotropy in EIT has yet been proposed. For anisotropic electrical impedance imaging see, e.g., Lionheart 1997. The obvious advantages of the method are speed of acquisition and relatively low-priced equipment.

Impedance plethysmography is closely related to tomography since both are concerned with the recording of local tissue immittivity variations or distributions. Impedance plethysmography is the measurement of immittance changes due to respiration or perfusion and is described in more detail in Section 8.3.

CHAPTER 6

Instrumentation and Measurement

6.1. General Network Theory, the Black Box

In many cases we will regard our biological material, together with the necessary electrode arrangements, as an unknown "black box". By electrical measurement we want to characterise the content of the box (we do not have direct access to the key to open the lid!). We want to use the data to *describe* the electrical behaviour, and perhaps even *explain* some of the physical or chemical processes going on in the box and discern the electrode and tissue contributions. The description must necessarily be based upon some form of *model*, for example in the form of an *equivalent* electric circuit, mimicking measured electrical behaviour. We may also want to link properties to distinct tissue parts or organ parenchyma behaviour. A basic problem is that always more than one model fits reasonably the measured electrical behaviour. The equivalent circuit is the tool of electronic engineers and facilitates their interpretation of the results, simply because they are trained and used to interpret such diagrams. As discussed in Section 7.1, the equivalent model may also go further and be of a more explanatory nature.

The black box may be assumed to "contain" the whole body, a part of the body, just an organ, or just a cell, together with the electrodes. It may also be assumed to contain not the real things, just the *equivalent model circuit* of the tissue of interest.

We will now give a very general description of the black box and how to characterise it electrically, irrespective of the box's content. The black box may be considered to contain the real tissue with electrodes for excitation and response measurement, or our model in the form of an electric *network* as a combination of *lumped* (discrete) electrical components. The network may have 2, 3 or 4 external *terminals* (cf. the number of electrodes used). A pair of terminals for excitation or recording is called a *port*. The treatment is so general that the content can be characterised with global variables not particularly linked with electrophysiology.

A very general box is the four-terminal type with two ports (Fig. 6.1a). This is the box corresponding to the four-electrode systems described in Section 5.4: two recording electrodes in the electric field of two current-carrying electrodes.

6.1.1. Admittance, impedance and immittance

The four external variables of a two-port black box are v_1 and i_1 (first port), v_2 and i_2 (second port). There are four possible v/i ratios: v_1/i_1, v_2/i_2, v_1/i_2 and v_2/i_1. These

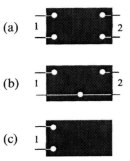

Figure 6.1. Black boxes. The two upper boxes allow for transfer parameters from one port to the other. The lowest box is a one-port, two-terminal box with only driving point parameters possible.

ratios may be inverted so actually there are eight possible ratios. If the signals are sine waves, most of the ratios have special names:

- If a chosen ratio is current to voltage, the ratio is called *admittance*. It is measured in siemens (S), and can be obtained directly by reading current when a constant-amplitude ac *voltage* is applied ($Y = i/v$). If the ratio is voltage to current, the ratio is called *impedance*. It is measured in ohm (Ω), and can be obtained directly by reading voltage when a constant amplitude current is applied ($Z = v/i$).
- *Driving point* immittance is defined with excitation and response at the same port (also as a one-port network); *transmittance* is defined with excitation and response at different ports. Transmittance may be transfer admittance or transfer impedance.
- *Immittance* is the general term covering the duality of admittance and impedance.

A *resistor* opposes and limits current flow. This ability is characterised by the resistance R measured in ohm, or the conductance measured in siemens. If the resistor is considered ideal, the resistance and the conductance are equal for dc and ac irrespective of frequency.

A *capacitor* also opposes and limits current flow. This ability is characterised by the impedance Z measured in ohm, or the admittance measured in siemens. If the capacitor is considered ideal, the capacitance C is considered frequency independent, but the impedance is frequency dependent and for sine waves is equal to $1/\omega C$. The impedance is therefore infinite at dc, but lower the higher the frequency.

Putting these two single components together in *series* in a two-component circuit, we obtain a more general expression for the impedance $\mathbf{Z}$ in complex form: $\mathbf{Z} = R + jX$. Here X is the reactance: for an ideal capacitor $X = -1/\omega C$. For the single component ideal capacitor, $R = 0$.

Putting the two single components together in *parallel* in a two-component circuit, we obtain a more general expression for the admittance $\mathbf{Y}$ in complex form: $\mathbf{Y} = G + jB$. Here B is the susceptance: for an ideal capacitor $B = \omega C$. For the single component ideal capacitor, $G = 0$.

6.1.2. The two-port network and the reciprocity theorem

Generally a two-port, 3–4-terminal network can be completely defined with four ratios (constants) characterising the network, and four variables. Here we will introduce two different equation sets.

The *admittance* equation set for a two-port network is

$$i_1 = Y_{11}v_1 + Y_{12}v_2 \qquad (6.1)$$

$$i_2 = Y_{21}v_1 + Y_{22}v_2 \qquad (6.2)$$

A driving point immittance Y_{11} or Y_{22} is admittance measured with excitation and response at the same port. The transmittances are in the form of the *transfer admittances* Y_{12} and Y_{21}; they are defined as

$$Y_{12} = i_1/v_2 \qquad (v_1 = 0, \text{ port 1 short-circuited}) \qquad (6.3)$$

$$Y_{21} = i_2/v_1 \qquad (v_2 = 0, \text{ port 2 short-circuited}) \qquad (6.4)$$

The variables v are independent, the i are dependent, and both transfer admittances are specified with one of the ports *short-circuited*.

The *impedance* equation set for a two-port network is

$$v_1 = Z_{11}i_1 + Z_{12}i_2 \qquad (6.5)$$

$$v_2 = Z_{21}i_1 + Z_{22}i_2 \qquad (6.6)$$

A driving point immittance Z_{11} or Z_{22} is impedance measured with excitation and response at the same port. The *transfer impedances* are defined as

$$Z_{12} = v_1/i_2 \qquad (i_1 = 0, \text{ port 1 open}) \qquad (6.7)$$

$$Z_{21} = v_2/i_1 \qquad (i_2 = 0, \text{ port 2 open}) \qquad (6.8)$$

The variables i are independent, the v are dependent, and both transfer impedances are specified with one of the ports *open*.

A somewhat different measuring principle is to record the voltage at a port, and record the reduced signal as a function of shunting the recording port with different load impedances (Mørkrid *et al.* 1980).

Many ratios may be formed between these variables. If the circuit is a current-to-voltage amplifier, the ratio of interest is v_o/i_i, which is a resistance. Such an amplifier is therefore called a *transresistance* amplifier. In a tube or a field-effect transistor the input voltage controls the output (anode or drain) current and the ratio i_o/v_i is a *transconductance*. Transmittance implies that a signal is sent from one port to another; v_2/v_1 is attenuation or gain, perhaps in decibel. Transmission is characterised by a *transfer function* of frequency $H(\omega)$ (time vector):

$$\mathbf{H}(\omega) = \mathbf{v}_2/\mathbf{v}_1 \qquad (6.9)$$

The transfer functions are of course related to the eqs (6.1) to (6.8) but they emphasise signal transmission and not tissue characteristics so directly.

Conditions

For the network theory presented, there are certain conditions to be met if the theory is to be valid. These conditions are that:

1. The network is *linear* if the immittances are independent of v or i, both the principles of *superposition* and *proportionality* must hold. Most of the systems of interest to us are not linear at dc but may have a linear amplitude range at ac: sometimes a broader range the higher the frequency.
2. The network is *passive* if the energy delivered to the network is positive for any excitation waveform, and if all currents or voltages are zero without excitation. As we know, tissue with electrodes does not fulfil this last requirement; we know for instance that it contains cells with endogenic ionic pumps.
3. The network is *reciprocal*[16] if the transmittances are equal, $Y_{12} = Y_{21}$ and $Z_{12} = Z_{21}$. Tissue is reciprocal if it is passive and linear. Then the ratio of the short-circuit current in electrode pair 1 to the voltage applied to electrode pair 2, is equal to the ratio of the short-circuit current in electrode pair 2 to the voltage applied to electrode pair 1. A network of passive components is reciprocal, but the insertion of a transistor makes it non-reciprocal. The transistor is a one-direction signal device.
4. The network is *causal* if its response is non-anticipatory, e.g. if there is no response before an excitation is applied. This is important in Fourier analysis, for instance the phase response of a capacitive network at the onset of a sine wave excitation. Often phase analysis presupposes that the sine wave has been there long before the time of analysis.

In this book all four of these conditions are considered to be satisfied when a system is characterised as *linear*, if not otherwise noted.

6.1.3. Extended immittance concepts

The classical immittance concept is linked to sine waves. The driving point admittance of a black box port is for instance defined as the ratio $Y = i/v$, where v is a sine wave as the independent variable. However, such ratios may be defined also for other waveforms than dc or sine waves. The immittance concept may be extended by the *Laplace*[17] *transform*.

By replacing the imaginary frequency variable $j\omega$ by an extended complex frequency variable $s = \sigma + j\omega$ (here σ is not conductivity), it is possible to define, e.g., impedance not only in the angular frequency ω-domain but also in the s-domain. The impedance of a capacitor of capacitance C is, for example, in the frequency domain: $Z(\omega) = 1/j\omega C$, and in the s-domain: $Z(s) = 1/sC$. The Laplace transforms of some very important excitation waveforms are very simple: for

---

[16] The *reciprocity* theorem originates from Helmholtz.
[17] Pierre Simon de Laplace (1749–1827), French mathematician. Famous for his work on differential equations and on probability, even so an extreme determinist.

example, for a unit impulse it is 1, for a unit step function $1/s$, for a ramp $1/s^2$, etc. That is why the excitation with, e.g., a unit impulse is of special interest when examining the response of a system. In the extended immittance definition, calculations with some non-sinusoidal waveforms become very simple. Even so, Laplace transforms are beyond the scope of this book.

6.1.4. The non-linear black box

The immittance concept may also be extended for *non-linear networks*, where a sine wave excitation leads to a non-sinusoidal response (cf. Section 4.4.4). Including an immittance value for each harmonic component of the response performs the necessary extension. In the linear region the principle of superposition is valid. This means, for example, that the presence of strong harmonics in the *applied* current or voltage will not affect immittance determination at the fundamental frequency or a harmonic (Schwan 1963). The lock-in amplifiers used today are able to measure harmonic components, making possible an analysis of non-linear phenomena and extending measurement to non-sinusoidal responses.

In the non-linear region, the principle of superposition is not valid. When measuring in the non-linear region, it is necessary to state whether the system uses constant-amplitude current (current clamp) or constant-amplitude voltage (voltage clamp). With, for example, constant-amplitude voltage, the voltage per definition is sinusoidal, but the current is not. If the measuring system is able to measure selectively sufficiently many current harmonic components (and a possible dc component, *rectification*), the actual current waveform in the time domain is defined. Such a steady-state non-linear analysis is very suitable in the low excitation energy end of the non-linear region. At the higher current densities used, for instance, in defibrillator shocks, steady state cannot be obtained (temperature rise and tissue destruction), and all measurements must be performed during one single pulse of energy.

6.1.5. The time constant

Immittance theory is based upon *sinusoidal* excitation and sinusoidal response. In relaxation theory (and cell excitation studies), a *step* waveform excitation is used, and the time constant is then an important concept. If the response of a step excitation is an exponential curve, the time constant is the time to reach 63% of the final, total response. Let us, for instance, consider a series *RC* circuit, excited with a controlled *voltage* step, and record the current response. The current as a function of time after the step is $I(t) = (V/R)e^{-t/RC}$, the time constant $\tau = RC$, and $I(\infty) = 0$.

However, if we excite the same series *RC* circuit with a controlled *current* step and record the voltage across the *RC* circuit, the voltage will increase linearly with time ad infinitum, corresponding to an infinite time constant. Clearly the time constant is dependent not only on the network itself but also on how it is excited. *The time constant of a network is not a parameter uniquely defined by the network itself.* Just as immittance must be divided between impedance and admittance depending on voltage- or current-driven excitation, there are two time constants depending on how the circuit is driven. The network may also be a three- or four-terminal network. The time constant is then defined with a step excitation signal at the first port, and the possibly exponential response is recorded on the second port.

The step waveform contains an infinite number of frequencies, and the analysis with such non-sinusoids is done with Laplace transforms.

6.1.6. Kramers–Kronig transforms

If the real part of a linear network function of frequency is known over the complete frequency spectrum, it is possible to calculate the imaginary part (and vice versa). There is a relationship between the real and imaginary parts of an immittance (or ε' and ε''), given by the Kramers–Kronig transforms (KKT). In theory there is no additional information in, e.g., the real data when the imaginary data are known. Of course a double data set increases accuracy and makes possible a control of data quality.

KKT are tools brought to network theory by the work of Kramers and Kronig on x-ray optics in the late 1920s (Kramers 1926; Kronig 1929). Just as the reciprocity theorem, they are purely mathematical rules of general validity in any passive, linear, reciprocal network of a *minimum phase shift* type. By minimum-phase networks we mean ladder networks that do not have poles in the right half-plane of the Wessel diagram. A ladder network is of minimum phase type; a bridge where signal can come from more than one ladder is not necessarily of the minimum phase type. The transforms are only possible when the functions are finite-valued at all frequencies. With impedance $\mathbf{Z} = R + jX$ the transforms are

$$R(\omega) - R(\infty) = \frac{2}{\pi} \int_0^\infty \frac{fX(f) - \omega X(\omega)}{f^2 - \omega^2} \, df \qquad (6.10)$$

If we seek dc values, ω is set to zero and we get

$$R(0) - R(\infty) = \frac{2}{\pi} \int_0^\infty \frac{X(f)}{f} \, df = \frac{2}{\pi} \int_{-\infty}^{+\infty} X(\ln f) d(\ln f) \qquad (6.11)$$

$$X(\omega) = -\frac{2\omega}{\pi} \int_0^\infty \frac{R(f) - R(\omega)}{f^2 - \omega^2} \, df \qquad (6.12)$$

$$\varphi(\omega) = \frac{2\omega}{\pi} \int_0^\infty \frac{\ln|\mathbf{Z}(f)|}{f^2 - \omega^2} \, df \qquad (6.13)$$

The frequency of integration f is from 0 to infinite. The resistance or reactance or modulus of impedance $|\mathbf{Z}|$ must therefore be known for the complete frequency spectrum. Dealing with one dispersion, the spectrum of interest is limited to that of the dispersion.

With *admittance* $Y = G + jB$ the transforms are

$$G(\omega) - G(\infty) = \frac{2}{\pi} \int_0^\infty \frac{fB(f) - \omega B(\omega)}{f^2 - \omega^2} \, df \qquad (6.14)$$

If we seek dc values, ω is set to zero and we get

$$G(0) - G(\infty) = \frac{2}{\pi} \int_0^\infty \frac{B(f)}{f} \, df = \frac{2}{\pi} \int_{-\infty}^{+\infty} B(\ln f) \, d(\ln f) \qquad (6.15)$$

The corresponding *permittivity* KKTs are

$$\varepsilon'(\omega) - \varepsilon'(\infty) = \frac{2}{\pi} \int_0^\infty \frac{f\varepsilon''(f)}{f^2 - \omega^2} \, df \qquad (6.16)$$

$$\varepsilon''(\omega) = \frac{2\omega}{\pi} \int_0^\infty \frac{\varepsilon'(f) - \varepsilon'(\infty)}{f^2 - \omega^2} \, df \qquad (6.17)$$

If we seek static values, ω is set to zero and we get

$$\varepsilon'(0) - \varepsilon'(\infty) = \frac{2}{\pi} \int_0^\infty \frac{\varepsilon''(f)}{f} \, df = \frac{2}{\pi} \int_{-\infty}^{+\infty} \varepsilon''(\ln f) \, d(\ln f) \qquad (6.18)$$

Consequently the area under one dispersion loss peak is proportional to the dielectric decrement and independent of the distribution of relaxation times. Equation (6.18) also represents a useful check for experimental data consistency.

6.2. Signals and Measurement

6.2.1. Dc, static values and ac

Dc (direct current) is a current flowing in the same direction all the time (*unidirectional* current). The abbreviation "dc" is so much used that it is common language to say dc current (tautology) and dc voltage (contradictory). A dc may be constant, but may also fluctuate or an ac may be superimposed, as long as the sum never changes direction. Any dc and ac signals may be added, but if the system is *non-linear* the response will not be equal to the sum of the individual signal responses. The dc current may be pulsed, but if the current changes direction in the cycle, it is an ac.

A *galvanic* current is the same as a dc current, and the term is used in particular for therapeutic applications and in electrochemistry. Anode and cathode are defined not from voltage polarity but from current direction. A galvanic (electrolytic) cell produces (passes) dc. If it does not, it is a dielectric cell and only displacement ac passes. Even so, an in-phase current may pass the cell, but it is due to dielectric losses and not to dc conductance. Thus in-phase components are not necessarily the same as dc components.

Stable values are constant values; the term *static* values corresponds to steady-state conditions and can by used for a dc potential or voltage, but not so well for a dc current, which does not fulfil *electrostatic* conditions. In dielectric relaxation theory the subscripts often refer to frequency, e.g. D_0 (Section 3.4.2) is the charge density at

$f = 0$, that is so long after the excitation step that the new equilibrium has been obtained and the charging current has become zero. With a single Debye dispersion this low-frequency value is called the *static* value (cf. Section 3.4.2). D_0 could therefore equally well have been called D_s, with s for "static". This is the case for the symbol for permittivity, where low-frequency permittivity is ε_s while ε_0 is the vacuum permittivity.

Ac (alternating current) is a current steadily changing direction. The abbreviation is so much used that it is common language to say ac current (tautology) and ac voltage (contradictory). We also say constant ac voltage (contradictory), by which we mean constant-amplitude ac voltage.

Dc compared with a sine wave ac with $f \to 0$
When a sine wave frequency approaches 0 Hz, corresponding to a period of, e.g., an hour or more, the signal may for a long time be regarded as a slowly varying dc. Strong dc polarisation effects may have time to develop at the electrodes, and capacitive susceptance is very small according to $B = \omega C$.

In order to maintain linear conditions in electrolytic systems, the signal amplitude must be reduced $\to 0$ as $f \to 0$ (cf. Section 2.5.4). Except in the bulk of an electrolyte (cf. eq. 2.2), *dc conditions are therefore virtual unobtainable in electrolytic systems*, cf. also the Warburg impedance concept described in Section 2.5. This is well illustrated with the logarithmic frequency scale, where both infinitely high and infinitely low frequencies are equally off-scale and unattainable. With electronic (not ionic) conduction and ordinary resistors, perfect dc conditions represent no difficulty, and these can therefore only be idealised *models* of electrolytic systems.

6.2.2. Periodic waveforms, Fourier series of sine waves

A *periodic* waveform repeats itself exactly at regular time intervals (the period *T*). It is predictive: at any moment in the future we can foresee the exact value. According to Fourier, any periodic waveform can be considered to be the sum of a *fundamental* sine wave of frequency $f_1 = 1/T$, and sine waves at certain discrete frequencies, the *harmonics* ($2f_1$, $3f_1$, $4f_1$ and so on). A periodic waveform is an idealised concept; the waveform is to have lasted and to last for ever. At the time we start and stop it, other frequency components than the harmonics appear as transients.

The sine wave is a very special periodic waveform in the sense that it is the only waveform containing just one frequency: the fundamental frequency. Why has just the *sine wave* such special qualities?

The sine wave is derived from the circle (Fig. 6.2); it is the projection of a rotating radius (cf. the phasor, Section 10.1). If the rotation is steady, the waveform is sinusoidal. A sinusoid is characterised by its *frequency f* (Hz, periods per second) or the *period T* $= 1/f$ (second). *Angular* frequency ω must be used for trigonometric functions and to emphasise the relationship with the angle of the rotating radius. $\omega = 2\pi f = 2\pi/T$ is the number of rotations (in radians or degrees) per second. *T* is the time of one complete rotation. $\varphi = \omega t$ is the angle of rotation during the time *t*.

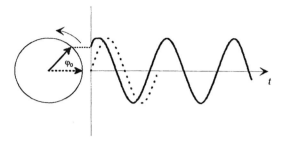

Figure 6.2. The sine wave, with a reference sine wave dotted.

A frequency-independent phase shift or a reference value φ_0 may be added: $\varphi = \omega t + \varphi_0$.

If the sine wave is symmetrical around 0, it has no dc component and is described by the equation:

$$v(t) = V_0 \sin(\omega t + \varphi_0) \tag{6.19}$$

$$\int v \, dt = -V_0 \cos(\omega t + \varphi_0) = -V_0 \sin(90° - \omega t - \varphi_0)$$

$$\frac{dv}{dt} = V_0 \cos(\omega t + \varphi_0) = V_0 \sin(90° - \omega t - \varphi_0) \tag{6.20}$$

The time derivative as well as the time integral of a sine wave is also a sine wave of the same frequency, but phase-shifted 90°. The relationship between a sine wave and the circle is seen more directly in the complex notation of a radius $\mathbf{r}$ rotating around the origin in the Wessel diagram:

$$\mathbf{r}(t) = r_0 \, e^{j\omega t + \varphi_0} = r_0 \left[\cos(\omega t + \varphi_0) + j \sin(\omega t + \varphi_0) \right]$$

As the time derivative of an exponential is the same exponential, then $\partial(e^{j\omega t})/\partial t = j\omega \, e^{j\omega t}$. That is why integration and derivation of sine waves in the equations describing the behaviour of electrical circuits can be replaced by algebraic operations with $j\omega$ instead of $\partial/\partial t$.

A phasor and a sine wave are given with respect to some reference sine wave. In Fig. 6.2 the reference is dotted, and the waveform of interest *leads* the reference by about 45°.

The peak value is called the *amplitude* V_p; φ is the *phase angle*. To define φ we must define a reference sine wave, for instance the known excitation signal. Although the mean value of a full period is 0, it is usual to quote the mean of half a period: $2V_p/\pi$. The *rms* value is $V_p/\sqrt{2}$.

The sum and product of two sine waves
To simplify the equations and the discussions we presuppose two sine waves of *equal amplitude*, symmetrical around zero.

The *sum* of two such sine waves of different frequencies is

$$f(t) = \sin \omega_1 t + \sin \omega_2 t = 2 \sin\left(\frac{\omega_1 + \omega_2}{2} t\right) \cdot \cos\left(\frac{\omega_1 - \omega_2}{2} t\right) \qquad (6.21)$$

The *product* of two sine waves of different frequencies is

$$f(t) = \sin \omega_1 t \cdot \sin \omega_2 t = \tfrac{1}{2}[\cos(\omega_1 - \omega_2)t - \cos(\omega_1 + \omega_2)t] \qquad (6.22)$$

Case 1: $\omega_1 = \omega_2 = \omega$
With a constant phase difference φ between the sine waves, $\omega_1 t = \omega t$ and $\omega_2 t = \omega t - \varphi$:

$$f(t) = \sin \omega t + \sin(\omega t - \varphi) = 2 \cos\frac{\varphi}{2} \sin\left(\omega t - \frac{\varphi}{2}\right) \qquad (6.23)$$

The *sum* is as a *pure sine wave* at the fundamental frequency, phase-shifted, with double amplitude when $\varphi = 0°$ and zero amplitude with $\varphi = 180°$. No new frequencies appear, in accordance with the law of superposition valid for a linear system.

$$f(t) = \sin \omega t \cdot \sin(\omega t - \varphi) = \tfrac{1}{2}[\cos \varphi - \cos(2\omega t - \varphi)] \qquad (6.24)$$

This is an important equation, proving the creation of a new frequency in a non-linear system where the laws of superposition are no longer valid. *A dc component and a pure sine wave of double the frequency (2nd harmonic component) have appeared, and the fundamental frequency has disappeared.* The equations are illustrated in Fig. 6.3.

Frequency spectrum analysis fundamentals
- The product of two sine waves of equal frequency contains a dc component.
- The product of a sine and cosine wave of equal frequency contains no dc component.
- Thus on multiplying a given signal function with a reference sine wave, the dc component of the product is proportional to the in-phase signal component at the reference frequency.
- The product of two sine waves of *different* frequencies contains no dc component.

The low-pass filtered result is a dc voltage proportional to the cosine of the phase difference between the signals; quadrature signals ($\varphi = 90°$) are thus cancelled. The peak-to-peak amplitude of the 2nd harmonic component is half the fundamental and independent of the phase. The equation formulates the basis for the lock-in amplifier and frequency spectrum analysis (Fourier analysis). The Fourier spectrum analysis is actually done this way, by multiplying the signal to be analysed with $e^{-j\omega t} = \cos \omega t - j \sin \omega t$, taking care of both the in-phase and quadrature component, and then sum up for the total time duration (eq. 6.33). In the lock-in amplifier

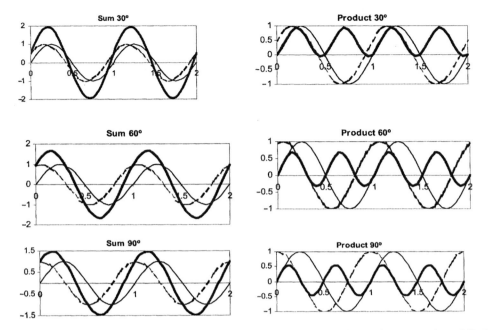

Figure 6.3. Sum and product of two equal-amplitude (= 1) sine waves of equal frequency, phase shifted by various amounts (30° (top), 60° (middle) and 90° (bottom)). Notice the dc component of the product (right hand side). Example is with amplitude = 1 (e.g. 1 volt), f = 1 Hz so that the time scale is in seconds.

one sine wave is the stable reference signal supplied by the experimental setup. The other sine wave is the measured response variable, usually also containing other non-synchronised signals and noise.

In tissue and at electrodes, the linear case corresponds to low-level excitation. The non-linear case and the creation of new frequencies correspond to high-level excitation (the non-linearity in the form of multiplication is of course only one possible form).

Case 2: $\omega_1 \neq \omega_2$

$\omega_1 \approx \omega_2$. If $\omega_1 \approx \omega_2 \approx \omega$, eq. (6.23) and eq. (6.24) can be used by introducing a slowly varying phase shift φ.

The most important effect for the *sum* (eq. 6.23) is the corresponding varying amplitude according to the $\cos(\varphi/2)$ factor. The sum is a sine wave of frequency ω with amplitude changing at the low beat frequency $\omega_1 - \omega_2$ from the double value to zero. The new waveform does not contain any new frequencies, and there is no dc component. This is illustrated in Fig. 6.4a. The curve is counterintuitive because it may easily be taken as a waveform containing the sum and beat frequencies, which it does not. It is the fundamental sine wave with a slowly varying amplitude at the beat frequency 0.125 Hz. The picture is different when the frequency difference is larger

(Fig. 6.4c). Now it is clearly seen that the high-frequency sine wave is simply superimposed on the slow sine wave.

In the *product* (eq. 6.24) waveform the dc level changes at the low beat frequency $\omega_1 - \omega_2$ according to the factor $\cos\varphi$. However, it has zero mean value over a beat frequency period, and thus does not contribute to a dc component after low-pass filtration. The second-harmonic component has constant peak-to-peak amplitude but a phase varying at the beat frequency.

Large difference between ω_1 and ω_2.   Now the discussion is based upon eq. (6.21) and eq. (6.22). It is clear from Fig. 6.4c that the *sum* waveform is just the 1 Hz and

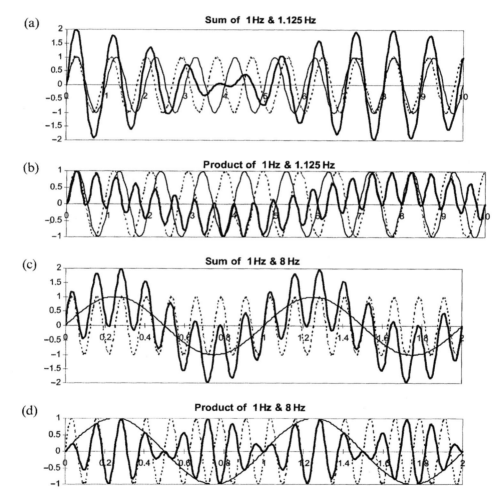

Figure 6.4. Sum and product of two equal amplitude ($= 1$) sine waves with unequal frequencies. (a,b) 1 Hz and 1.125 Hz sine waves ($\omega_1 \approx \omega_2$); (c,d) 1 Hz and 8 Hz sine waves.

8 Hz signals with no phase shifts. However, the *product* waveform has a new higher frequency (the sum $8 + 1 = 9$ Hz). When studying Fig. 6.4d, the waveform is amplitude-modulated with an envelope frequency of 1 Hz. The frequency difference component of 7 Hz ($8 - 1 = 7$ Hz) is difficult to see.

In radio communication systems the signal with frequency ω_1 is called the carrier and ω_2 the modulation, and $\omega_1 \gg \omega_2$. The amplitude-modulated signal from a perfect multiplier under these conditions does not contain the low-frequency signal ω_2, just the upper and lower sideband frequencies $\omega_1 + \omega_2$ and $\omega_1 - \omega_2$. They are very near to the carrier frequency and can therefore be transmitted through the ether as radio waves.

It is clear from these discussions that the waveforms in the time domain may be difficult to interpret correctly, and it is the mathematical treatment that gives the correct answers.

The sum of a fundamental sine wave and its harmonic components: Fourier series
The only waveform containing just one frequency is the sine wave. A periodic waveform can be created by a sum of sine waves, each being a harmonic component of the sine wave at the fundamental frequency determined by the period. This is illustrated in Fig. 6.5a,b, showing the sum of a fundamental and its 3rd and 5th harmonic components. It indicates that odd harmonic components may lead to a square wave, with a precision determined by the number of harmonic components included.

Figure 6.5c shows the frequency spectrum of the waveform. It is a *line* or *discrete* spectrum, because it contains only the three discrete frequencies. Continuously repetitive waveforms have line spectra; their periodicity is composed only of the fundamental and its harmonic components.

Fourier formulated the mathematical expression for the sum of the fundamental and its harmonics. The condition is that a fundamental period of a waveform $f(t)$ can be determined, and that the waveform $f(t)$ is extended outside its defined interval so that it is *periodic* with period $T = 2\pi/\omega_1$.

$$f(t) = \frac{a_0}{2} + \sum_{n=1}^{\infty}(a_n \cos n\omega_1 t + b_n \sin n\omega_1 t) \tag{6.25}$$

where a_n and b_n are the amplitudes of each harmonic component n, $a_0/2$ is the dc component, and ω_1 the angular fundamental frequency defining the period $T = 2\pi/\omega_1$.

According to the Fourier series eq. (6.25), any periodic waveform is the sum of a fundamental sinusoid and a series of its harmonics. Notice that in general each harmonic component consists of a sine and cosine component. Of course either of them may be zero for a given waveform in the time domain. Such a *waveform synthesis* (summation) is done in the time domain, but each wave is a component in the frequency domain. The *frequency spectrum* of a periodic function of time $f(t)$ is

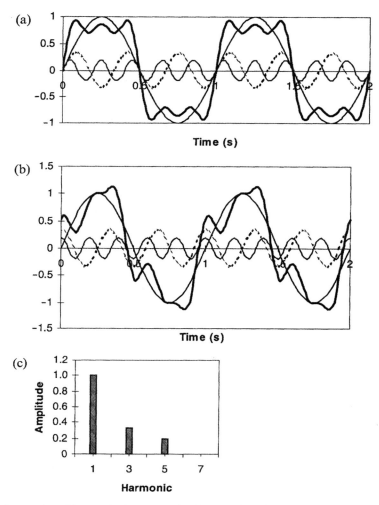

Figure 6.5. Summation of harmonic sine waves, showing waveform dependence on the phase relationships. Amplitude of fundamental sine wave = 1. Time domain: (a) in-phase harmonics, (b) phase-shifted harmonics. (c) Amplitude line frequency spectrum, equal for both cases.

therefore a *line* spectrum. The amplitude of each discrete harmonic frequency component is

$$a_n = \frac{1}{T} \int_{-T/2}^{T/2} f(t) \cos(n\omega_1 t)\, dt \qquad (6.26)$$

$$b_n = \frac{1}{T} \int_{-T/2}^{T/2} f(t) \sin(n\omega_1 t)\, dt \qquad (6.27)$$

$$\mathbf{A}_n = a_n + jb_n \tag{6.28}$$

$$A_n = \sqrt{a_n^2 + b_n^2} \tag{6.29}$$

$$\varphi = \arctan(b_n/a_n) \tag{6.30}$$

Because the waveform is periodic, the integration can be limited to the period T as defined by ω_1. However, the number n of harmonic components may be infinite. The presentation of a signal in the time or frequency domain contains the same information and there is a choice of how data is to be presented and analysed.

Figure 6.5 illustrates how two rather different waveforms in the time domain may have the same amplitude magnitude A_n frequency spectrum. The amplitude magnitude frequency spectrum does not contain all necessary information since the phase information is lacking. For each harmonic component both the sine and cosine (eqs (6.26) and (6.27)), or the magnitude and phase (eqs (6.29) and (6.30)) must be given, a magnitude and a *phase spectrum*. The amplitude $\mathbf{A}_n$ is a vector, therefore amplitude magnitudes A_n cannot just be added as scalars. A given waveform is the sum of only one unique set of sine and cosine harmonics.

An infinite number of harmonics must be added in order to obtain, for instance, a true square wave. The Fourier series for a periodic square wave of unit amplitude is (cf. Fig. 6.5):

$$f(t) = \frac{4}{\pi} \sum_{n=1,3,5...} \frac{1}{n} \sin n\omega_1 t \tag{6.31}$$

The square wave can therefore be realised as the sum of only sine components.

Any waveform with sharp ascending or descending parts, like the square wave or sawtooth, contain large amplitudes of higher harmonic components. The triangular pulses contain more of the lower harmonics. No frequencies lower than the fundamental, corresponding to the repetition rate, exist. The waveform may contain a dc component. If the waveform is symmetrical around zero, the dc component is zero. However, if a periodic waveform is started or stopped, it is no longer periodic. During those non-periodic intervals the Fourier series approach is no longer valid.

Using non-sinusoids as excitation waveforms, a system is excited at several frequencies simultaneously. If the system is linear, the response of each sine wave can be added. If the system is non-linear, new frequencies are created that influence the frequency spectrum. The square wave to the left of Fig. 6.6 has no dc component. One of the ramps in the middle waveform is used in scanning devices such as polarographs. As drawn, the waveform has a dc component. The pulse to the right has a dc component dependent on the repetition frequency.

6.2.3. Aperiodic waveforms

These are the waveforms of bioelectricity, because the repetition rate of the respiration or the heart, for example, is not perfectly periodic.

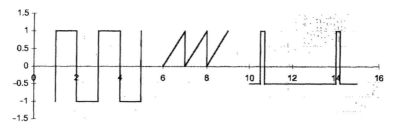

Figure 6.6. Periodic waveforms: square, ramp and pulse.

Single pulse or step

A single pulse or a step function excitation is the basis of relaxation theory. Power dissipation and temperature rise may, for instance, impede the use of repetitive waveforms, and single pulse excitation becomes necessary. A single pulse is a pulse waveform with repetition interval → ∞, it has a *continuous* frequency spectrum as opposed to a line spectrum. The *unit impulse* (delta) waveform is often used as excitation waveform. It is obtained with the pulse width → 0 and the pulse amplitude → ∞, keeping the product = 1. The frequency spectrum consists of equal contributions of all frequencies. In this respect it is equal to white noise (see below). Also, the infinite amplitude of the unit impulse automatically brings the system into the non-linear region. The unit impulse is a mathematical concept, a practical pulse applied for the examination of a system response must have a limited amplitude and a certain pulse width.

What then is the frequency content of a single rectangular pulse? It can be found from the periodic waveform by letting the period → ∞. The frequency spectrum $F(\omega)$ of a positive pulse of amplitude A and duration T is

$$F(\omega) = \frac{2A}{\omega} \sin \omega \frac{T}{2} \qquad (6.32)$$

Note that the frequency spectrum as defined by $F(\omega)$ in eq. (6.32) is *signal amplitude per angular frequency increment (spectral density)*. Equation (6.32) defines a continuous frequency spectrum: *all frequencies* are present except the discrete frequencies n/T where $n = 1, 2, 3$, etc. This may be regarded as another way of saying that the periodicity of a single pulse does not exist—it has no characteristic harmonics. All these frequency components must also be a function of time: all components must be zero before the single pulse has arrived (cf. the causality criterion, Section 6.1). During and after the pulse the frequency spectrum components build up and decay, with time constants depending on the filters used to record them. Another illustration of a *frequency spectrum as a function of time* is the frequency analysis of speech or music. Biological events do not occur strictly periodically, and there is therefore a general need for frequency analysis as a function of time. In order to have good resolution in the time domain, the time interval used should be short. But a short time interval makes it impossible to analyse low frequencies. This is the basis for the special short time Fourier transform (STFT) presented by Gabor (1946), cf. Section 6.2.11.

White noise waveform

Whereas the sine wave contains only one frequency, a white noise signal is the other extreme, containing all frequencies of equal amplitudes. Like the sine wave or unit impulse, white noise is an idealised concept. It is a fractal curve: any enlargement will just bring up similar curves. It is an interesting excitation waveform because the system is examined at all frequencies simultaneously. White noise is a curve whose value in the future can not be predicted: there is an equal probability of any amplitude at any moment. Both the unit impulse and white noise are represented by a flat frequency spectrum. However, the ideal unit impulse is of infinitesimally short time duration, while the ideal white noise is of infinitely long time duration.

However, because amplitude in any system is limited, we do not have absolute white "colour" of the noise found in a system. Also, our total sampling time is limited. The curve in Fig. 6.7 is calculated for a certain time interval Δt. Based on Fig. 6.7, we can evidently say nothing about a possible periodicity for times $> \Delta t$ (frequencies below $1/\Delta t$). In general any amplitude is possible; noise must therefore be described by a value averaged over a defined time interval. Usually the root-mean-square (rms) value is used. This is practical because noise is related to energy according to Boltzmann, Einstein and the interpretation of Brownian motion (random walks).

Figure 6.7 shows a computer-generated curve, for which the computer was programmed to generate 1000 random numbers (periodic samples at the x-axis) of values between 1 and 1000 (amplitude at the y-axis). These choices are related to the graphical limitations of the illustration. Figure 6.7 therefore does not illustrate ideal white noise, because both the number of samples and the amplitudes are limited. The waveform is somewhat counterintuitive, because visual inspection may easily give the impression of periodicities.

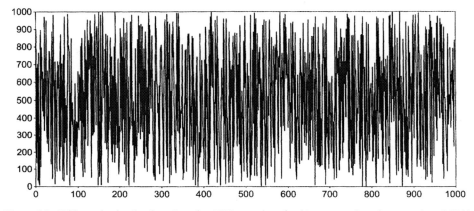

Figure 6.7. White noise in the time domain. 1000 samples of arbitrary numbers between 1 and 1000. Mean (dc) value = 500.

6.2.4. Spectrum analysis, Fourier transforms

Frequency spectrum (Fourier) analysis
The *time domain* is well known from everyday experience; it is, for example, the way a signal is recorded as a function of time as an ECG waveform. From the waveform in the time domain the components can be found in the *frequency domain*: a search for periodicity. In the frequency domain we have seen the line spectra of *periodic* signals. But the heart does not beat regularly. The heart rate varies both in a noisy way and by the way it is controlled by nervous and biochemical systems. Making a frequency analysis of an ECG waveform will therefore not give a *line* spectrum, the spectrum will have a more *continuous* character. Regular heartbeats during sleep correspond to more pronounced line spectra than those obtained during variable physical activity. The frequency spectrum of an electronically driven pacemaker however, is a line spectrum.

There are several ways of finding the frequency spectrum of a time domain waveform, periodic or aperiodic. If the signal is input to a filter bank, the output of each filter represents the signal content as a function of time (both frequency and time domain!) within the frequency passband of the filter. The result is not optimal because the output of each filter is an amplitude magnitude, the phase information is lost. In Section 6.2.2 we showed that on multiplying the waveform to be analysed by a sine wave, the dc value of the result indicates the amplitude content at that frequency. By multiplying also by a 90° phase-shifted signal, information about the phase relationships in the waveform may also be obtained. This must be repeated at each frequency of interest, and therefore is a slow procedure. Instead of these analogue methods, the signal can be digitised and treated by a mathematical algorithm called the fast Fourier transform (FFT). In such ways the frequency content can be extracted, and this is the Fourier[18] or spectrum analysis. Such an analysis is a search for periodicities in a waveform.

The Fourier *series* with discrete harmonics were introduced in Section 6.2.2. Mathematically the Fourier *transform* of a function (periodic or non-periodic) in the time domain, $f(t)$, to the corresponding function in the frequency domain, $f(\omega)$, is described by

$$\mathbf{F}(\omega) = \int_{-\infty}^{\infty} f(t) e^{-j\omega t}\, dt \qquad \text{(to frequency spectrum from time domain)} \qquad (6.33)$$

The machinery behind this equation is shown in Section 6.2.2 and Fig. 6.3. The inverse Fourier transform is then

$$f(t) = \int_{-\infty}^{\infty} \mathbf{F}(\omega) e^{j\omega t}\, d\omega \qquad \text{(to time domain from frequency spectrum)} \qquad (6.34)$$

Note that by multiplying by the complex expression $e^{j\omega t} = \cos \omega t + j \sin \omega t$, both the in-phase and quadrature components, and thus the phase information, are taken

---

[18] Joseph Fourier (1768–1830), French mathematician. Participated as scientist in the Napoleonic military expedition in Egypt 1798–1801.

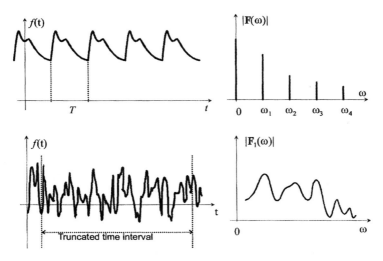

Figure 6.8. Frequency spectra (right) of periodic and non-period waveforms (left). Upper part (right): amplitude $|\mathbf{F}(\omega)|$ in volt. Lower part (right) continuous spectrum: amplitude spectral density $|\mathbf{F}_1(\omega)|$ in volt/$\sqrt{\text{Hz}}$.

care of (cf. eq. (6.24)). When dealing with signals defined, e.g., as voltage, $F(\omega)$ represents the signal distributed in the frequency spectrum: the *spectral distribution of signal amplitude per frequency bandwidth*. The unit may be, e.g., $\mu V/\sqrt{\text{Hz}}$. In general there will be two spectra: one in-phase spectrum and one quadrature spectrum, or one amplitude spectrum and one phase spectrum. Thus $\mathbf{F}(\omega) = F'(\omega) + jF''(\omega)$, or $\mathbf{F}(\omega) = |\mathbf{F}(\omega)|\, e^{j\varphi(\omega)}$, and $|\mathbf{F}| = (F'^2 + F''^2)^{1/2}$. Because of the phase dependence, the amplitudes at each frequency are not simply additive.

Equations (6.33) and (6.34) are the integral, and thus more general forms of eq. (6.25). The integral form can, for example, be used for periodic signals when the pulse interval $\to \infty$, that is for a single pulse (eq. 6.32). Note also the non-realistic integration over the complete frequency spectrum from $-\infty$ to $+\infty$, and that the frequencies are not limited to the harmonics. As we have seen, the integration interval is not a problem with periodic waveforms, the waveform can easily be extended without limitation. With aperiodic waveforms, approximations that introduce errors must be made, see below.

According to eq. (6.33), there may be a non-zero signal amplitude density at any frequency, and thus a *continuous spectrum* is possible. With periodic waveforms only *line* spectra are possible; this is illustrated in Fig. 6.8.

The *energy* of a waveform pulse is proportional to the square of voltage or current in a time interval. According to Plancheral's theorem (a special case of the more general Parseval's theorem[19]), the *energy* corresponding to an ideal resistor and a

---

[19] Parseval des Chênes (1755–1836), French mathematician. Forced to flee France after writing poems critical of the Napoleonic government.

voltage or current waveform $f(t)$ computed in the time domain is equal to the energy computed in the frequency domain:

$$\int_{-\infty}^{\infty} f^2(t)\, \mathrm{d}t = \int_{-\infty}^{\infty} |\mathbf{F}(\omega)|^2\, \mathrm{d}\omega \qquad (6.35)$$

For steady-state conditions the integral diverges and *power* (energy per time interval) spectra are used. Note that the magnitude $|\mathbf{F}(\omega)|$ used in eq. (6.35) implies that the information contained in the phase relationships is lost in the power spectrum. Stated more positively: power spectra are not sensitive to phase relationships and values may simply be added.

When dealing with power spectra related to, e.g., rms voltage, v_{rms}, according to $W = v_{rms}^2/R$, $|\mathbf{F}(\omega)|^2$ represents the distributed power spectrum: the *spectral density of signal power per frequency bandwidth*. The unit for $F(\omega)$, e.g. when dealing with noise spectra, may be $\mu V_{rms}^2/\mathrm{Hz}$. When the spectrum is plotted with amplitude squared per Hz on the y-axis and frequency on the x-axis and scaled so that the area under the $F(\omega)$ curve is equal to the total rms value in the time domain (Plancheral's theorem), the spectrum is called a *power density spectrum*. A less stringent definition is simply that a spectrum is called a *power spectrum* when the function is squared before analysis.

Aperiodic signal in a limited time interval
With a recorded waveform, we must generally assume that it represents the sum of non-synchronised aperiodic signals, e.g. from exogenic sources and endogenic activities such as respiration, peristaltic movements, heart beats and nerve activity. In addition there may be wideband noise and noise at discrete frequencies, e.g. from the power line 50 or 60 Hz fundamentals.

In a practical case, the waveform to be analysed must be of a limited time duration. This is particularly clear when the analogue signal has been digitised for computer analysis. Thus a long-lasting waveform must be *truncated* with a finite sampling time, Fig. 6.8. Errors are introduced when such a waveform is analysed, because the Fourier transform (eq. 6.33) presupposes that the integration interval is infinite. When the end value (trailing edge) is not equal to the start (leading) value, the abrupt change of level corresponds to high-frequency components introduced by the truncation. Generally truncation results in sharp discontinuities in the time domain, and the additional frequency components in the frequency spectrum are called *leakage*.

To reduce the leakage effect, the signal can be amplitude-weighted around the leading and trailing edges, so that the signal starts and ends near zero value. A Hanning or Blackman truncation function is often used for this purpose.

The truncated time interval also defines the lowest frequency that can be analysed; nothing can be said about sine wave components with half-periods longer than the time analysed.

In conclusion, the truncation introduces errors due to leakage, and the (totally sampled) interval limits the lowest frequency analysed. In addition, the limited sampling frequency may also introduce errors when an analogue waveform is to be digitised. The sampling frequency must be higher than twice the highest frequency to be analysed (*Nyquist criterion*). If this is not the case, *aliasing* errors (frequency folding) are introduced. If the sampling frequency cannot be increased, the signal must be low-pass filtered before analysis, so that the Nyquist criterion is met for all signal components reaching the analyser.

Correlation and convolution

In EEG waveforms, it may not be easy to estimate visually whether there is interdependence between the waveforms from two different leads. *Correlation* analysis is used to find common periodicities of two functions (waveforms) $f_1(t)$ and $f_2(t)$. We have seen that if we multiply two sine waves of the same frequency, the dc value of the product is proportional to the cosine of the phase difference φ between them (eq. (6.24)). We can therefore calculate the product as a function of the delay (τ) of one of the waveforms with respect to the other, and look for maxima corresponding to $\varphi = 0$. Mathematically, for each time t, the correlation value $c_{cor}(t)$ can be found by summing the products of one of the waveforms and a time displaced version of the other:

$$c_{cor}(t) = \int_{-\infty}^{\infty} f_1(\tau) f_2(t + \tau)\, d\tau \tag{6.36}$$

The correlation will be maximal if one signal can be displaced with respect to the other until they fluctuate together. The correlation function $c(t)$ will be a more or less noisy sine wave symmetrical around $t = 0$. The decay of the amplitude envelope from $t = 0$ indicates the degree of correlation: the slower the decay, the higher the correlation. If $f_1(t) = f_2(t)$, *autocorrelation* is done by delaying a copy of the function itself and performing the integration of eq. (6.36). The process will be much the same as a Fourier analysis, a search for periodicity.

For the sake of completeness, the *convolution* transform will also be mentioned because it is so closely related to cross-correlation:

$$c_{con}(t) = \int_{-\infty}^{\infty} f_1(\tau) f_2(t - \tau)\, d\tau \tag{6.37}$$

The plus/minus sign difference in the integrand is the only difference between the integrals of cross-correlation and convolution. Convolution is a powerful mathematical tool also strongly related to the Fourier transform. By performing a usual logarithmic transformation, a multiplication is simplified to a summation, and then the anti-logarithmic transformation brings up the result. In a similar way, *convolving* two functions in the time domain corresponds to a *multiplication* of the same functions in the frequency domain. The convolving action implies a folding in the frequency domain, represented by the term $(t - \tau)$. In the integrand of the correlation transform there is no folding process in the term $(t + \tau)$.

Signal averaging

If a synchronisation signal is available from the stimulus source, special noise-reducing techniques are available. The response of an organism to a stimulus is

called an *evoked* potential or *event-related* signal. The technique is somewhat similar
to the principle of a lock-in amplifier (Section 6.3). By recording the response as a
function of time after a stimulus, storing it, repeating the stimulus many times, each
time summing the response with the previous sum, we gradually increase the signal
to noise ratio. Non-synchronised waveforms will cancel out in the long run. It can be
shown that the signal to noise ratio increases with $\sqrt{N}$, where N is the number of
stimuli (cf. eq. (6.59) in Section 6.3.2). The signal to noise ratio limits are related to
the variability of the responses, both with respect to amplitude and time. Applications are found, for example, in hearing, brain stem evoked potentials, EEG with
visual stimuli and electrodermal response (EDR).

Other forms of signal processing
- An important *time domain* analysis method is the *probability density function* of
  finding a certain signal amplitude value. This involves histograms of the
  amplitude window values versus the number of counts in each amplitude
  window.
- Recording the number of *zero crossings* per time interval.
- *Bispectral* analysis. In stead of phase spectra obtained in Fourier analysis where
  the phase relates to the start of the epoch, the bispectrum correlates the phase
  between different frequency components. It is used in EEG.
- *Wavelet analysis*, treated in Section 6.2.11.

6.2.5. Signal generators

The constant-amplitude voltage output (voltage clamp)
From the electric mains and our use of batteries we are well acquainted with the
constant-amplitude voltage supply. The ideal voltage supply has zero internal
resistance. It supplies the set voltage from no load (load resistance ∞, open circuit)
to full load (minimum load resistance and maximum current). Two ideal voltage
sources can not be coupled in parallel. In series the voltages are added.

   With V = constant, Ohm's law $I = V/R = VG$ shows that the current is proportional to G, not to R.

The constant-amplitude current output (current clamp)
There are no constant-amplitude current supplies in our daily surroundings, so we
are not so well acquainted with them. This sort of supply may be constructed in two
ways: either by electronic circuitry or by a voltage supply with a large series
resistance.

   The ideal current supply has infinite internal resistance. It supplies the set current
from no load (load resistance 0, short circuit) to full load (maximum load resistance
and maximum voltage). Leaving a current supply open-circuited is the same as
leaving a voltage supply short-circuited. Two ideal current sources can not be
coupled in series. In parallel the currents are added.

   With I = constant, Ohm's law $V = RI = I/G$ shows that the voltage is proportional to R, not to G.

Choice of supplies

The power dissipated as a function of load resistance R is

$$W(R) = RI^2 = V^2/R \qquad (6.38)$$

We realise from these equations the important difference between using a constant-amplitude current or a constant-amplitude voltage supplying a variable load resistance. The current–voltage characteristic of the black box may decide the choice of constant-amplitude voltage or current.

6.2.6. The basic measuring circuit

Let us consider a black box containing some series combination of ideal components: one resistor, one or no battery and one or no capacitor. From external measurements we shall find the component values in three different cases.

First, we are told that there is only one resistor in the black box. To find the resistance value R we apply a known dc voltage V (Fig. 6.9a). From measured current I the resistance is calculated: $R = V/I$. Now let somebody go into the black box and add the ideal battery B in series with the resistor (Fig. 6.9c). The dc current will of course change. How can we from the outside know whether the dc current change was due to a change in resistance, or an additional emv in the black box? *We can not.*

In order to find out, we must add something to the measurement. For instance, by varying V, the current I also varies. By applying $V = 0$, we still have a current flowing, which must be $I = V_B/R$. We can then assume that there must be an internal battery in the circuit.

To obtain this information, we actually had to superimpose a varying signal, or an ac voltage, on the dc voltage. In addition, this manual method can not be used if we want to continuously monitor changing values of V_B and R.

If R and B are non-ideal with current-dependent values, as in an electrolytic electrode system, the above approach can not be used. A better approach is to superimpose a *small, continuous* sine wave voltage on the applied dc voltage. Our current-measuring device must then be able to measure both ac currents with phase, and dc. The battery (being ideal with zero internal resistance) will not influence the ac current, and we consequently measure the resistance of R at ac, but a different "R" at dc. As there is no phase shift, we then know that the battery is in the circuit. If

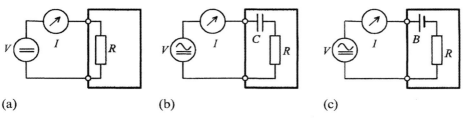

(a)   (b)   (c)

Figure 6.9. Basic black-box measuring problems.

we repeat the measurement at many frequencies and the results are identical, we know that there is no capacitor inside the black box.

What if the battery B is replaced by a capacitor C? With the sine wave superimposed, the current will be phase shifted. From a single frequency measurement, we can find the R–C values by the impedance formula $\mathbf{Z} = R + 1/j\omega C$.

The dc current will decay from the value found when the applied voltage was switched on. The voltage V_c on the capacitor depends on the charge and the capacitance according to $V_c = Q/C$. Just like the battery, it stores energy and it can give this energy back. The capacitor voltage represents a voltage that changes the current I in the same way as a battery. The ac current will not change during the charging of the capacitor, so constant impedance reveals that the dc current decay must be caused by a gradually increased capacitor charge.

If the battery also is inserted, it can not necessarily be detected from the outside because it is blocked by the capacitor. A battery *may be* regarded as a very large capacitor with a nearly constant charge and voltage, so large a capacitance that its reactance $(1/\omega C)$ is negligible. Also, an ideal battery has zero internal resistance (reactance). However, the battery generates an electromotive voltage (emv) from a chemical reaction. A dry capacitor is a system with energy stored in the dielectric, and a voltage according to $V = Q/C$. It is a matter of definition whether the capacitor voltage is called a counter-emv; in this book only emv from electrolytic net charge distributions will be regarded as such.

Measurement of immittance with an endogenic signal source
It is possible to determine the immittance of an electrode system without the use of exogenic signal sources. An endogenic signal is recorded, and the electrode system is loaded with a known admittance in parallel. The *reduction* in signal amplitude is measured as a function of the admittance load. The source immittance can then be calculated. This method has been used for checking the influence of a limited input impedance of ECG amplifiers (Geddes and Valentinuzzi 1973), and the estimation of signal source impedance of implanted pacemaker electrodes (Mørkrid *et al.* 1980).

6.2.7. Operational amplifiers and filters

The operational amplifier
The operational amplifier (op-amp) is an amplifier with two input terminals (inverting and non-inverting), always used with negative feedback, and having so large an amplification that the voltage difference between the input terminals is negligible. It may be used as single-ended, non-inverting voltage amplifier (negligible input current) (Fig. 6.10d), as a current amplifier (negligible voltage drop (b), called a *transresistance* amplifier). It is not suited to measuring a differential voltage without loading the measured circuit; this is done by a special circuit (often composed of three operational amplifiers) called an *instrumentational amplifier* (same circuit drawing as Fig. 6.10a).

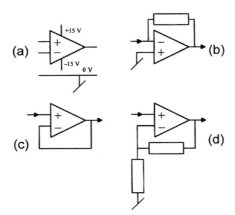

Figure 6.10. Operational amplifier circuits. The details shown in (a) are always present, but are usually omitted in circuit diagrams.

These amplifiers are active devices, and the inputs need a certain bias current and must be within the voltage limits of the power supply. The power supply leads and the reference lead (0 V) as illustrated in Fig. 6.10a are omitted on most circuit diagrams. Notice that the inputs are galvanically separated neither from the output, nor from the power supply. Transformers or opto-couplers are needed if galvanic separation is necessary.

Figure 6.10c is the voltage follower; this is just a buffer with amplification very near unity. It is also the principle of a constant amplitude voltage supply. The purpose is to read a voltage without loading (drawing current) from the measured point. Often the op-amp may be brought as near the recording electrode as possible. Then the output lead is not critical and need not necessarily be shielded. A shield for the input lead may preferably be connected to the output instead of ground, because the capacitance between inner lead and shield is then eliminated (bootstrapping). By adding two resistors (as in (d)) it is possible to obtain amplification.

The current-measuring circuit is very attractive instead of introducing a current-reading shunt resistor with the necessary (even if small) voltage drop. The voltage drop in this circuit is virtually zero.

The circuit of (b) can be used also as a constant-amplitude current circuit. A constant-amplitude voltage and a resistor are used to supply the input current. The load is the feedback resistor; the constant input current will pass the resistor for any resistance value up to the voltage limit of the operational amplifier output circuit (a few volt or less below the power supply voltage to the op-amp).

Frequency filtering
High-pass filter
To eliminate dc from a signal, a blocking capacitor C is inserted. Together with a resistor R they form a high-pass filter (Fig. 6.11a). The time constant is RC, and the

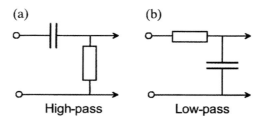

(a)   (b)

High-pass   Low-pass

Figure 6.11. High-pass and low-pass filters.

so-called 3 dB *corner frequency* f_0 is $1/2\pi RC$. At that frequency the phase shift is 45°, and the amplitude has dropped to 63%. This is clear from the transfer function:

$$\mathbf{H}(\omega) = v_0/v_i = \frac{\omega^2 R^2 C^2 + j\omega RC}{1 + \omega^2 R^2 C^2} \qquad \varphi = \arctan(1/\omega RC) \qquad (6.39)$$

At f_0 the phase shift $\varphi = 45°$, at $10f_0$ $\varphi = 5.7°$, and at $100f_0$ $\varphi = 0.57°$. The phase shift in a filter is thus substantial even far away from the corner frequency in the passband, and the frequency must be much higher than the corner frequency to ensure negligible phase shift.

If a repetitive signal is applied to the tissue electrodes via a high-pass filter, no dc polarisation is possible.

Low-pass filter
The low-pass filter (Fig. 6.11b) passes low frequencies and dc. The transfer function is

$$\mathbf{H}(\omega) = v_0/v_i = \frac{1 - j\omega RC}{1 + \omega^2 R^2 C^2} \qquad \varphi = \arctan(-\omega RC) \qquad (6.40)$$

The same precaution holds for the phase shift: to ensure low phase shift, the frequency must be much lower than the corner frequency. Such low-pass filtering effects are an important source of error when reading signals through high-impedance systems such as microelectrodes owing to inevitable stray capacitance (cf. Section 8.1).

The low-pass filter is also easily realised with a capacitor in parallel with the feedback resistor of the op-amp circuits of Fig. 6.10b and d.

6.2.8. Ground, reference and common mode voltage

In a building there are materials forming the ceiling, floor and walls of rooms. The materials used may have a certain electrical conductivity. We must remember that the dimensions are very large, so that even a small conductivity may result in appreciable conductances according to $G = \sigma A/L$. These materials form a Faraday cage around the room, and it is of interest to have electrical access to this cage. That is one of the functions of the household *ground wire*. In our context the importance of the ground wire is not that it is connected to the earth, but to the building and the room we are in. The ground wire is of interest with respect to noise (*functional grounding*), but also with respect to safety (*safety grounding*). Therefore, the net plug of electromedical class I equipment contains three wires: the two power line wires and the ground wire with the colour yellow/green according to the international standard IEC-60601. Electromedical equipment may also be of class II with *double insulation*, when there is no safety ground wire in the power cable.

For safety reasons, we ideally wish to have a floating (not grounded) patient (person), because then if the patient by accident has a potential with respect to ground, no current flows. Therefore modern electromedical equipment is usually designed with a *floating applied part* of type BF or CF (cf. Section 8.14). The applied part is the part of the equipment that by intention is in physical contact with the patient. B means body, C cardiac and F floating. The F-type equipment is designed with a *galvanic separation* between the applied part and the rest of the equipment. Some equipment still grounds the patient, type B.

Electronic circuitry with operational amplifiers is usually supplied from a symmetrical power supply, e.g. ± 12 volt. This implies three wires to the amplifier: $+ 12$ V, 0 V and -12 V. The 0 V wire is also called the *reference* wire, or chassis wire. The reference wire is used for connecting local shields of cables or chassis, for example. The reference wire is often floating, that is not connected to ground. The symbols for reference and ground are shown in Fig. 6.12d. It may also be grounded to reduce noise (functional grounding) or to increase safety (safety grounding).

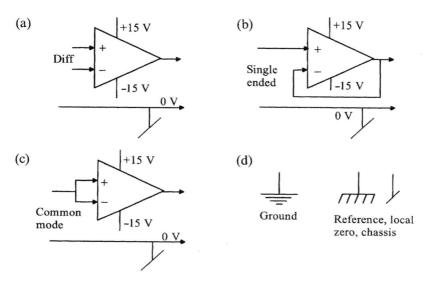

Figure 6.12. (a) Differential (instrumentational amplifier) input. (b) Single-ended input. (c) Common mode coupled input. (d) Symbols for ground (protective earth) and local reference.

The signal input of electronic amplifiers may be designed single-ended or differential. *Single-ended* input (Fig. 6.12b) is an asymmetrical two-wire input circuit, one wire for the signal plus the reference wire. The reference wire, grounded or not, is common for the input, output and power supply. The circuit is a two-port, three-terminal device.

Differential input (a) is a symmetrical three-wire input circuit, two wires for the input signal (+ wire non-inverting, − wire inverting) plus the reference wire. The signal between the differential wires v_i is the differential, wanted signal. The voltage between the differential input wires (together) and the reference wire, is the *common mode voltage* (CMV, Fig. 6.12c). The amplifier should be as insensitive to CMV as possible, and this is expressed by the amplifier's *common mode rejection ratio* (CMRR), usually given in decibel (dB). A CMRR of 100 dB ($= 10^5$) means that, e.g., a CMV input signal of 1 volt (between the two input wires connected together and the reference wire) is equivalent to a differential input signal of 10 μV. The CMRR is always strongly frequency dependent, with the highest value at dc. In the circuit of Fig. 6.12b an ordinary operational amplifier is used, in the circuit of Fig. 12a the special *instrumentational amplifier* is used. That circuit is a two-port, five-terminal device. However, two of the terminals are connected together.

The CMV must usually be within limits set by the power supply of the amplifier. If the supply is ± 12 volt, the CMV input range is perhaps ± 9 volt. For this reason the patient/test person must usually have a third electrode connected to the reference wire of the input amplifier. Without this third wire, the input amplifier's CMV range may easily be exceeded. Both dc and ac must be considered in this respect. In BF and CF equipment the third electrode is a floating reference electrode; in B equipment it is a ground electrode.

An operational or instrumentational amplifier of the applied part is usually *galvanically coupled* to the patient. If it has infinitely high CMRR, then with respect to CMV noise cancellation it will be the same as if input and output are galvanically separated. But with respect to safety it is not ideal, because the CMV input range is restricted to less than the power supply voltage to the amplifier. Outside that linear range, the junctions of the input transistors may enter a nonlinear breakdown region. To keep it inside the CMV range, a third electrode is usually necessary. An input circuit is therefore only *galvanically separated* if its energy supply is from batteries or transformers, and if the signal output is by optical, transformer or radio signal (telemetry) coupling. Then the allowed CMV may be in the kilovolt range. A transformer winding input is the nearly perfect input both for safety and dc CMV cancellation, but unfortunately not for broadband amplification and high input impedance.

Power line noise reduction methods

Figure 6.13 shows three situations of power line capacitatively coupled noise voltage. Suppose first (left) that leakage current is passing through a grounded electrode on the patient. The skin impedance under such an electrode may easily attain 100 kΩ at 50/60 Hz. With a leakage current of 1 μA, the voltage on the patient

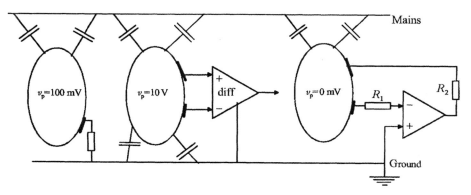

Figure 6.13. Three noise reduction approaches. Left: grounded patient. Middle: floating patient with differential (instrumentational) amplifier. Right: ground-clamping circuit. v_p is typical patient ac voltage with respect to ground. v_p is in practice non-sinusoidal: high-pass filtering effects will expand the harmonics of the mains supply waveform.

with respect to ground is 100 mV. With a single-ended signal amplifier (two electrodes on patient: signal and ground), the 100 mV is superimposed on the signal of interest. To reduce the influence of the noise voltage, one must reduce either the coupling to the mains (increase distance to source, shield, Faraday cage) or the impedance of the ground electrode (dominated by skin impedance).

Another approach (middle) is to convert the noise voltage to a common voltage by applying an instrumentational amplifier (two measuring electrodes on patient). With a floating patient the voltage may be about 10 V, so a common mode rejection factor of 120 dB (10^6) would be needed to reduce the noise contribution to 10 μV. In addition, the common mode voltage (CMV) must be within the linear range of the instrumentational amplifier. By *also* grounding the patient (three electrodes on patient), the CMV range can be better controlled.

A third approach (right) is to clamp the patient to ground potential by an active operational amplifier circuit (two electrodes on patient). We have already introduced this technique in Section 5.4. The reference electrode picks up the voltage of the deeper skin layers (skin impedance independent: no current flow and no voltage drop in the stratum corneum). The operational amplifier sets up a current in the current-carrying electrode to just counterbalance the capacitatively coupled noise current; the reference electrode is virtually grounded. It is also possible to put in safety resistors (R_1 and R_2) without reducing the effect of the circuit. In that way the patient is not directly grounded even if the semiconductor circuits break down. Of course this is also possible in the instrumentational amplifier input leads. The signal amplifier may be single ended—ground referenced (a total of three electrodes on the patient), or differential (a total of four electrodes on the patient).

The patient has been considered equipotential in this analysis. From the human body segmental resistances described in Section 4.3.6, the chest has a segmental resistance of the order of 10 ohm. With 1 μA flowing the voltage difference is of the order of 10 μV. In the limbs the segmental resistances are much higher. If this is critical, care must be taken about where to locate the reference and current-carrying electrodes of the clamping circuit, and where to locate measuring electrodes with respect to noise current flow paths.

6.2.9. Chaos theory and fractals

An exciting development is currently seen in the use of chaos theory in signal processing. A chaotic signal is not periodic, it has random time evolution and a broadband spectrum, and is produced by a deterministic non-linear dynamical system with an irregular behaviour. The evolution in state space (also called phase space) of a chaotic signal must take place on a strange *attractor* of volume zero, which requires sensitive dependence to initial conditions. An attractor is a set of values to which all nearby values converge in an iterative process. In order to classify the attractor as strange, this set of values must be a fractal. Consider for example the sequence produced by the iterative process

$$S_n = (\cos x)^n \tag{6.41}$$

The sequence appears when you, e.g., repeatedly press the cosine button on a pocket calculator and it converges to 0.739 if x is an arbitrary number in radians. Other iterative processes do not converge to a specific number, but produce what seems like a random set of numbers. These numbers may be such that they always are close to a certain set, which may be a *fractal*. This set is called a *fractal attractor* or *strange attractor*.

Fractal patterns have no characteristic scale, a property that is formalised by the concept of *self-similarity*. The complexity of a self-similar curve will be the same regardless of the scale to which the curve is magnified. A so-called fractal dimension D may quantify this complexity, which is a non-integer number between 1 and 2. The more complex the curve, the closer D will be to 2. Other ways of defining the fractal dimension exist, such as the Haussdorf–Besicovitch dimension.

Self-affine curves resemble self-similar curves, but have weaker scale-invariant properties. Whereas self-similarity expresses the fact that the shapes would be identical under magnification, self-affinity expresses the fact that the two dimensions of the curve may have to be scaled by different amounts for the two views to become identical (Bassingthwaighte et al. 1994). A self-affine curve may hence also be self-similar; if it is not, the curve will have a local fractal dimension D when magnified a certain amount, but this fractal dimension will approach 1 as an increasing part of the curve is included. Brownian motion plotted as particle position as a function of time gives a typical example of a self-affine curve.

The theory of fractal dimensions may be used in bioimpedance signal analysis, for example, for studying time series. Such analysis is often done by means of *Hurst's rescaled range analysis* (R/S analysis), which characterises the time series by the so-called *Hurst exponent* $H = 2 - D$. Hurst found that the rescaled range often can be described by the empirical relation

$$R/S = (N/2)^H \tag{6.42}$$

where R is the peak-to-peak value of the cumulative deviation from the average value of the signal, and S is the corresponding standard deviation over a time period of N sampling intervals. Natural time series such as temperature and rainfall figures are found to have Hurst exponents more or less symmetrically distributed around 0.73 ($SD = 0.09$) (Feder 1988).

6.2.10. Neural networks

A neural network is a system of interconnected processing elements called neurons or nodes. Each node has a number of inputs and one output, which is a function of the inputs. There are three types of neuron layers: input, hidden and output layers. Two layers communicate via a weight connection network. The nodes are connected together in complex systems, enabling comprehensive processing capabilities. The archetype neural network is of course the human brain, but there is no further resemblance between the brain and the mathematical algorithms of neural networks used today.

A neural network performs parallel and distributed information processing that is learned from examples, and can hence be used for complex bioimpedance signal processing. The "learning" capabilities of neural networks are by far their most fascinating property. The processing may be simulated in a computer program, but, because of the sequential nature of conventional computer software, the parallel feature of the neural network will be lost and computation time will increase. However, simulation on a computer gives the great advantage of full control over the network algorithm at any time. Physical neural networks are most often implemented as VLSI (very large scale integration) electronic circuits, and sometimes a hybrid solution is chosen.

Neural networks may be constructed in many ways, and those presented so far in the literature can be classified as follows:

1. *Feedforward networks:* These networks are usually used to model static systems. Data from neurons of a lower layer are propagated forward to neurons of an upper layer, and no feedback is used.
2. *Recurrent networks:* These networks include feedback, and are usually used to model dynamic systems.
3. *Associative memory networks:* A type of recurrent network whose equilibrium state is used to memorise information.
4. *Self-organising networks:* The neurons are organised on a sort of dynamic map that evolves during the learning process, in a way that is sensitive to the history and neighbouring neurons. Clustering of input data is used to extract extra information from the data.

The most commonly chosen approach is the feedforward network using a so-called back-propagation algorithm. The back-propagation algorithm can be thought of as a way of performing a supervised learning process by means of examples, using the following general approach. A problem, i.e. a set of inputs, is presented to the network, and the response from the network is recorded. This response is then compared to the known "correct answer". The result from this comparison is then fed back into the system in order to make the network adapt according to the information inherent in the examples. This adaptation is accomplished by means of adjustable parameters that control the behaviour of the network. The act of repeatedly presenting inputs to the network, and providing it with feedback regarding its performance, is called *training*, and the network can be said to be *learning*.

6.2.11. Wavelet analysis

It is difficult to give an introduction to wavelet analysis without reference to Fourier analysis which is discussed earlier in this chapter.

Consider the following signal:

$$f(t) = \sin(\omega_0 t) = \sin\left(\frac{2\pi}{T}t\right) = \begin{cases} \sin(\pi t), 0 \leq t \leq 2 \\ \sin(2\pi t), 2 < t \leq 3 \end{cases} \tag{6.43}$$

which is the sum of two single-period sine signals, as shown in Fig. 6.14. The Fourier transform for this signal will be

$$\mathbf{F}(\omega) = \int_0^2 \sin(\pi t)e^{-j\omega t}dt + \int_2^3 \sin(2\pi t)e^{-j\omega t}dt \tag{6.44}$$

which solves to

$$\mathbf{F}(\omega) = \frac{\pi}{\omega^2 - \pi^2}(e^{-j2\omega} - 1) + \frac{2\pi}{\omega^2 - 4\pi^2}(e^{-j3\omega} - e^{-j2\omega}) \tag{6.45}$$

The amplitude spectrum of $\mathbf{F}(\omega)$ is shown in Fig. 6.15.

Figure 6.15 shows that you need an infinite number of sine waves to reconstruct the original signal perfectly, or at least a large number of sine waves to get close to the original.

Looking at Fig. 6.14, it is obvious that the signal could be represented by only one or two frequency components if the transformation was performed within a moving time window. This kind of time–frequency analysis is available in the so-called short-time Fourier transform (STFT), cf. Section 6.2.3. The problem with STFT is that the resolution in time and frequency is restricted once you have selected the kind of window you want to use for your analysis.

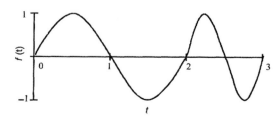

Figure 6.14. Sequence of two single-period sine wave signals.

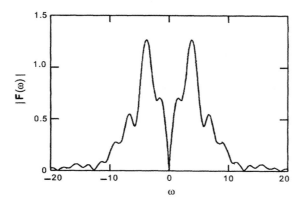

Figure 6.15. Amplitude spectrum of the signal in Fig. 6.14.

An intuitive solution to this problem would be to allow for sine waves with finite duration to appear as building blocks in the transformed data. This is the basis of wavelet analysis. The wavelet transform is based on such building blocks or elementary functions, which are obtained by dilatations, contractions, and shifts of a unique function called the wavelet prototype (or mother wavelet) $\psi(t)$.

There are four different types of wavelet transforms:

1. *The continuous wavelet transform.* This transform is given as

$$CWT(a, \tau) = \frac{1}{\sqrt{a}} \int_{-\infty}^{\infty} f(t) \psi\left(\frac{t - \tau}{a}\right) dt \qquad (6.46)$$

It has a parallel in the Fourier transform, and the variable t, scale a and shift τ are all continuous.

2. *The discrete parameter wavelet transform (or wavelet series).* Here the parameters a and τ are discretised to $a = a_0^m$ and $\tau = n\tau_0 a_0^m$. The parameters m and n are integers and the function $f(t)$ is still continuous. The transform is then given as

$$DPWT(m, n) = a_0^{-m/2} \int_T f(t) \psi(a_0^{-m} t - n\tau_0) dt \qquad (6.47)$$

The wavelet prototype $\psi(t)$ is hence shifted in time by increasing n or in frequency by increasing m. An increased m will reduce the time duration and thereby increase the centre frequency and frequency bandwidth of $\psi(t)$.

3. *The discrete time wavelet transform.* This is a version of the DPWT with discrete time where $t = kT$ and the sampling interval $T = 1$, given as

$$DTWT(m, n) = a_0^{-m/2} \sum_k f(k) \psi(a_0^{-m} k - n\tau_0) \qquad (6.48)$$

4. *The discrete wavelet transform.* This is defined as

$$DWT(m,n) = 2^{-m/2}\sum_k f(k)\psi(2^{-m}k - n) \qquad (6.49)$$

The discrete wavelet $\psi(k)$ can be, but is not necessarily, a sampled version of a continuous counterpart.

Let us now see how discrete parameter wavelet transform can be applied to the simple example given above. The transform coefficients were given as

$$DPWT(m,n) = a_0^{-m/2}\int_T f(t)\psi(a_0^{-m}t - n\tau_0)\mathrm{d}t \qquad (6.50)$$

Considering the signal in Fig. 6.14, one could choose the mother wavelet to be a single period of a sine wave, e.g.

$$\psi(t) = \sin\left(\frac{2\pi}{T}t\right) = \sin(\pi t) \qquad (6.51)$$

and furthermore choose, $a_0 = \frac{1}{2}$ and $\tau_0 = 2$, giving transform coefficients on the form:

$$DPWT(m,n) = 2^{m/2}\int_T f(t)\psi(2^m t - 2n)\mathrm{d}t \qquad (6.52)$$

where the baby wavelets are

$$\psi_{m,n}(t) = 2^{m/2}\psi(2^m t - 2n) \qquad (6.53)$$

The procedure is then to compare the value of this wavelet with $f(t)$ for one period of the wavelet, i.e. $0 \le t < 2$, which gives you DPWT(0,0) = 0. Then increase m and n individually in positive unit steps, to get DPWT(0,0) equal to

	$n = 0$	$n = 1$	$n = 2$
$m = 0$	1	0	0
$m = 1$	0	0.707	0
$m = 2$	0	0	0

Hence, $f(t)$ can be represented by two terms only:

$$f(t) = \psi_{0,0}(t) + \frac{1}{\sqrt{2}}\psi_{1,1}(t) \qquad (6.54)$$

An in-depth description of wavelets and their use in biomedical signal processing is given by Akay (1998).

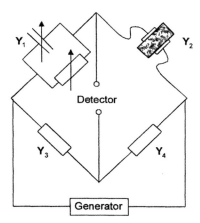

Figure 6.16. Admittance measurement with an ac bridge.

6.3. Bridges, Impedance Analysers, Lock-in Amplifiers

Circuitry for impedance measurements has changed dramatically since the first measurements were performed on biological tissue. The development of digital electronics and the incorporation of computer power in almost all instruments have had an important impact on the possibilities of studying the passive electrical behaviour of biomaterials over a wide frequency range, and with a speed which was not feasible only a few years ago. Only a brief summary of some of the techniques will be given in this section. In low-frequency measurements on tissue with a large dc conductivity from body fluids, high-resolution methods are of particular interest.

6.3.1. Bridges

A variety of bridges have been used for immittance measurements for a long time. The general principle of an ac bridge is illustrated as an admittance bridge in Fig. 6.16, where $Y_1 = 1/Z_1$ is a parallel circuit of a variable resistor and a variable capacitor, and $Y_2 = 1/Z_2$ is the measured sample. By balancing the bridge so that the signal measured by the detector is equal to zero, the unknown admittance Y_2 can be calculated using the relation $Y_1Y_4 = Y_2Y_3$.

The advantage of bridges is their high-resolution capabilities, a feature very important for extending dielectric measurements to low frequencies in tissue (cf. Section 3.8). Blood, for instance, has a conductivity of about 1 S/m and ε_r' of 1500, frequency independent up to 100 kHz. A conduction resolution of about 10^5 is necessary for a precision of 10% at 1 kHz (Schwan 1963). Schwan (1963) discussed both low- and high-frequency bridges, and Schwan et al. (1968) discussed high-resolution tetrapolar bridges. Hayakawa et al. (1975) have further increased the precision of bridge instrumentation.

However, manual bridges are slow, and not suited for measurements on dynamic systems. Although automated bridges are commercially available, they have given way to other methods for bioimpedance measurements, such as lock-in amplifiers.

The lower practical frequency limit for an ac bridge is about 10 Hz if it is transformer coupled, and down to dc if it is directly coupled. The upper frequency limit may be high in the MHz region with special constructions.

6.3.2. Digital lock-in amplifiers

Lock-in amplifiers are commonly used to detect minute signals buried in noise. This can only be accomplished, however, if the signal of interest appears as an amplitude modulation on a reference frequency. The ideal lock-in amplifier will then detect only the part of the input signal having the same frequency and phase as the reference signal. This technique is ideal for immittance measurements, since admittance appears as an amplitude modulation on the measured current, and impedance appears as an amplitude modulation on the measured voltage.

Lock-in amplifiers can be basically digital or analogue. Analogue amplifiers will be treated in the next section.

Digital lock-in amplifiers are based on the multiplication of two sine waves, one being the signal carrying the amplitude-modulated information of interest, and the other being a reference signal with the chosen frequency and phase. Many commercial lock-in amplifiers can also generate reference signals with frequencies equal to a multiple of the base frequency and hence enable harmonic analysis. If the reference signal is given (assuming an amplitude of unity) by

$$v_r = \sin(\omega_r t) \tag{6.55}$$

and the input signal is

$$v_i = v_1 \sin(\omega_i t + \varphi) \tag{6.56}$$

the output signal will be (cf. eq. (6.22))

$$\begin{aligned} v_o &= v_1 \sin(\omega_i t + \varphi) \sin(\omega_r t) \\ &= \frac{v_1}{2} [\cos((\omega_i - \omega_r)t - \varphi) - \cos((\omega_i + \omega_r)t + \varphi)] \end{aligned} \tag{6.57}$$

A low-pass filter will follow this multiplier module, and the right-hand cosine expression may therefore be ignored. In fact the only dc signal that will appear at the low-pass filter output will be the one corresponding to $\omega_i = \omega_r$, which gives

$$v_o = \frac{v_1}{2} \cos \varphi \tag{6.58}$$

A lock-in amplifier is perfect for immittance measurements. Two amplifiers should be used, one with a reference signal identical to or in phase with the excitation signal, and one with a reference signal 90° out of phase with the excitation signal. If the excitation signal is a voltage, the measured current should be converted to a voltage using, e.g., a transresistance amplifier. In that case the dc outputs from the lock-in amplifiers will be proportional to the parallel admittance values of the measured object, i.e. conductance and susceptance, respectively. Using a current as excitation will correspondingly produce a measured voltage that can be separated into signals proportional to the series impedance values, which are resistance and reactance.

Total suppression of noise in this system is of course only possible if the integration time is infinite, i.e. the multiplication is carried out over an infinite number of signal cycles. Gabrielli (1984) shows that the suppression of random (white) noise is given by an equivalent filter function with a bandwidth Δf given by

$$\Delta f = \frac{f_r}{N} \tag{6.59}$$

where f_r is the analysed frequency and N is the number of cycles. Hence, in a 10 Hz measurement, the bandwidth is 0.1 Hz when integrating over 100 cycles, but only 2 Hz when integrating over only five cycles of the signal. The corresponding transfer function is given by

$$|\mathbf{H}(\omega)| = \frac{2}{\pi N} \left[\frac{\sin(N\pi\omega/\omega_0)}{1 - (\omega/\omega_0)^2} \right] \tag{6.60}$$

where ω/ω_0 is the normalised angular frequency. The transfer function is also shown in Fig. 6.17 as a function of number of integration cycles.

Digital lock-in amplifiers have virtually no low-frequency limitations. Commercial amplifiers typically operate down to 1 mHz, but, for example, the Solartron 1260 is constructed for measurements down to 10 μHz (corresponding to a period of about 28 hours!). The upper frequency limit is about 100 kHz mainly limited by the conversion time in the analogue to digital converters. Above this frequency, heterodyne sampling can be used as described in Section 6.3.5 which extends the range of Solartron 1260 to 32 MHz. Another example of a radiofrequency digital lock-in amplifier is the SR844 from Stanford Research Systems. It has a frequency

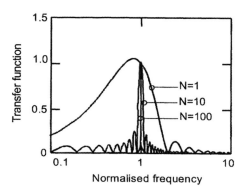

Figure 6.17. Absolute value of transfer function for a digital lock-in amplifier as a function of normalised frequency and number of integration cycles.

range from 25 kHz to 200 MHz with an *absolute* phase error less than 2.5° below 50 MHz, increasing to a maximum of 10.0° at 200 MHz. The *relative* phase error is less than 2.5°, however. This can be utilised in a practical experimental situation, for example by measuring before and after biomaterial is added to an electrolyte solution. Those interested in constructing their own digital lock-in amplifier should read Vistnes *et al.* (1984), for example.

6.3.3. Analogue lock-in amplifiers

The heart of the analogue lock-in amplifier is the synchronous rectifier that includes a phase-sensitive detector (PSD) and a low-pass filter.

The PSD in Fig. 6.18 comprises an inverter and an analogue switch. The switch is connected to the in-signal in the positive half-periods of the reference signal, and to the inverted in-signal in the negative half-periods. Hence, if the in-signal and reference signals are in phase, the detector will act as a full-wave rectifier, providing a maximum dc signal out from the low-pass filter. If the in-signal and reference signal are 90° out of phase, the resulting signal out from the PSD will be without any dc component as shown in Fig. 6.18, and the low-pass filter will produce no output signal. A phase shift between 0 and 90° will accordingly produce a dc signal lying between these two limits. It can easily be shown that the output signal from the synchronous rectifier is given by the equation

$$V_{out} = v_{in}(av) \cos \varphi \qquad (6.61)$$

where V_{out} is the dc voltage out from the low-pass filter, $v_{in}(av)$ is the average value of the ac in-signal and φ is the phase shift between the in-signal and the reference.

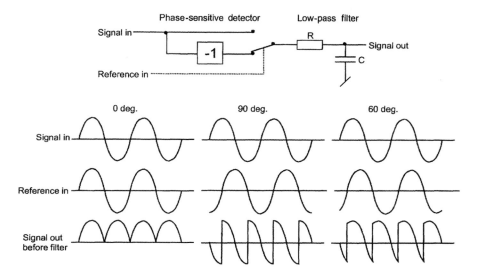

Figure 6.18. Synchronous rectifier with 0°, 90° and 60° phase shift between signal and reference.

As described in the previous section for digital lock-in amplifiers, the analogue PSD can mathematically be described as a multiplication of the in-signal and the reference signal. Since in this case the reference signal controls a switch, it must be regarded a square wave. We should then take into account the Fourier components of this signal, which include the frequency of the square wave itself as well as all odd harmonic frequencies. The synchronous rectifier will hence be sensitive both to the reference frequency and to all odd harmonic components, but with sensitivity proportional to $1/n$ where n is the odd harmonic number (cf. eq. (6.31)).

Analogue lock-in amplifiers have a practical lower frequency limit of about 1 Hz, owing to the necessity of ac coupled inputs. Since any dc signal will influence the results, the method requires pure ac signals.

The upper frequency limit of the analogue lock-in technique is about 100 kHz, and is mainly due to problems of stray capacitance.

6.3.4. Current mode lock-in amplifiers

Most signal processing has traditionally been confined to manipulating electrical voltages, and current signals are typically transformed into voltages in most electronic instruments. However, in semiconductor devices currents are fundamentally controlled, and it has proved to be advantageous to use current mode signal processing in some applications.

The basic idea for using current mode analogue lock-in technique, e.g. for RF measurements, is that currents may be driven with relatively low voltages. This reduces the charging time of stray capacitance, and hence shortens signal propagation time, enabling high-precision measurements at higher frequencies. A low-cost current mode lock-in amplifier has been presented by Min and Parve (1996, 1997).

As shown in Fig. 6.19, it comprises a programmable oscillator that provides a symmetrical controlled current through the specimen of interest, Z_x, by means of a current-to-current converter (CCC). The voltage across the same specimen is again converted into a current by a differential input voltage-to-current converter (VCC),

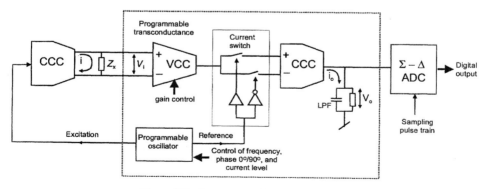

Figure 6.19. A current mode lock-in amplifier.

which is also termed a transconductance amplifier. The current from the transconductance amplifier is then decomposed into its in-phase and quadrature component by means of current mode switches that are driven by the reference signal from the programmable oscillator. Hence, to measure the quadrature component, the oscillator must provide also a second reference signal that is 90° out of phase with the excitation signal. A differential input current-to-current converter ($\pm$CCC) produces the output current i_o, which is filtered and converted into an output voltage V_o by the low-pass filter (LPF).

To avoid aliasing, the measured signal, which is synchronously rectified, is filtered by the LPF before reaching an analogue-to-digital converter (ADC). The final averaging of the result is performed using digital signal processing. The system has an upper practical frequency limit of about 10 MHz.

To achieve full vector operation (detection of both the in-phase and quadrature components simultaneously), two identical lock-in channels are needed, one driven by the 0° reference signal and the other by the 90° reference.

6.3.5. Impedance analysers and LCR-meters

There are several different kinds of automated instruments for impedance measurements, which can be divided into two general categories: LCR-meters and impedance analysers. LCR-meters use the auto-balancing bridge technique to measure the impedance of materials. They can generally be used in a frequency range roughly between 5 Hz and 1 GHz. Low-cost meters are also available, like the SR720 or HP4263B which measure at five pre-defined frequencies from 100 Hz to 100 kHz.

Impedance analysers are also called *frequency response analysers* and are in most cases combined instruments employing two or more different techniques to cover a larger frequency range. One example is the HP4194A, which covers a frequency range from 10 Hz to 100 MHz. It can be used for impedance (100 Hz–40 MHz) or gain-phase (full frequency range) measurements and has four terminals for impedance measurements. Proper four-electrode measurements are not available because the voltage and current electrodes must be connected together at the device under test. However, Gersing (1991) designed a special pre-amplifier for these analysers that gives separate connections for the voltage and current electrodes. Such front-end amplifiers for increased measurement accuracy were also later designed by Yelamos et al. (1999), for example.

Another example is the Solartron 1260, which uses digital correlation techniques for measurements in the range 10 μHz to 32 MHz. The analogue to digital converters (ADCs) used have a resolution of 16 bits and the basic accuracy is 0.1% for magnitude and 0.1° for phase measurements. The 1260 has three channels measuring in parallel (two voltage and one current measurement channel) and ordinary digital lock-in technique is used after the ADCs. For the highest frequency range (65.5 kHz to 32 MHz) an analogue phase-locked loop system is used that generates both the high-frequency output waveform to the sample under test and an internal high-frequency reference signal, which is arranged to be at a slightly different frequency from the output waveform. The input waveform is mixed (heterodyned) by this reference waveform to produce sum and difference frequencies. The sum frequency is filtered, leaving the low-frequency waveform, which can then be analysed using the ADC. A digital heterodyning process is used for measurements in the mid-frequency band (300 Hz to 65.5 kHz) to mix the input signals down to low frequency. Low-frequency measurements up to approximately 300 Hz are measured directly by the ADC.

The 1260 has a generator, a current measurement input and two differential voltage inputs. This allows the instrument to be used either in two- or four-terminal connection mode. In order to further improve the instrument's performance in four-electrode bioimpedance measurements, a new interface, the 1294 impedance interface, which is designed to be operated as a pre-amplifier for the 1260 impedance analyser (or any other frequency response analyser) has been developed by Solartron. The 1294 makes use of driven shields, balanced generator and high input impedance voltage sense inputs in order to minimise stray currents to ground and hence maximise accuracy when measuring in difficult four-terminal conditions. In addition, this interface has been developed to comply with the IEC-60601 standard for connection of electrical equipment to live subjects. The 1294 can also be used for two-electrode measurements.

Frequency-domain techniques such as *spectrum analysers* and *Fourier analysers* can also be used for impedance measurements at audio and higher frequencies. A variety of excitation functions may be chosen and the transfer function of the measured object can hence be found by analysing the response spectrum. Spectrum analysers are typically swept-tuned, superheterodyne receivers that display amplitude versus frequency. However, unlike the Fourier analysers, spectrum analysers do not provide phase information. Fourier analysers use digital sampling and mathematical algorithms to form a Fourier spectrum of a signal, and they can be used for both periodic and transient signals.

Another type of instrument that can be used for impedance measurements is that of *network analysers*, which are typically used in a frequency range between 100 kHz and 100 GHz. Measurements of the transmission through and/or reflection from a material are used together with information about the physical dimensions of the sample to characterise the electrical properties. Careful calibration and correction procedures must be utilised when measuring in this frequency range since even minor changes in the measuring setup may largely influence the results.

CHAPTER 7

Data and Models

7.1. Models, Descriptive and Explanatory

We will see in Section 7.2.2 that when measuring on a parallel circuit with controlled voltage, the real part of the resulting current will be proportional to the conductance, and the imaginary part will be proportional to the susceptance. And furthermore, that if the physical reality is a series circuit, this simple proportionality will be absent, and the values must be mathematically calculated in each case. The same proportionality is also present for controlled current measurements on a series circuit. Values for conductance and susceptance of the skin are thus always related to an opinion on whether these phenomena exist electrically in series or in parallel.

The problem when trying to make an electrical model of the physical or chemical processes in tissue is that often it is not possible to mimic the electrical behaviour with ordinary lumped, physically realisable components such as resistors, capacitors, inductors, semiconductor components and batteries. Let us mention three examples: (a) The constant phase element (CPE), not realisable with a finite number of ideal resistors and capacitors. (b) The double layer in the electrolyte in contact with a metal surface. Such a layer has capacitative properties, but perhaps with a capacitance that is voltage dependent or frequency dependent. (c) Diffusion-controlled processes, cf. Section 2.4.2. Distributed components such as a CPE can be considered composed of an infinite number of lumped components, even if the mathematical expression for a CPE is simple.

In this chapter we will use electrical models for human skin as examples in a general discussion on the use of electrical models. An electrical model of the skin with only two components will obviously not be able to simulate the frequency response measured on skin—it is certainly too simple compared to the complex anatomy of human skin. We suggest three different solutions to this problem, which would result in the following archetypes.

- *Model 1:* One solution is to make complete distinction between the different structures of the skin, and hence make an electrical model as complex as the skin itself. Lipid membranes would then typically be replaced by capacitors, electrolytes by resistors, and semipermeable membranes by voltage sources. The model will, without compromise on the cellular level, aim to reflect the electrical properties of the various *microanatomical* structures.
- *Model 2:* A different solution would be to focus on those structures of the skin that make the largest contributions to the electrical properties, decide whether each of these structures should be represented by a resistor or a capacitor, and

finally find the simplest electrical equivalent that comprises all these structures. This *organ-oriented* model is thus a simplified version of Model 1.

- *Model 3:* A third solution would be to consider the skin as a *black box*, make extensive measurements on it, and find the electrical circuit that matches the admittance levels and frequency response measured. The anatomy of the skin is not considered in this model.

These three models will appear very different, but they are all electrical models of the skin. Is one of them more "correct" or "better than" the other models—and in that case, why?

The model concept

The word *model* is ambiguous. A model can be a three-dimensional representation of an object, but it can also be a person presenting fashion clothing, an object used to produce casting moulds, a person who is portrayed, an instrument for physical model experiments, a substitute for something, an illustration of a mathematical expression, etc. The word *model* has essentially two different interpretations; *prototype/picture* and *equivalent/substitute*. Mathematical models should also be mentioned, which are a sort of predictive substitute for reality. A characteristic of models is that they represent selected properties of the reality they reflect.

Models that serve as visual pictures of reality have a geometric resemblance to the original, but can be scaled compared to it. The substitute models depict other properties than the visual. They can, with regard to these properties, act as substitutes for the original. (As an alternative, one could claim that the pictorial models serve as substitutes with regard to visual properties, and include them also in this group.) Their predictive function is of great interest, because they are often used to investigate a specific side of the functionality or structure of the original, or its response to given stimuli. Such models might be flight simulators, models of the hull of a ship, miniature railroads, mathematical models, electrical models or maps. Hence, for each of these models, one has chosen to emphasise one or more attributes from the original, and has thus made them selective acting substitutes. The models to be discussed here are substitute models, where the electrical properties are important.

Ian Hacking (1983) writes: "Suppose there are theories, models and phenomena. A natural idea would be that the models are doubly models. They are models of the phenomena, and they are models of the theory. That is, theories are always too complex for us to discern their consequences, so we simplify them in mathematically tractable models." The simplification can always be done in alternative ways, depending on the choice of theories and differences in the amount of stress put on theories and phenomena. We shall now take a closer look at how these choices influence the models, and how the models appear as more or less *descriptive* or *explanatory*.

Some examples of models

Tregear presents a model of the skin comprising 12 resistors and 12 capacitors in his book *Physical Functions of Skin* (Tregear 1966) (Fig. 7.1). The capacitors have equal

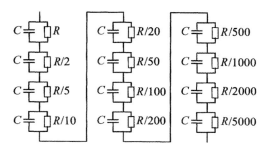

Figure 7.1. Tregear's model.

values, and represent cell membranes in the stratum corneum. The decreasing resistance values represent intercellular conductance, considering that the stratum corneum is drier towards the surface.

The transcellular admittance should, according to Tregear, actually be represented by two capacitors in series with a resistor, where each capacitor is a cell membrane and the resistor represents intracellular fluid. In the stratum corneum, however, the cells have collapsed, and the series resistance is thus small (short-circuited), so that the cells can be represented by a resulting capacitor. Tregear measured the impedance and phase angle in the frequency range 1 Hz–10 kHz, both in the model and in human skin, and found acceptable correlation between the results. The circuit is a good example of Model 1.

Neuman's circuit (Fig. 7.2) is presented in Webster's book *Medical Instrumentation* (Neuman 1992), and is more oriented towards organised tissue than is Tregear's model. The components E_{se}, C_e and R_e are related to the electrical properties of the epidermis itself, while E_p, C_p and R_p represent the sweat ducts. The voltage sources E_{se} and E_p are DC potentials generated by the respective organs. The resistance R_u in the viable layers of the skin is in most cases regarded as

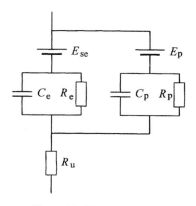

Figure 7.2. Neuman's model.

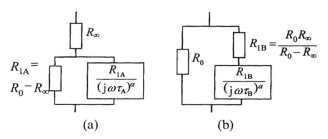

(a)   (b)

Figure 7.3. Salter's models.

negligible. The frequency response of the model is not compared to measurements on skin, and the model hence comes naturally under Model 2.

Salter's two circuits in Fig. 7.3 are presented in Rolfe's *Non-invasive Physiological Measurements* (Salter 1979). They model with a certain precision the frequency response measured on human skin, and are closely related to the Cole equations, which will be treated in more detail in the next section. The model in Fig. 7.3a is probably the most frequently used, and it originates from fundamental work done by, among others, Cole, Fricke and Schwan (Foster and Schwan 1989). The model is based on measurements, and reveals little about the anatomy of the skin. One can certainly empirically relate variations in the frequency response to changes in skin function, but the model gives no explanations for the changes. The model corresponds to our Model 3.

The last model in this selection is taken from a paper by Kontturi *et al.* (1993). R_1in Fig. 7.4 is described as the ohmic resistance of the skin and CPE as a capacitive impedance given by

$$Z_{CPE} = \frac{1}{Y(j\omega)^\alpha} \qquad (7.1)$$

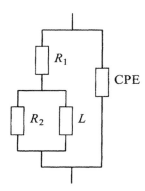

Figure 7.4. Kontturi's model.

Alpha (α) in this equation, has a value between 0.5 and 1, and the following relation to the fractal dimension D_F of the measured surface is suggested: $\alpha = 1/(D_F - 1)$. The fractal dimension is somewhere between 2 and 3, depending on the roughness of the surface. It is furthermore pointed out by the authors that while a simple RC-circuit has a single time-constant $\tau = RC$, a CPE-element has a continuous distribution of time-constants. This constant phase element is further discussed in the next section. The components R_2 and L only exist after treatment with so-called penetration enhancers, i.e. substances that increase the transdermal transport of other substances, e.g. medical drugs. The inductor L is particularly interesting, the hypothesis put forward by Kontturi *et al.* (1993) is that enhancers increase the mobility of large dipolar molecules in the skin, and that their relaxation at low frequencies is of a nature that causes induction. The strength of the model is that it simulates skin *in vitro* with good precision, and it resembles our Model 3.

Description or explanation?
Let us return to the question of which model is the best or most correct. The question as posed is obviously not precise enough to be given an exact answer. What is required from a good or correct model?

The term "electrical equivalent" is as frequently used as "electrical model" in the literature. The more rigorous word "equivalent" has a more precise meaning than "model". An electrical equivalent is a circuit that electrically behaves exactly like the original when studied from predefined terminals. Model 3 is in this respect the best electrical equivalent, and consequently also the best model, if the overall object is to describe the electrical properties of the skin. The descriptive models should therefore in this context be separated from the explanatory.

The descriptive models characterise the skin electrically by means of both known electric components and algorithms. The models reflect primarily the phenomena, i.e. the measured values and time courses, and the theories are not to any great extent connected to the microanatomy of the skin. The entities of the model do not necessarily exist as isolated biological structures, and even though the model includes known electric components, they do not necessarily resemble corresponding electrophysiological processes in the skin. An instrumentalist, who will claim that the ontological reality is only a heuristic construction that is used to organise experience in a suitable way, may say that these models are good despite their consisting of inductors and boxes with equations. They simulate the electrical behaviour of skin, and are in that sense proper models. They make good instruments for their particular use, e.g. for calculating the frequency response of skin.

The explanatory models are based on the basic concepts of electrical theory—potential, conductance, polarisation, induction, etc. Knowledge about the physical mechanisms behind these phenomena is used to provide understanding of similar phenomena in biological materials, and the models are largely influenced by theories concerning the relationship between microanatomy and fundamental electrical properties. It is vital that these models include only discrete electrical components for which the essential mode of operation is known. The models are explanatory because one believes that the components of the model represent isolated anatomical structures or physical processes, such that the dominating electrical property can be explained by means of the properties of the component. The reality of entities, i.e. belief that the components of the model exist as structures in the skin, is consequently a vital prerequisite in these models.

Tregear's and Neuman's models are essentially explanatory. They provide a fundamental understanding of the dominating electrical mechanisms of the stratum corneum, and they are appropriate for predicting the effect of, e.g., increased ambient humidity, "stripping" of the skin, increased sweat activity, increased measurement frequency, and so on. Their weak point, however, is that they are far from simulating the electrical frequency response of human skin as well as Salter's and Kontturi's models. Achieving something like a Cole plot of a circular arc with depressed locus by means of ideal, passive components requires a very complex model, where the simple heuristic analogy to the electric components will be lost.

One can easily be lead to believe that the descriptive and explanatory models express pure instrumentalism and realism, respectively. As archetypes they presumably do, but it is important to understand that the explanatory models are approximations in which the heuristic content is substantial. Research on skin admittance has revealed electrical phenomena that are not easily represented by the traditional components, but the models have undisputed value in that to a certain extent they illustrate the dominating electrical properties of substructures of the skin. Their limitations are that they are too simple, because they do not take into account the difference between an electronic and an ionic medium. The biological charge carriers can be large molecules that cause relaxation processes, and hence frequency-dependent capacitance and conductance values. This is not easily modelled by means that are compatible with the simple analogy to the passive, basic components of electronics. The explanatory models are thus simplifications, but can at least be regarded as representing the electrical properties of skin at one single frequency.

The most correct model
The most correct model will take all recent knowledge about relaxation processes, frequency dispersion, diffusion, fractals, etc. into account. It will be an electrical equivalent to the skin, i.e. it has the same frequency response. It should be simple and use symbols in a way that makes it easy to understand the outlines of the electrical properties of the different substructures of the skin. Such a model takes care of all requirements of an electrical model of the skin—but unfortunately it does not exist! The most correct of the existing models is therefore the one best adapted for the target group.

The realism of the lumped element models is not easy to assert in view of modern theories and measurement techniques, and these models can be regarded as pragmatic explanatory models. They are, however, the only models that differentiate between substructures of the skin, and they have clinical value in that they aid in choosing measuring technique.

The pure instrumentalism of the descriptive models is also questionable. There is an obvious need to correlate the parameters τ and α with well-defined mechanisms in the skin, particularly regarding the use of skin admittance measurements as a diagnostic tool. The descriptive models can therefore be considered as essential intermediate stages in the development of better explanatory models. The new explanatory models will comprise discrete, passive components, but there will be a change in the use of symbols compared to the descriptive models, as the relationship between anatomical structure and electrophysiological mechanisms are revealed. These new models should be able to explain the so far empirically disclosed connections between skin diseases and the frequency response of the skin, by showing how changes within a substructure in the skin influence one or more of the components of the model, and hence how this can be detected by choosing the appropriate measuring technique.

7.2. Equations and Equivalent Circuits

A mathematical equation is also a model, and a very idealised one. There is an important interaction between these two model forms—the mathematical equation and the equivalent circuit diagram—with respect to the phenomena to be described.

7.2.1. Maxwell's equations

Maxwell's four equations may be written in differential or integral form and with different variables involved. Equations (7.2) to (7.5) show one example set. The differential form relates to a *point* in space, the integral form to a defined finite *volume*. Ideal charge distributions are often discontinuous, and so not differentiable, *therefore the integral form is a more generally applicable form.*

Eq.	Differential form	Explanation by integral
(7.2)	$\nabla \cdot \mathbf{E} = q_v/\varepsilon_0$	Flux of $\mathbf{E}$ through a closed surface $=$ net enclosed charge$/\varepsilon_0$
(7.3)	$\nabla \times \mathbf{E} = -\partial \mathbf{B}/\partial t$	Line integral of $\mathbf{E}$ around a loop $=$ $-$rate of change of flux of $\mathbf{B}$ through the loop
(7.4)	$\nabla \cdot \mathbf{B} = 0$	Flux of $\mathbf{B}$ through a closed surface $= 0$
(7.5)	$\nabla \times \mathbf{H} = \mathbf{J}_f + \partial \mathbf{D}/\partial t$	Line integral of $\mathbf{H}$ around a loop $=$ current density of free charges through the loop $+$ rate of change of flux of $\mathbf{D}$ through the loop

Local symbols used in this chapter are: $\mathbf{B}$ = magnetic flux density, $\mathbf{H}$ = magnetic field strength, $\mathbf{J}_f$ = current density of free charges.

∇ is the differential *vector operator* called nabla or del: $\nabla = \mathbf{i}\,\partial/\partial x + \mathbf{j}\,\partial/\partial y + \mathbf{k}\,\partial/\partial z$, where $\mathbf{i}$, $\mathbf{j}$ and $\mathbf{k}$ are unity vectors in a Cartesian coordinate system.

$\nabla \Phi$ is the *gradient* of Φ (the gradient of a scalar is a vector), $\nabla \cdot \mathbf{B}$ is the *divergence* of $\mathbf{B}$ (the divergence of a vector is a scalar), and $\nabla \times \mathbf{B}$ is the *curl* of $\mathbf{B}$ (the curl of a vector is a vector).

These four equations go back to Maxwell[20] himself and the year 1864. Maxwell's four differential equations describe almost all known electrical effects in one complete, concentrated, powerful set. They were very much based upon Faradays discoveries; for example, eq. (7.3) is a more general formulation of Faraday's law of induction. They state that a variable electrical field can only exist together with a magnetic field, and vice versa. They are based upon the presupposition that a charge can not be created anywhere, charges can only be moved. They apply to any inertial reference frame: the *charge* of a particle moving at an increasing velocity is constant, while the *mass* increases. The Maxwell equations are linear: they do not contain products of two or more variables. They are valid also for non-homogeneous, non-linear and anisotropic materials.

Maxwell's equations do not directly use the force field concept. Maxwell was not familiar with it, and Coulomb's law of mechanical forces between charges is not explicitly expressed. Maxwell also did not use the ∇-operator, but that is more a question of compactness. It was Oliver Heaviside (1850–1925) who first expressed them in the form we know today. It was also he who introduced the concept of impedance around 1880. Note that older literature uses CGS units implying e.g. a factor 4π in eq. (7.2).

Most other equations describing electrical effects are derivable from the Maxwell equations. Equation (7.2) is also called *Gauss law*. It is important for us, describing how the *E*-field is changed by the existence of net space charges. It is equally valid for free (q_{vf}) and bound (q_{vb}) charge densities, $q_v = q_{vf} + q_{vb}$. However, $\nabla \cdot D = q_{vf}$ is related to free charges, meaning that the flux of **D** through a closed surface = net enclosed free charges inside. If we replace E by $-\nabla\Phi$, we obtain the *Poisson equation*:

$$\nabla^2\Phi = -q_v/\varepsilon_0 \left(= \frac{-q_{vf}}{\varepsilon} \right) \quad \text{or} \quad \frac{\partial^2\Phi}{\partial x^2} + \frac{\partial^2\Phi}{\partial y^2} + \frac{\partial^2\Phi}{\partial z^2} = -q_v/\varepsilon_0 \quad (7.6)$$

The special case of the volume charge density q_v being zero is called the *Laplace equation*:

$$\nabla^2\Phi = 0 \quad (7.7)$$

Other equations derivable from Maxwell's are, for example, the law of conservation of charges, Ohm's law $J = \sigma E$, the power density (Joule effect) $W_v = E \cdot J$ (watt/m^3) and the energy density $E_v = \frac{1}{2}E \cdot D$ (joule/m^3).

Maxwell introduced the term *displacement current* for the current through free space and the local displacement of the *bound* charges in the dielectric, as opposed the movement of *free* charges.

From Maxwell's equations the fundamental difference between low frequency near fields (electromagnetic *fields*) and radiation (electromagnetic *waves*) is demonstrated. In Section 5.2.3 we treated the dipole. Consider that the dipole moment varies as a sine function of time. Now if the frequency is very high, the time delay

---

[20] James Clerk Maxwell (1831–79), British mathematician and physicist, founder of the electromagnetic field theory and the kinetic theory of gases.

will be noticeable if we were at a distance from the dipole longer than the wavelength $\lambda = c/f$. In the *near field* (distance to the source $\ll \lambda$), electric and magnetic fields are independent of each other, it is possible to have, e.g., a pure 50 Hz electric field with a negligible magnetic field. In the *far field* ($\gg \lambda$), the laws of *radiation* dominate. Then the **E**- and **B**-fields are strongly interdependent, with their directions mutually perpendicular. The wave is a plane wave, and the field falls as $1/r$ and not as $1/r^2$ or faster. The terms *radiation* and *wave* are reserved for such far-field cases.

7.2.2. Two-component equivalent circuits, ideal components

We will now discuss the simplest equivalent circuits mimicking the immittance found in tissue measurements. In this subsection the R–C components are considered *ideal*, that is frequency independent and linear. Immittance values are typically examined with sine waves, relaxation times with step functions. A sine wave excitation results in a sine wave response. A step excitation results in a single exponential response with a simple R–C combination.

The two-component model with one resistor and one capacitor is a one-port network, and is the simplest and most important model because every measurement on a specific frequency is reduced to such a circuit. The results are given as complex immittance values with two figures corresponding to the two components: one resistor and one capacitor. Inductive properties and corresponding resonance phenomena are possibilities that are found, for example, in membranes, but here we limit the treatment to capacitive systems.

Admittance and the parallel model

In the parallel two-component model circuit it is direct access to both components from the external terminals. Therefore the voltage can be externally controlled, but not the current division. Accordingly, a *constant amplitude voltage* v is applied across the model parallel circuit, and the current i is measured. The admittance **Y** has a direct relationship with a parallel G–C circuit, the real part Y' is G, and the imaginary part Y'' is $B = \omega C$. The parallel values are measured directly because it is proportionality between admittance **Y** and measured i. v is the independent reference sine wave, with zero phase shift per definition, and therefore here designated as a scalar.

$$\mathbf{Y} = i/v \tag{7.8}$$

$$\mathbf{Y} = G + j\omega C_p \tag{7.9}$$

$$\begin{aligned} \varphi &= \arctan (\omega C_p/G) \\ \varepsilon' &= C_p \qquad \text{(unit cell)} \\ \varepsilon'' &= G/\omega \qquad \text{(unit cell)} \end{aligned} \tag{7.10}$$

$$\tau = C_p/G \tag{7.11}$$

The circuit has some very characteristic properties:

1. The admittance diverges at higher frequencies (eq. 7.9); here the model is not in agreement with biomaterial properties.
2. It lets dc-current pass.
3. The phase angle is positive, meaning the current (dependent variable) leads the voltage. Since **i** is the dependent variable, for causality it can not lead at the start of the sine wave, only after steady-state conditions have been obtained.
4. The time constant τ of eq. (7.11) is not found with a controlled step voltage excitation as shown in Fig. 7.5: the capacitor will be charged in zero time during the voltage step. The characteristic time constant of the parallel circuit alone *can only be found* with a *constant-amplitude current* ($R_i = \infty$) excitation.

Time constant of the parallel G–C circuit
The time constant τ of the parallel circuit depends on how it is driven. With a controlled step *current* across the circuit (not constant-amplitude voltage as shown in Fig. 7.5), the voltage across the G–C combination increases exponentially from zero to $V = I/G$. The time constant is $\tau = C_p/G$. Driven by a controlled *voltage* step V, the starting current is infinitely high, and the time constant is zero.

If this parallel model is characterised with impedance, we may express the impedance **Z** of the parallel circuit using the parallel values G and C_p, $Y^2 = G^2 + B^2$ and $\tau_Y = C_p/G$:

$$\begin{aligned}
R &= G/Y^2 \\
X &= -B/Y^2 \\
\mathbf{Z} &= 1/\mathbf{Y} = (1/G)(1 - j\omega\tau_Y)/(1 + \omega^2\tau_Y^2) \\
Z' &= (1/G)/(1 + \omega^2\tau_Y^2) \\
Z'' &= -(1/G)j\omega\tau_Y/(1 + \omega^2\tau_Y^2) \\
C_s &= C_p(1 + 1/\omega^2\tau_Y^2)
\end{aligned} \qquad (7.12)$$

Equations (7.12) illustrate the extreme importance of choosing the best model circuit:

- Z' will be frequency dependent, even if G is not.

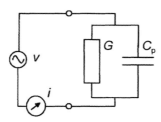

Figure 7.5. Parallel circuit driven from an ideal voltage source ($R_i = 0$), the response recorded by an ideal current reading device ($R_i = 0$).

- The series capacitance C_s *diverges towards infinite capacitance values at low frequencies.* This will, for instance, be experienced with a measuring bridge if an unknown parallel combination is to be zeroed by a series combination in the balancing arm (Section 6.3.1).

Impedance and the series model

In the series two-component model circuit there is no direct access to both components from the external terminals. Since the two components are in series, the current can be externally controlled, but not the voltage division. Accordingly, a *constant amplitude current i* is applied across the model series circuit, and the voltage **v** is measured. The impedance **Z** has a direct relationship with a series R–C circuit, the real part Z' is R, and the imaginary part Z'' is $X = -1/\omega C$. The series values are measured directly because it is proportionality between impedance **Z** and measured **v**. i is the independent reference sine wave, with zero phase shift per definition, and therefore here designated as a scalar:

$$\mathbf{Z} = \mathbf{v}/i \tag{7.13}$$

$$\mathbf{Z} = R - j/\omega C_s \tag{7.14}$$

$$\varphi = \arctan(-1/\omega\tau) \tag{7.15}$$

$$\tau = RC_s \tag{7.16}$$

Important properties of the series circuit are:

1. The impedance converges at higher frequencies (eq. 7.14).
2. It does not allow the passage of dc-current.
3. The phase angle is negative. This means that the voltage as dependent variable lags the current (cf. current leading in the parallel circuit).
4. The time constant τ of eq. (7.16) is not found with constant-amplitude current excitation as shown in Fig. 7.6; the capacitor will be charged ad infinitum during the current step. The characteristic time constant of the parallel circuit alone *can only be found with a constant-amplitude voltage ($R_i = 0$) excitation.*

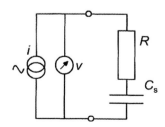

Figure 7.6. Series circuit driven from an ideal current source ($R_i = \infty$), the response recorded by an ideal voltage reading device ($R_i = \infty$).

Time constant of the series R–C circuit
The time constant τ of the series circuit depends on how it is driven. With a controlled step *voltage* V across the circuit (not controlled current as shown in Fig. 7.6), R is effectively in parallel with C_s, and the current decays exponentially from an initial value of $i = V/R$ to $i = 0$ with a time constant $\tau = RC_s$. The voltage across the capacitor increases exponentially with time from $v = 0$ to $v = V$ with the same time constant. Driven by a controlled *current* step I, the total series resistance is effectively infinite, and the voltage across the circuit (after a transient period) increases linearly with time with a velocity $\Delta v/\Delta t = I/C_s$. The time constant is therefore infinitely large.

If this series model is characterised with admittance, we may express the admittance **Y** of the series circuit with the series values R and C_s, $Z^2 = R^2 + X^2$ and $\tau_Z = RC_s$:

$$
\begin{aligned}
G &= R/Z^2 \\
B &= -X/Z^2 \\
\mathbf{Y} &= 1/\mathbf{Z} = (\omega^2 C_s \tau_Z + j\omega C_s)/(1 + \omega^2 \tau_Z^2) \\
Y' &= (1/R)\omega^2 \tau_Z^2 (1 + \omega^2 \tau_Z^2) \\
Y'' &= C_s \omega/(1 + \omega^2 \tau_Z^2) \\
C_p &= C_s/(1 + \omega^2 \tau_Z^2) \\
\varepsilon' &= C_s/(1 + \omega^2 \tau_Z^2) \qquad \text{(unit cell)} \\
\varepsilon'' &= \omega C_s \tau_Z/(1 + \omega^2 \tau_Z^2) \qquad \text{(unit cell)}
\end{aligned}
\tag{7.17}
$$

Equations (7.17) illustrate the extreme importance of choosing the best model circuit:

- Y' will be frequency dependent, even if R is not.
- The parallel capacitance C_p will be frequency dependent, even if C_s is not. C_p converges to zero at high frequencies. C_p is now the capacitance as seen from the outside (C_{ext}), different from C_s because we do not have direct access to C_s from the model terminals.

Figure 7.7 shows an example of how a model series circuit will look in a Bode diagram. Most of the phase shift is within a frequency range of 2 decades, centred on the characteristic frequency 1.592 Hz where $\varphi = 45°$. $|\mathbf{Z}|$, however, has hardly started to increase at the characteristic frequency when passing towards lower frequencies.

The parallel and series models in the Wessel diagram
Figure 7.8 illustrates the *parallel* circuit in the Wessel diagram. In a Y-plot representation the locus is a *straight line*. Because of the frequency scale direction, it is reasonable to interpret this as an infinitely high characteristic frequency or an infinitely small time constant, as if the circuit were voltage driven. However, *in the impedance plane* (Z-plot), *the parallel circuit has a semi-circle locus*. Then the characteristic frequency of the apex is $f_c = G/2\pi C_p$, as if the circuit were current driven. Notice the *narrow and different* frequency ranges presented in each digram.

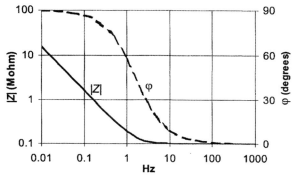

Figure 7.7. Impedance of the series model of Fig. 7.6 presented as a Bode diagram: impedance and frequency scales logarithmic, phase scale linear. $R = 0.1$ Mohm, $C_s = 10^{-6}$ F ($\tau = 0.1$ s, $f_c = 1.6$ Hz).

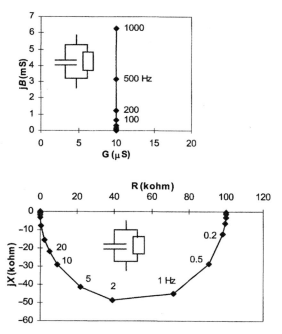

Figure 7.8. Wessel diagrams for the parallel model circuit. $G = 10$ μS, $C_s = 10^{-6}$ F ($\tau = 0.1$ s, $f_c = 1.6$ Hz).

Figure 7.9 illustrates the *series* circuit in the Wessel diagram. In a Z-plot representation the locus is a *straight line*. Because of the frequency scale direction, it is reasonable to interpret this as zero characteristic frequency or an infinitely high time constant, as if the circuit were current driven. However, *in the admittance plane* (Y-plot), *the series circuit has a semi-circle locus*. Then the characteristic frequency of the apex is $f_c = 1/2\pi RC_s$, as if the circuit were voltage driven.

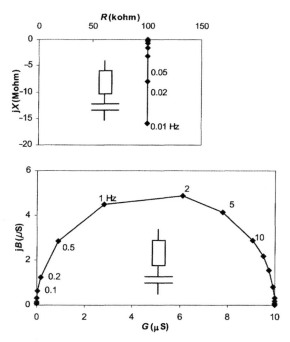

Figure 7.9. Wessel diagrams for the series model circuit. $R = 100$ kohm, $C_s = 10^{-6}$ F ($\tau = 0.1$ s, $f_c = 1.6$ Hz).

In conclusion, the parallel and series two-component circuits are *complementary models* and very different from each other. A choice of model is unavoidable when electrical data is to be analysed and presented. The choice must be based upon a presumption on the actual physical arrangement in the biomaterial/electrode system to be modelled.

Parallel- and series-equivalent circuits contain the same information in ideal systems. *The choice is often actually made in the measurement setup*: by choosing constant-amplitude voltage or current mode from the signal generator, or by choosing parallel or series coupling in the bridge arm.

Of course it is also possible to use non-ideal sources and to measure both voltage and current at the terminals of the system under examination. With special feedback circuits it is also possible to control, for example, the power dissipation in the unknown (e.g. iso-watt mode). It must still be decided whether admittance or impedance values are to be calculated and presented.

If the system under examination is non-linear, the information in the parallel and series models is not the same. Series values can not be calculated from parallel values and vice versa, and the actual measuring circuit determines the variables and the model chosen.

7.2.3. Constant phase elements (CPEs)

The two-component equivalent circuit models presented in the last subsection are of course too simple to mimic the admittance found with real biomaterials at all frequencies. A much better agreement can often be obtained by allowing the components to be non-ideal, that is frequency dependent. Such elements are not physically realisable with ordinary lumped electric components. In particular, the frequency dependence can be modelled so that the *phase* of the immittance is *independent* of frequency.

The constant phase element (CPE)
Let us calculate the characteristic properties of a general constant phase element (CPE) as a conductor and a susceptor in parallel, both frequency dependent. The frequency dependence of the admittance $Y = G + jB$ is sought so that the phase angle ($\varphi_{cpe} = \arctan B/G$) becomes frequency *independent*:

$$B_{cpe}/G_{cpe} = k \tag{7.18}$$

From eq. (7.18) it is clear that, in order to keep the phase angle constant, both G and B must be dependent on the frequency in the same way:

$$G_{cpe} = G_{\omega=1}\,\omega^m, \qquad B_{cpe} = B_{\omega=1}\,\omega^m \tag{7.19}$$

The dimension of G_{cpe} and B_{cpe} is siemens (S), so the dimensions of $G_{\omega=1}$ and $B_{\omega=1}$ is more complex: S/ω^m or $S \cdot s^m$.

Here the introduction of a parameter τ with the dimension of time and forming the product $\omega\tau$ is of interest for two reasons:

1. τ represents a useful frequency scale factor.
2. The product $\omega\tau$ is dimensionless (an angle).

Equation (7.19) is then changed to

$$G_{cpe} = G_{\omega\tau=1}(\omega\tau)^m, \qquad B_{cpe} = B_{\omega\tau=1}(\omega\tau)^m \tag{7.20}$$

The introduction of τ has the interesting consequence that it *changes the dimension of* $G_{\omega\tau=1}$ *and* $B_{\omega\tau=1}$ *to be that of an ideal conductance: siemens.*

$$Y_{cpe} = (\omega\tau)^m(G_{\omega\tau=1} + jB_{\omega\tau=1}) \tag{7.21}$$

$$\varphi_{cpe} = \arctan B/G \tag{7.22}$$

Equations (7.21) and (7.22) define a general CPE, determined by $\omega\tau$, m, $G_{\omega\tau=1}$ and $B_{\omega\tau=1}$. *There is no correlation between m and φ_{cpe}, and there are no restrictions to the value of m. The parameter m does not determine the constant phase value, but it defines the frequency dependence of the Y_{cpe} and the frequency scale together with τ.

If the m-values are restricted to positive values in the range $0 \leqslant m \leqslant 1$, then the admittance Y_{cpe} increases with frequency, in accordance with what is found for tissue. As can be seen from eq. (7.21), this also implies the intrinsic property of zero dc conductance and susceptance. By also specifying the susceptance to be capacitative, obeying the equation $B = \omega C$ and therefore $B_{\omega\tau=1} = (1/\tau)C_{\omega\tau=1}$, we get for the *admittance parallel* version:

$$\mathbf{Y}_{cpe} = (\omega\tau)^m G_{\omega\tau=1} + j\omega^m \tau^{m-1} C_{\omega\tau=1}, \qquad 0 \leqslant m \leqslant 1 \qquad (7.23)$$

$$\varphi_{cpe} = \arctan(C_{\omega\tau=1}/\tau G_{\omega\tau=1}) \qquad (7.24)$$

$$C_{cpe} = C_{\omega\tau=1}(\omega\tau)^{m-1} \quad \text{(falling with increasing frequency when } 0 \leqslant m \leqslant 1) \quad (7.25)$$

Let us examine the extreme values of frequency. When $f \to \infty$, $\mathbf{Y}_{cpe} \to \infty$ (eq. 7.23), even when $C_{cpe} \to 0$! (eq. 7.25). When $f \to 0$, $\mathbf{Y}_{cpe} \to 0$, even when $C_{cpe} \to \infty$!

Let us examine the extreme values of $m = 0$ and $m = 1$. For $m = 0$: $G_{cpe} = G_{\omega\tau=1}$, an ideal (frequency independent) resistor. $C_{cpe} = C_{\omega\tau=1}/\omega\tau$, a very special capacitor. For $m = 1$: $G_{cpe} = G_{\omega\tau=1}\omega\tau$, a very special resistor. $C_{cpe} = C_{\omega\tau=1}$, an ideal (frequency independent) capacitor. In both cases the constant phase character is valid with $\varphi_{cpe} = \arctan(C_{\omega\tau=1}/\tau G_{\omega\tau=1})$, not necessarily 0° or 90°. But with $G_{\omega\tau=1} = 0$, $\varphi_{cpe} = 90°$; and with $C_{\omega\tau=1} = 0$, $\varphi_{cpe} = 0°$.

The *impedance, series* version (note that $R_{\omega\tau=1} \neq 1/G_{\omega\tau=1}$ and $C_{\omega\tau=1}$ is not the same as in eq. 7.25) is similar:

$$\mathbf{Z}_{cpe} = (\omega\tau)^{-m}(R_{\omega\tau=1} + jX_{\omega\tau=1}) \qquad (7.26)$$

$$\varphi_{cpe} = \arctan(-\tau/C_{\omega\tau=1}R_{\omega\tau=1}) \qquad (7.27)$$

$$C_{cpe} = C_{\omega\tau=1}(\omega\tau)^{m-1} \quad \text{(falling with frequency)} \qquad (7.28)$$

When $f \to \infty$, $\mathbf{Z}_{cpe} \to 0$ (eq. 7.26), even when $C_{cpe} \to 0$ (eq. 7.28). When $f \to 0$, $\mathbf{Z}_{cpe} \to \infty$, even when $C_{cpe} \to \infty$.

The equivalent circuits are shown in Fig. 7.10.

In the Wessel diagram of Fig. 7.11, the loci of the $\mathbf{Y}_{cpe}$ and $\mathbf{Z}_{cpe}$, both the parallel and series circuits, are straight lines, with no circular arcs as seen in Figs 7.8 and 7.9. *The choice between parallel or series representation is not important for a CPE, and using a series or parallel circuit as shown in Fig. 7.11 is arbitrary.* In particular, inverting, e.g., the admittance of a parallel CPE into an impedance Wessel diagram, the locus will still be a line (Fig. 7.11), not a circular arc.

A CPE comprises a frequency-dependent capacitor and resistor. From an electronic point of view, a CPE is a two-component descriptive model. If the

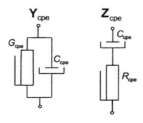

Figure 7.10. Equivalent circuits of parallel and series CPEs. The symbols are for non-ideal, frequency dependent components.

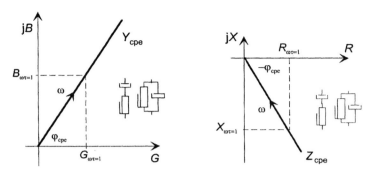

Figure 7.11. Immittance of an isolated CPE. There is no difference between series and parallel representation, in contrast to Figs 7.8 and 7.9.

mechanism behind it is seen as one process, it is a one-component explanatory model, the Warburg process is an example, Section 2.5.3.

CPE in accordance with Fricke's law (CPE$_F$)

According to Fricke's law, there is a correlation between the frequency exponent m and the phase angle φ in many electrolytic systems: when experiment shows that $C = C_1 f^{m-1}$, then $\varphi = m\pi/2$. As pointed out by Fricke (1932), m is often also found to be frequency dependent. However, it can be shown that Fricke's law is not in agreement with the Kramers–Kronig transforms *if m is frequency dependent* (Daniel 1967). If we therefore also presuppose m and φ to be frequency independent, we have a constant phase element (CPE). For such a Fricke CPE$_F$ it is usual practice to use the exponent symbol α $(m = \alpha)$. The immittance of a Fricke CPE$_F$ can then be written in a very simple way in complex notation (remembering that $j^\alpha = \cos\alpha\pi/2 + j\sin\alpha\pi/2$, cf. eq. 10.6):

$$
\begin{aligned}
\mathbf{Y}_{cpeF} &= G_{\omega\tau=1}(j\omega\tau)^\alpha = \omega^\alpha\tau^\alpha G_{\omega\tau=1}(\cos\alpha\pi/2 + j\sin\alpha\pi/2) \\
\mathbf{Z}_{cpeF} &= R_{\omega\tau=1}(j\omega\tau)^{-\alpha} = \omega^{-\alpha}\tau^{-\alpha}R_{\omega\tau=1}(\cos\alpha\pi/2 - j\sin\alpha\pi/2)
\end{aligned}
\tag{7.29}
$$

The factor α intervenes *both* in the constant phase expression ($\cos\alpha\pi/2 + j\sin\alpha\pi/2$) *and* the frequency exponent, in accordance with Fricke's law. The dimension of Y_{cpeF} and $G_{\omega\tau=1}$ is siemens (S) and of $R_{\omega\tau=1}$ is ohm (Ω). The values of Y_{cpeF} at extreme frequencies are like the general CPE: zero dc conductance and no admittance limit at very high frequencies. The admittance at the extreme values of $\alpha = 0$ and $\alpha = 1$ must not be confused with the extreme value of m given in the general discussion of eqs (7.23) to (7.28) above:

$\alpha = 0$: $\varphi_{cpeF} = 0°$. $Y_{cpeF} = G_{\omega\tau=1}$: the Fricke CPE$_F$ is an ideal conductance

$\alpha = 1$: $\varphi_{cpeF} = 90°$. $Y_{cpeF} = j\omega\tau G_{\omega\tau=1}$: the Fricke CPE$_F$ is an ideal capacitor, with $C = \tau G_{\omega\tau=1}$

A CPE$_F$ has the locus of a straight line through the origin in the Wessel diagram, just as the general CPE (Fig. 7.11). Notice that the factor $(j\omega)^\alpha$ implies a Fricke compatible CPE_F, but the factor $j^\alpha \omega^m$ ($m \neq \alpha$) implies a Fricke (and Cole) non-compatible CPE.

In conclusion: For a general CPE there is no correlation between the frequency exponent m and the constant phase angle φ_{cpe}. For a Fricke CPE_F, $\varphi_{cpeF} = m\pi/2 = \alpha\pi/2$.

7.2.4. Augmented Fricke CPE_F

Freely chosen augmenting ideal resistor
We may *augment* the Fricke CPE_F by adding a resistor G_{var} (an ideal, frequency-independent dc conductance) to a CPE_F. Let us consider the case with G_{var} in parallel with the CPE_F (eq. 7.30) (with an ideal resistor augmenting a CPE in *series*, the case will be similar).

The *admittance* locus will still be a line, but not through the origin (Fig. 7.12). But the *impedance* locus corresponds to an arc in the Wessel diagram according to the equation:

$$Z_{cpeFA} = 1/[G_{var} + G_{\omega\tau=1}(j\omega\tau_z)^\alpha] \qquad (7.30)$$

In the admittance plane the parallel combination will not have a finite characteristic time constant, and τ has the character of being just a frequency scale factor. Inverted to the *impedance* plane *the added conductor G_{var} causes the locus to be a circular arc* with a characteristic time constant $\tau_c = \tau_Z$ and frequency ω_c corresponding to the apex (the point of maximum mangitude of reactance) of the arc. At the apex, $\omega_c\tau_Z = 1$. In general τ_Z will be dependent on G_{var}; this is easily shown by

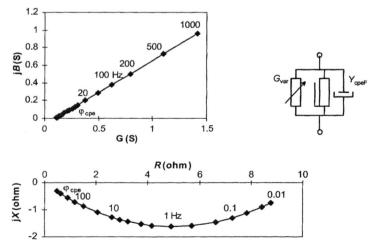

Figure 7.12. Y-plot and Z-plot for an augmented CPE_F with a freely chosen $G_{var} = 0.1$ (S). $\alpha = 0.4$ ($\varphi_{cpe} = 36°$), $G_{\omega\tau=1} = 0.1$ (S), $\tau = 0.1$ (s), according to eq. (7.30).

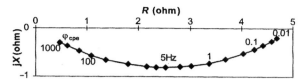

Figure 7.13. Demonstration of τ_Z being dependent on the freely chosen augmenting conductor. Parameters as for Fig. 7.12 except that $G_{var} = 0.2$ (S). Higher parallel conductance leads to higher characteristic frequency.

programming these equations in a spreadsheet and an example is illustrated in Fig. 7.13.

This model with a free variable conductance in parallel with a CPE or CPE_F is often found in the literature and is analysed as if in agreement with the Cole model (cf. Section 7.2.6) which it is not.

Cole-compatible augmented CPE_F
Restricting the value of $G_{var}(= 1/R)$ so that it is equal to the $G_{\omega\tau = 1}$ of the CPE_F element, gives a special case with different properties, the impedance of such an element is

$$Z_{elem} = \frac{R}{1 + (j\omega\tau_Z)^\alpha} \qquad (7.31)$$

Equation (7.31) describes a parallel Cole element (cf. eq. 7.33): G_{var} is now a dependent variable. With ideal components the time constant of a parallel circuit is $\tau = C/G$, and thus dependent on the parallel conductance. In eq. (7.30) the characteristic time constant is dependent on the parallel conductance G_{var}. However, in the special case of eq. (7.31), the parallel conductance *does not* influence the characteristic time constant. τ_Z and the characteristic frequency ω_c corresponding to the apex of the ZARC are determined by the equation $\omega_c\tau_Z = 1$ and are independent of $G_{var} = G_{\omega\tau = 1}$ and α. This is the Cole case and has been obtained by linking the dc conductance value G_{var} to the magnitude of the immitance value of the CPE, implying that they must be due to the same physical mechanism. The case is illustrated in Fig. 7.14.

Equation (7.31) is a part of the Cole equation (cf. Section 7.2.6). *If the characteristic frequency ω_c is found to vary during an experiment and there is reason to believe that this is due to an independent parallel conductance, the process cannot be modelled with eq. (7.31) and the process is Cole incompatible. However, eq. (7.30) or (7.43) can be used.*

In conclusion, a circular arc locus in the immitance Wessel diagram

1. Can not be due to a CPE alone.
2. Can be due to a general CPE augmented by an ideal resistor in parallel or series (not Fricke, not Cole compatible).
3. Can be due to a Fricke CPE_F augmented by an ideal resistor of freely chosen value in parallel or series (not Cole compatible).

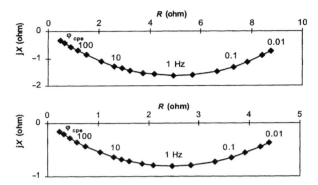

Figure 7.14. Demonstration of τ_Z being independent of the augmenting conductor (Cole case). Z-plot for a CPE$_F$ with a Cole-compatible parallel conductor. $G_{var} = G_{\omega\tau=1} = 0.1$ (S) (upper); 0.2 (S) (lower). Other parameters as for Fig. 7.12. No change of characteristic frequency, the only changes are the scales of the axes.

4. Can be due to a Fricke CPE$_F$ augmented by an ideal resistor of a particular value in parallel or series (Cole compatible).

7.2.5. Three- and four-component equivalent circuits, ideal components

Usually the two-component circuit is too simple to mimic with sufficient precision the frequency dependence of the variables measured. The three-component model combines features of both the series and parallel models. It may consist of two capacitors and one resistor (dielectrics), or two resistors and one capacitor (conductors). Detailed equations can be found in Section 10.2.

Two resistors–one capacitor
As illustrated (in Fig. 7.15) only two versions are of interest. Both allow dc-current, and both guarantee current limitation at high frequencies. The detailed equations are found in the appendix, Section 10.2. The two circuits are very

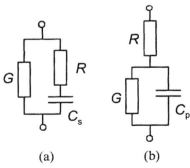

(a)   (b)

Figure 7.15. The two alternative equivalent circuits with two resistors and one capacitor. (a) Parallel version, (b) series version. Ideal components.

similar because it is possible to obtain the same immittance values for all frequencies with two matching sets of component values. As we have seen, this was not possible with the two-component series and parallel circuits. Their descriptive powers are therefore identical, and a choice must be made on the basis of their explanatory possibilities. However, the choice is important because the component values for the same frequency dependence are not identical. It is easy to see that the limiting cases at very low and very high frequencies determine the resistor values (Fig. 7.16).

The *parallel* version is best characterised by admittance because the *time constant then is uniquely defined* (Section 10.2). It has been used for cells and living tissue, with C for cell membranes, R for intracellular fluids and G for extracellular fluids.

The *series* version is best characterised by impedance because the *time constant then is uniquely defined* (Section 10.2). It has often been used as a skin electrical equivalent, with R for deeper tissue in series and the skin composed of G and C in parallel.

Two capacitors–one resistor
Detailed equations for such circuits with ideal components are found in Section 10.2. As illustrated in Fig. 7.17, two versions are of interest. Neither allows dc-current, and neither guarantees current limitation at high frequencies. The two circuits are similar because it is possible to obtain the same immittance values for all frequencies with two matching sets of component values. With one set of component

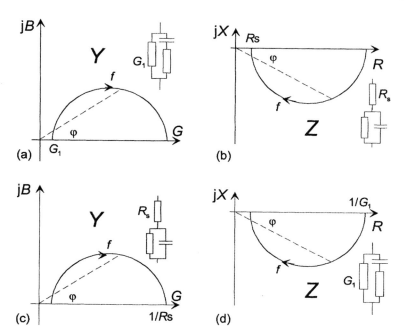

Figure 7.16. ZARC and YARC in a Wessel diagram for the equivalent circuits of Fig. 7.15, Debye case.

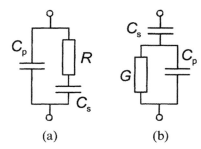

Figure 7.17. The two alternative equivalent circuits with two capacitors and one resistor. (a) Parallel version (cf. Fig. 3.4), (b): series version.

values, they can not be made equal to the one capacitor, two resistors models at more than one frequency.

Their descriptive powers are identical, and a choice must be made on the basis of their explanatory possibilities. However, the choice is important because the component values for the same frequency dependence are not identical. It is easy to see that the limiting cases at very low and high frequencies determine the capacitor values.

The models are best suited for lossy dielectrics with negligible parallel dc conductance. The parallel version is an important model for a simple relaxation process, and corresponds to the Debye model. It is the preferred version because it can be defined *with one unique time constant* (Section 10.2), and is treated according to the Debye equation (eq. 3.26) in Section 3.4 on relaxation.

It should be pointed out, however, that the simple Debye model most often is not in accordance with experimental findings. The discharge current $i(t)$ of a capacitor more often follows the non-exponential Curie–von Schweidler power law; $i(t) \sim t^{-n}$ (Jonscher 1983).

The effect of additional parallel conductors and capacitors
The permittivity locus for a Debye dispersion in the Wessel diagram is a complete semi-circle with the centre on the real axis (Fig. 7.18a). An ideal resistor in parallel destroys the circle at low frequencies (Fig. 7.18b; cf. Fig. 3.8). The conductivity locus is equally sensible for an ideal capacitor in parallel at high frequencies (Fig. 7.18d).

7.2.6. Cole equations

Immittance is the dependent variable in the Cole equations. For most biological systems it is observed that the centre of the impedance circular arc locus is situated above the real axis in the Wessel diagram. This was clear from the late 1920s, and Cole and Fricke gave some diagrams and equations based upon a frequency-independent phase angle. But in 1940 Kenneth S. Cole proposed the empirical equation $z = z_\infty + (r_0 - r_\infty)/(1 + (j\omega\tau)^\alpha)$ to describe tissue impedance, based on findings he presented in 1928 and onwards. In 1940 Cole explained the frequency dependence as membrane capacitive effects, and not relaxation as he did a year later

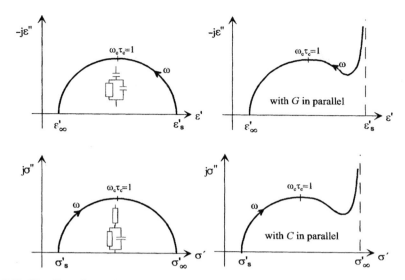

Figure 7.18. The effect of a parallel conductor on a two-capacitor–one-resistor circuit (Fig. 7.17) (upper). The effect of a parallel capacitor in parallel with a two-resistor–one-capacitor circuit (Fig. 7.15) (lower).

together with his brother. The term z_∞ is misleading and is replaced by an ideal resistor R_∞ in the usually quoted version of the *Cole$_Z$ equation*:

$$\mathbf{Z} = R_\infty + \frac{\Delta R}{1 + (j\omega\tau_Z)^\alpha}, \qquad \Delta R = R_0 - R_\infty \qquad (7.32)$$

The subscripts for the resistances relate to frequency. The simplest equivalent circuit for this equation is with two ideal resistors and one CPE immittance. As shown in Section 7.2.5, there are two possible variants that may be adapted over the whole frequency range. Examining the equations in Section 10.2, we see that the *series* version has the advantage that $R = R_\infty$ and $\Delta R = \Delta R$ directly and that the circuit is characterised with only one time constant. The simplest equivalent circuit for the Cole$_Z$ equation (7.32) is therefore the series version shown in Fig. 7.15, redrawn in Fig. 7.19. Here $\Delta G = 1/\Delta R$; the symbol ΔG is preferred to G_0 because it is not the dc value of the system.

If we omit the series resistance R_∞ in eq. (7.32), we find the admittance of the remaining parallel part of the circuit to be $\mathbf{Y}_{\text{Cole}} = \Delta G + \Delta G(j\omega\tau)^\alpha$. This admittance is difficult to handle because it has no characteristic time constant of finite value, the admittance locus of $\mathbf{Y}_{\text{Cole}}$ is a line. τ is merely a frequency scaling factor (cf. Section 7.2.3). Using the impedance form, the locus becomes a circular arc with a characteristic time constant τ_Z:

$$\mathbf{Z}_{\text{elem}} = \frac{1}{\Delta G + \Delta G(j\omega\tau_Z)^\alpha} \qquad (7.33)$$

Equation (7.33) describes an ideal conductance in parallel with a Fricke CPE$_F$; we

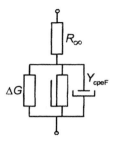

Figure 7.19. The complete Cole$_Z$ system.

may call it the *parallel Cole element*. Together with the series resistor R_∞ they form a *complete Cole series system*. The admittance and capacitance of the Fricke constant phase element CPE$_F$ alone is

$$Y_{cpeF} = \Delta G(j\omega\tau)^\alpha$$
$$C_{cpeF} = \omega^{\alpha-1}\tau^\alpha \Delta G \sin \alpha\pi/2 \qquad (7.34)$$

The Cole element is a very special combination of a Fricke CPE$_F$ and a parallel ideal (dc) conductance ΔG so that ΔG also controls the magnitude of the CPE$_F$ admittance. A change in a parallel conductance influences the magnitude (not the α) of the Z_{cpeF}, so the parallel conductance *is not a separate mechanism*, but a part of the Cole dispersion mechanism. The Cole$_Z$ equation does not allow an *independent* variable dc conductance in parallel with the CPE$_F$ without disturbing the components of the CPE. This may limit the applicability of the Cole$_Z$ equation in many real systems, because parallel processes are often independent of the CPE$_F$. For example, in the skin, with sweat duct conductance in parallel with the capacitive properties of the stratum corneum, the series model has serious limitations. However, the series resistance R_∞ of the Cole$_Z$ equation is not correlated with the parameters of the Z_{elem}: ΔG, τ or α. R_∞ may therefore be freely regarded as an independent *access resistance* to the parallel Cole element.

In the *impedance* (Z-plot) Wessel diagram the series resistance R_∞ of the Cole system moves the arc of the Cole element to the right along the real axis a distance equal to the value of R_∞ (Fig. 7.20) .

Cole equation in admittance form

By examining the equations for the equivalent circuits in Section 10.2, we find that the *parallel* circuit with two resistors and one capacitor is best suited. Then the components in the equivalent circuit are directly equal to the parameters in the equation (G_0 and ΔG). Accordingly the Cole$_Y$ equation in admittance form is

$$Y = G_0 + \frac{\Delta G}{1 + (j\omega\tau_Y)^{-\alpha}}, \qquad \Delta G = G_\infty - G_0 \qquad (7.35)$$

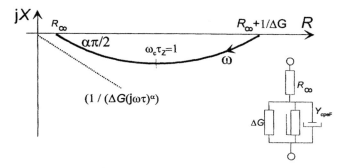

Figure 7.20. The impedance locus (ZARC) of the Cole$_Z$ system.

The subscripts for the conductors relate to frequency. The best equivalent circuit is shown in Fig. 7.21. The resistor $\Delta R \ (= 1/\Delta G)$ and the CPE$_F$ are now in series, and it is better to operate with the impedance of the CPE$_F$ alone:

$$\mathbf{Z}_{\text{cpeF}} = \Delta R (j\omega\tau)^{-\alpha} = \Delta R (\omega\tau)^{-\alpha} (\cos \alpha\pi/2 - j \sin \alpha\pi/2)$$

$$C_{\text{cpeF}} = \frac{\omega^{\alpha-1}\tau^{\alpha}}{\Delta R \sin \alpha\pi/2} \tag{7.36}$$

In the Cole parallel model, the capacitance of the Fricke element C_{cpeF} is dependent on ΔR, α, ω and τ_Y. A variable G_0 will not disturb the locus curve form in the admittance Wessel diagram but will move the arc along the real axis (Fig. 7.22). C_{cpeF}, α, ΔR and the frequency scale on the arc will not change, but G_∞ will change.

Because $\mathbf{Y} = \sigma A/d$, and A/d is equal for all the components, eq. (7.35) may also be written

$$\sigma = \sigma_0 + \frac{\Delta\sigma}{1 + (j\omega\tau_Y)^{-\alpha}} \tag{7.37}$$

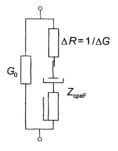

Figure 7.21. The complete Cole$_Y$ system.

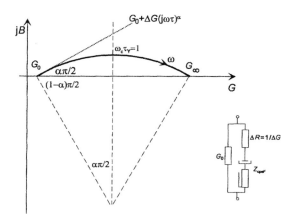

Figure 7.22. The admittance locus (YARC) of the Cole$_Y$ system.

If we omit the parallel conductance G_0 in eq. (7.35), we find the impedance of the remaining part of the circuit to be $\mathbf{Z}_{\text{elem}} = \Delta R + \Delta R(j\omega\tau)^{-\alpha}$, or

$$\mathbf{Y}_{\text{elem}} = \frac{1}{\Delta R + \Delta R(j\omega\tau_Y)^{-\alpha}} \tag{7.38}$$

This is a Fricke CPE$_F$ in series with an ideal resistor; we may call it the *series Cole element*. Together with the parallel conductance G_0 it forms the *complete Cole parallel system*.

The term ΔR is found *both* in the Cole CPE$_F$ *and* the series resistance of eq. (7.38). A change in a series resistance influences the value of the CPE$_F$, so the series resistance *is not a separate mechanism*, but a part of the Cole polarisation mechanism. The Cole$_Y$ equation does not allow an *independent* variable dc resistor in series with the CPE$_F$. This limits the applicability of the Cole$_Y$ equation in many real systems, because often series processes are processes independent of the CPE$_F$. However, the parallel conductance G_0 of the Cole$_Y$ equation is not correlated to ΔR, nor to τ or the α of the $\mathbf{Y}_{\text{elem}}$. G_0 may therefore be freely regarded as an *independent* parallel conductance to the series Cole element. For example, in the skin, with sweat duct conductance in parallel with the capacitive properties of the stratum corneum, the parallel Cole model may therefore be a better choice than the series model.

Figure 7.23 shows the admittance locus of a Cole$_Y$ system with the u and v lengths defined (to be used for control of Fricke compatibility, cf. Section 7.2.8). Measured phase angle φ is smaller than $\alpha\pi/2$; φ_{max} and the corresponding angular frequency are shown on the figure.

In conclusion

For Cole-compatible systems, the immittance arc locus is completely defined by three parameters, e.g., G_0, ΔG and α. Vice versa: a *given* arc completely defines three parameters: G_0, ΔG and α.

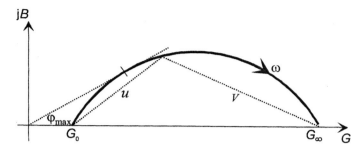

Figure 7.23. YARC plot showing the u–v variables and the maximum possible measured phase angle φ_{max}.

The fourth parameter is the characteristic time constant τ_c, which defines the frequency scale. Vice versa: a given frequency scale defines the characteristic time constant.

These four parameters are independent as variables. τ is so only because one of the two ideal resistors of a Cole system is correlated with the CPE.

If a Cole model is to be used not only for descriptive but also for explanatory purposes, then it is necessary to discuss the relevance of the equivalent circuit components with respect to the physical reality that is to be modelled. The existence of possible dc paths is of special importance in this respect. If the characteristic frequency is found to vary, and dc paths with independent conductance variables can not be excluded, the Cole$_Z$ equation is for instance not the best model to use.

7.2.7. Cole–Cole equations

Permittivity is the dependent variable in the Cole–Cole equations. In 1941 Cole and Cole proposed the following version:

$$\varepsilon = \varepsilon_\infty + \frac{\Delta\varepsilon}{1 + (j\omega\tau_c)^{1-\alpha}}, \qquad \Delta\varepsilon = \varepsilon_s - \varepsilon_\infty \qquad (7.39)$$

Here ε_s is used for static values (and not ε_0 which is reserved for vacuum permittivity). This is in agreement with the complex permittivity as $f \to 0$: in this model (Fig. 7.24) there is no dc conduction and accordingly electrostatic conditions.

As $C = \varepsilon A/d$, the Cole–Cole equation (eq. 7.39) may equally well be written as a capacitance equation:

$$C = C_\infty + \frac{\Delta C}{1 + (j\omega\tau_c)^{1-\alpha}} \qquad (7.40)$$

By inspection of the equations for the two-capacitor models found in Section 10.2, we find that the parallel version is best adapted. According to eq. (10.26) it can be characterised by a single time constant. The equivalent circuit is shown in Fig. 7.24. Notice that when $\omega \to \infty$, $C_{cpeF} \to 0$ (eq. 7.25/28/34/36) thus decoupling ΔC (even if

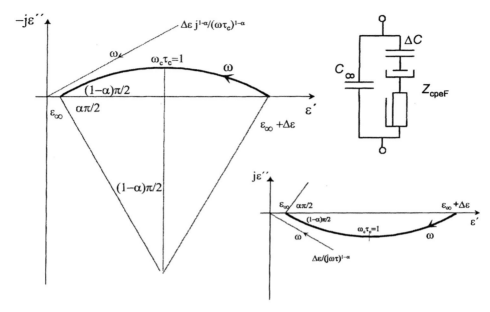

Figure 7.24. Equivalent circuit and Wessel diagrams for the Cole–Cole equation.

the fact that $Z_{cpeF} \rightarrow 0$ could lead to the false conclusion that ΔC actually is in the circuit).

A *Cole–Cole plot* is permittivity plotted in a Wessel diagram. If the permittivity is according to the Cole–Cole equation, the locus will be a circular arc. The permittivity used in the Cole–Cole equations implies that the model is changed from regarding tissue as a conductor (Cole equations) to regarding tissue as a dielectric with only bound charges and dielectric losses. The equivalent circuits contain basically two capacitors and one resistor, *without opening for dc conductance.*

Now, in living tissue there is a substantial dc conductance. Such a conductance in parallel will disturb the circular arc locus, and *must be subtracted for circular arc analysis* (cf. Fig. 7.18).

Vice versa: if the admittance values are represented by a circular arc locus, the permittivity derived from it will *not* be a circular arc. Admittance is linked with a *two-resistor* equivalent circuit, and permittivity is linked with a *two-capacitor* circuit. With fixed component values, these equivalent circuits can not be made to have the same frequency dependence.

7.2.8. Control of Fricke compatibility

For a data set, one procedure is to plot the data in a Wessel diagram and then do a best fit to adapt a circular arc to the data. The extension of this arc gives, for example, G_0, G_∞ and α. τ_c is found from the characteristic frequency f_c corresponding to the apex of the arc, from the equation $\omega_c \tau_c = 1$. However, this is under the assumption that the data set is in accordance with Fricke's law and a frequency-independent phase element. As shown in Section 7.2.3, there are CPEs that do not have the link between the frequency exponent m and the constant phase angle φ_{cpe}. If the data is in disagreement with Fricke's law, the parameters found from the arc in the Wessel diagram do not correspond to the parameters G_0, G_∞, α and τ of the Cole equation. A perfect circular arc locus is not proof of accordance with the law of Fricke or the equations of Cole, as shown, e.g., by eq. (7.30) and Fig. 7.12. The data *must be checked* for Fricke compatibility. This can be done by plotting, e.g., R or X or C_s or G or B or C_p as a function of log-frequency and determining the frequency exponent m, and then checking whether m corresponds to the α found in the Wessel diagram. α, m and τ_c can be determined graphically. With reference to Fig. 7.23 and the ratio u/v, it can be shown that $u/v = (\omega\tau)^m$ for immittance and $(\omega\tau)^{1-m}$ for permittivity. $\log(u/v)$ plotted as a function of $\log(f)$ is therefore a straight line for a circular arc. m can be found from the slope, and α can be found from arc analysis in the Wessel diagram. If $\alpha = m$ the system is Fricke compatible. The characteristic time constant τ_c can be found from the intercept of the straight line and the vertical axes at $\log(f) = 0$; this $\log(u/v)$ value is equal to $m \log(\tau_c)$. An example is given in Fig. 7.41.

7.2.9. The α parameter

It is common observation that the value of Z for tissue decreases with frequency, Y increases with frequency, and ε decreases with frequency. Choosing α as exponent in the Z equation (Cole 1940), $-\alpha$ in the Y equation, and $1 - \alpha$ (equivalent to the loss factor of a capacitor, the phase angle of an ideal capacitor is 90°, but the loss angle is 0°) in the ε equation (Cole 1940; Cole and Cole 1941); the correct frequency dependence is taken care of with α *always positive*: $1 \geqslant \alpha \geqslant 0$. It is possible to regard the parameter α in several ways:

1. As a measure of a distribution of relaxation times (DRT).
2. As not due to a distribution of relaxation times, but based upon the theory of many-body interactions between clusters in the material. Such a non-empirical model has been developed by Dissado and Hill (1979).
3. As a measure of the deviation from an ideal resistor and capacitor in the equivalent circuit.
4. According to energy models, e.g. charge carriers trapped in energy wells (Jonscher 1983), or continuous-time, random walk charge carrier translocation (Salter 1981).
5. Physical processes like Warburg diffusion.

According to the first interpretation, the spread of relaxation times may be due to: (a) different degrees of molecular interaction (minimal interaction corresponds to $\alpha = 1$ for immittance and $\alpha = 0$ for permittivity), (b) cellular interactions and properties of gap junctions, (c) anisotropy, (d) cell size, or (e) fractal dimensions.

Table 7.1. The exponent α in the Cole equations

Equation	Exponent	Equivalent circuit components	
		$\alpha = 0$	$\alpha = 1$
Z	α	Ideal resistors	Ideal resistors and capacitor
Y	$-\alpha$	Ideal resistors	Ideal resistors and capacitor
ε	$1 - \alpha$	Ideal resistor and capacitors (Debye)	Ideal capacitors

For immittance $\alpha = 0$ corresponds to the purely resistive case, for the permittivity $\alpha = 1$ corresponds to the no loss case. α is therefore analogous to the *phase* angle, $1 - \alpha$ to the *loss* angle (cf. Section 3.3).

Table 7.1 shows a summary according to the first and third interpretations. $\alpha = 0$ corresponds to the lossy case; $\alpha = 1$ for the permittivity corresponds to the no-loss case.

In Section 2.1 it was pointed out that semiconductor theory is based upon a theory of local energy wells. With reference to point (5) above, Salter (1981, 1998) proposed a model in which hopping between a pair of energy wells is also possible not just with neighbouring but also with more distant energy wells, thus allowing for dc conductance. He claimed that by applying the energy well theory together with a continuous-time, random walk model, it is possible to explain the Cole equations not from empirical or DRT theory but from more basic physical laws. According to Salter this is true without having to define whether the conduction is electronic or ionic.

Characterising tissue with the parameter α has the advantage that the parameter is not dependent on the geometry of the tissue sample as long as the measured tissue volume is constant and not a function of frequency. α is a material constant like ε.

7.2.10. Symmetrical DRT (distribution of relaxation times)

Perfect circular arcs, Cole-compatible DRTs
The characteristic time constant τ_c in the form of τ_Z and τ_Y of eqs (7.32) and (7.35) deserves some explanation. A two-component RC circuit with ideal components has a time constant $\tau = RC$. A step excitation results in an exponential response. With a CPE the response will not be exponential. In Section 7.2.3 the time constant was introduced simply as a frequency scale factor. However, the *characteristic* time constant τ_c may be regarded as a mean time constant due to a distribution of relaxation time constants (DRT). When transforming the Cole$_Z$ impedance Z to the Cole$_Y$ admittance Y or vice versa, it can be shown that the αs of the two Cole equations (7.32) and (7.35) are equal, but the characteristic time constants τ_c are not. In fact:

$$\tau_Y = \tau_Z (G_0 R_\infty)^{1/\alpha} \tag{7.41}$$

Usually $R_\infty < 1/G_0$, then $\tau_Y < \tau_Z$.

It is possible to determine the relationship between the DRT and the two parameters α and τ_c in the Cole–Cole equation (Cole and Cole 1941):

$$F(\tau) = \frac{1}{2\pi} \frac{\sin \alpha\pi}{\cosh[(1-\alpha)\ln(\tau/\tau_c)] - \cos \alpha\pi} \qquad (7.42)$$

$F(\tau)$ is the relaxation time distribution or density of time constants (Fig. 7.25). This Cole–Cole distribution is the only distribution corresponding to a Fricke CPE_F. The distribution may be broad, allowing for a considerable density two decades away from the mean (characteristic) time constant (cf. Fig. 7.25). The $\alpha = 0$ case corresponds to the Debye case with one single time constant and $F(\tau_c) \to \infty$.

Perfect circular arcs, Cole-noncompatible
Based upon eq. (7.30) it is possible to set up a Fricke compatible but non-Cole-compatible, e.g., impedance equation as an alternative to the $Cole_Z$ eq. (7.32):

$$Z_{arcF} = R_\infty + \frac{1}{G_{var} + G_{\omega\tau=1}(j\omega\tau_Z)^\alpha} \qquad (7.43)$$

The impedance equivalent circuit is shown in Fig. 7.26. This is a model often found in the literature. It is not Cole compatible. With G_{var} as an independent variable, the value of the τ_Z of the CPE_F will depend on it. However, α will be invariant, it is determined by the CPE_F and its $\varphi_{cpe} = \alpha\pi/2$. Of course it is possible to use the model also for a more general Fricke-noncompatible CPE, with the necessary adjustments in eq. (7.43). The locus will still be a perfect circular arc.

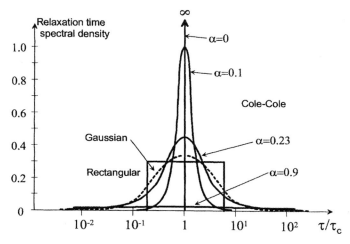

Figure 7.25. Relaxation time distribution for four different Cole–Cole permittivity $(1-\alpha)$ cases, as well as the log-normal (Gaussian) and rectangular distributions. They are all symmetrical around the characteristic time constant τ_c, and correspond to perfect circular arcs (Cole–Cole) or near-circular arcs.

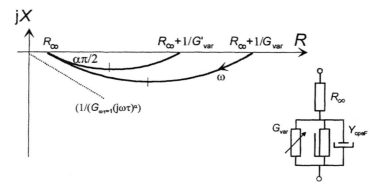

Figure 7.26. Cole-incompatible impedance model with an independent conductance G_{var} in parallel with a Fricke-compatible CPE$_F$. In certain aspects indistinguishable from the ZARC of Fig. 7.20.

Quasi-circular arcs
Other distributions than the Cole–Cole of eq. (7.42) are possible, for instance a constant function extending over a limited range of time constants (Fig. 7.25). Such different distributions result in Wessel diagram loci surprisingly similar to circular arcs (Schwan 1957). High-precision measurements, calculations and plotting are necessary in order to determine the type of a DRT. A plot in the complex plane is much more sensitive to whether the distribution is logarithmically symmetrical. *An estimation in the Wessel diagram therefore depends more on whether the time constants are logarithmically symmetrical distributed than whether the data actually are in agreement with Cole models.* By presupposing an equally plausible distribution such as the Gaussian, the mathematical treatment and models become more complex. The popularity of the Cole models is due to the fact that on *choosing* the arc as the best regression curve to the data points found, the mathematical equations become apparently simple (the Cole equations), and the corresponding equivalent circuits also become simple.

7.2.11. Non-symmetrical DRT

The characteristic relaxation time in suspensions is dependent on particle size, larger particles correspond to lower dispersion frequencies (Takashima and Schwan (1963), eq. (3.61)). If the particle size distribution is non-symmetrical to a mean size, the DRT must accordingly be skewed. Davidson and Cole (1951) proposed another version of the Cole–Cole equation:

$$\varepsilon = \varepsilon_\infty + \Delta\varepsilon/(1 + j\omega\tau)^{1-\beta} \tag{7.44}$$

Havriliak and Negami (1966) and Williams and Watts (1970) proposed even more general versions of the form:

$$\varepsilon = \varepsilon_\infty + \Delta\varepsilon/[(1 + (j\omega\tau)^{1-\alpha}]^{1-\beta} \tag{7.45}$$

These equations do not correspond to circular arcs in the complex plane but to a non-symmetrical DRT with a larger density of relaxation times on the low-frequency side. The models are compared by Lindsey and Patterson (1980).

With measured data showing a non-symmetrical DRT, the choice of model is important: e.g., a model according to this subsection, or a model with multiple Cole systems (see next subsection).

Parameters similar to those of the Cole equation, for example, can also be computed without choosing a model. Jossinet and Schmitt (1998) suggested two new parameters that were used to characterise breast tissues. The first parameter was the distance of the 1 MHz impedance point to the low-frequency intercept in the complex impedance plane and the second parameter was the slope of the measured phase angle against frequency at the upper end of the spectrum (200 kHz–1 MHz). Significant differences between carcinomas and other breast tissues were found using these parameters.

7.2.12. Multiple Cole systems

With a given data set obtained from measurements it must be decided how the data is to be interpreted and presented (cf. Section 7.1). This immediately leads to the question of which model should be chosen. The same data can be interpreted using Cole variables like α and τ, but also with many non-Cole models as explained in earlier sections. As we shall see, many data sets can be explained with multiple dispersions, e.g. multiple Cole systems. As we now present multiple Cole systems, it must always be kept in mind that the tissue data set certainly could have been made to fit other models. The ability to fit the multiple Cole model to the data set *is not a proof* that this model is the best one for describing, or explaining, the data found. Skin, for instance, is used as an example in this chapter, and we know that skin data is not necessarily in agreement with the Cole models.

As stated in Section 7.2.6, the four-component equivalent circuit may be used as a model for the Cole equation. The Cole equation represents a single electrical dispersion with a given distribution of relaxation times, and can hence be an interesting model for a macroscopically homogeneous tissue within a limited range of frequencies. A practical experimental setup, however, will in most cases involve a measured object that can be divided in two or more separate parts, where each of these parts can be modelled by an individual Cole equation. Furthermore, when measuring over a broader frequency range, different dispersion mechanisms will dominate in different parts of the frequency range, and each of these mechanisms should also be ascribed to an individual Cole equation. We suggest using the phrase *Cole system* for any system comprising one predominating dispersion mechanism that is decided to be adequately represented by a single Cole equation. This section deals with the necessity of and problems with identifying more than one Cole system in a measured object.

Consider an assembly of Cole systems like the one shown in Fig. 7.27. This electrical equivalent corresponds with the object under investigation when making impedance measurements on human skin. Measured data may represent contribu-

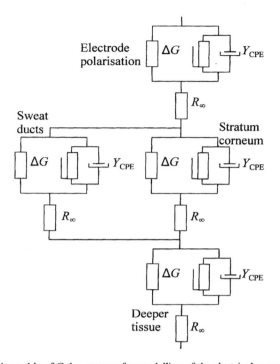

Figure 7.27. Assembly of Cole$_Z$ systems for modelling of the electrical properties of skin.

tions from electrode polarisation, stratum corneum, sweat ducts and deeper tissue, and also several dispersions of some of these components. Only one Cole system is shown for the electrode polarisation, although two dispersions have been found in some studies (Onaral and Schwan 1982). The stratum corneum is dominated by one broad dispersion (Yamamoto and Yamamoto 1976) and the sweat ducts may exhibit dispersion due to counterion relaxation (Martinsen *et al.* 1998). The viable skin will probably have several dispersions, as explained in Section 4.3.5.

This rather incomprehensible equivalent circuit may of course be simplified on the basis of existing knowledge about the different parts of the circuit. The Cole system representing the sweat ducts will, for example, be reduced to a simple resistor since the polarisation admittance is most probably negligible (Martinsen *et al.* 1998). This corresponds to α being close to zero in the Cole equation. The β and γ dispersions of deeper tissue may also be dropped if the measurements are made at sufficiently low frequencies, and there may likewise be rationales for neglecting the electrode polarisation in a given set of data. It is nevertheless of great importance to recognise all relevant Cole systems that may have influence on the interpretation of measured data, and then to carefully make the necessary simplifications. (We choose to use the Cole system as the basic model for any separate part of the measured object, since, for example, a pure resistor can readily be achieved by choosing $\alpha = 0$ in the Cole equation.)

Table 7.2. Parameters for the two Cole systems in series

	R_0	R_∞	τ_Z	α
System 1	1.7 kΩ	200 Ω	20 μs	0.55
System 2	10 kΩ	300 Ω	5 ms	0.75

The succeeding discussion will be restricted to the case of two Cole systems in series or in parallel, and how the measured data in those two cases will appear in the complex admittance (Y-plot) or impedance (Z-plot) plane.

Consider two Cole systems in series having the arbitrarily chosen parameters given in Table 7.2 (τ is given for the impedance plane). Each of these two Cole systems will produce a circular arc in the complex impedance plane as shown in Fig. 7.28, and, when impedance measurements are done on the total system, complex values like those indicated by circles in the diagram will be obtained.

Measurements will hence in this case reveal the presence of two Cole systems if the measured frequency range is broad enough. Measurements up to about 1 kHz, however, will only disclose one circular arc, corresponding to system 2 shifted to the right by approximately the value of R_0 in system 1.

The same data are plotted in the complex admittance plane in Fig. 7.29. The small arc representing system 1 in Fig. 7.28 has now become a large arc owing to the low value of R_∞. This illustrates the importance of using descriptions like "small" or "large" dispersion in connection with Cole plots with considerable care. The existence of two dispersions will be revealed in this plot also, provided the frequency range exceeds 1 kHz.

The analysis will become more troublesome if the relaxation times of the two Cole systems are closer. Consider now two new Cole systems, system 3 and system 4, having almost identical parameters to the systems in Table 7.2, but with the relaxation time of system 1 increased from 20 μs to 2 ms as shown in Table 7.3.

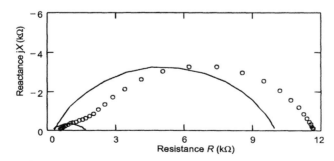

Figure 7.28. Cole plot in the impedance plane (Z-plot) for values given in Table 7.2 showing system 1 (small solid arc), system 2 (large solid arc) and measured values (circles).

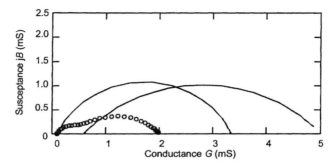

Figure 7.29. Cole plot in the admittance plane (Y-plot) for values given in Table 7.2 showing system 1 (right solid arc), system 2 (left solid arc) and measured values (circles).

Table 7.3. New parameters for the two Cole systems in series

	R_0	R_∞	τ_Z	α
System 3	1.7 kΩ	200 Ω	2 ms	0.55
System 4	10 kΩ	300 Ω	5 ms	0.75

Dotted arcs in Fig. 7.30 represent each of the two systems. The corresponding measured values show that the existence of two Cole systems can not be discovered in the diagram. The natural thing will thus be to attempt to fit one circular arc to the measured values, as indicated by the solid line. The parameters of this fitted arc are given in Table 7.4. The values of R_0 and R_∞ are of course the sum of the corresponding values for the two isolated systems, but the most interesting features are the values of τ_Z and α. The value of τ_Z will be close to the value of the dominating system, i.e. the largest arc in the impedance plane. (For two Cole systems in parallel, this will be the largest arc in the admittance plane.) The value of α will be closer to

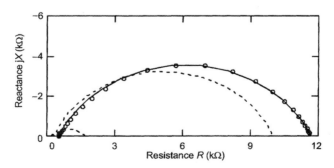

Figure 7.30. Cole plot in the impedance plane (Z-plot) for values given in Table 7.3 showing system 3 (small dotted arc), system 4 (large dotted arc), measured values (circles) and fitted circular arc (solid line).

Table 7.4. Parameters for arcs fitted to measured values for the two Cole systems in series

	R_0	R_∞	τ_Z	α
Fitted impedance arc	11.7 kΩ	500 Ω	~5 ms	0.72
Fitted admittance arc	11.7 kΩ	500 Ω	~5 ms	0.67
Fitted admittance arc (low frequency only)	11.7 kΩ	500 Ω	~5 ms	0.72

the value of the dominating system, but may still be significantly different from either of the systems.

The same data are presented in the complex admittance plane in Fig. 7.31. In this plot, the arcs are about equal in size, but they produce total values that coincide well with a single circular arc. The parameters of this arc are given in Table 7.4, and the most interesting thing is the value of α, which is roughly the mean of the two original values. However, when plotting only the low frequency part as in Fig. 7.32 the

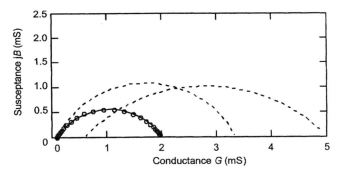

Figure 7.31. Cole plot in the admittance plane (Y-plot) for values given in Table 7.3 showing system 3 (right dotted arc), system 4 (left dotted arc), measured values (circles) and fitted circular arc (solid line).

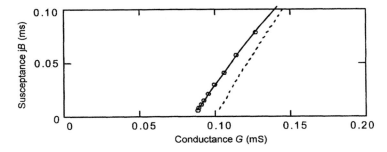

Figure 7.32. Low-frequency part of Fig. 7.31, showing system 4 (dotted arc), measured values (circles) and fitted circular arc (solid line).

calculated value of α will be closer to that of system 4, as shown in Table 7.4, and, in this case, equal to the value found in the impedance plane.

The preceding examples demonstrate how two Cole systems in series influence the measured electrical data. Two parallel Cole systems will behave in a similar manner, and we leave to the reader to investigate how systems with two parallel dispersions behave. We encourage the use of simulations like this to acquire a general sense of how a change of one parameter or part of a system comes out in different plots.

7.3. Data Calculation and Presentation

7.3.1. Measured data, model data

Measured data and model data: the one can not live without the other. A clear distinction is necessary between measured "raw" data, *calculated measured* data, and *derived* data. All these types of data are both obtained and analysed according to models, equations included, as described in Sections 6.1, 7.1 and 7.2.

The measuring setup determines what data is *raw* data. If a constant-amplitude current is applied to a set of electrodes, and the corresponding voltage is measured, this voltage is directly proportional to the impedance of the unknown. Then the raw data and the measured variable are impedance.

Raw data has to be presented in some chosen form. Some of the data is collected as a function of time, some as a function of frequency. If data is collected as a function of both frequency and time, the need for a three-dimensional presentation arises.

By a simple mathematical operation the inverse values of impedance may be *calculated*; the admittance values found contain no additional information, and the calculations do not contain any apparent conditions with respect to any disturbing influence such as temperature, atmospheric pressure, etc.: it is a purely mathematical operation. Therefore, the admittance data is correct to the extent that the impedance data is correct; these are just two different ways of presenting the measured data.

The problem of *derived* data is much larger: the variable of interest may be non-electrical, e.g. tissue water content, derived from electrical variables. Then a new question arises: what *other* variables influence the correlation between the water content and the measured electrical variable? This is a *selectivity* problem, and a *calibration* problem.

If the measured volume can be defined (not always possible in measurements *in vivo*), admittivity may be calculated, and also permittivity. Permittivity can not be measured, it can only be calculated; capacitance can be measured. If instead an index is calculated (see next section), knowledge of dimensions is not needed, but it must be ascertained that measured data is from the same tissue volume.

The interpretation of the measured data is always linked to some sort of model: an equation or a black-box concept. Choosing a model is a major decision, e.g. deciding whether it is to be descriptive or explanatory (cf. Section 7.1). The task is to choose a productive model, e.g. as simple as possible, still allowing conclusions and predictions.

7.3.2. Indexes

An index is the ratio between two measurements made under different conditions but *on the same sample volume*. The index can be calculated from measurement results and be compared with the results from a "gold standard", or can be based for instance upon the same parameter measured at different energies or frequencies. Indexes can also be made by a mathematical operation on the combination of measurement results and other parameters such as height or weight of the measured person. The phase angle $\varphi = \arctan(B/G)$ is an index.

An index is often useful because it is a relative parameter that becomes independent on, e.g., sample dimensions (as long as they are invariable) or some other invariable material property. By calculating several indices one may be able to choose one that is particularly well correlated with clinical judgements. An index may be based on a clear understanding of the underlying physical process, for instance the admittance ratio of high-frequency and low-frequency values referring to the total and the extracellular liquid volumes. But indices may also be a "blind" route when they are used without really knowing why the correlation with clinical judgements is high. They have, however, a long and very positive tradition in medicine, as for instance in the interpretation of EEG waveforms.

7.3.3. Presentation of measured data, example

We have already discussed the use of multiple Cole system models for the interpretation of measured data. Here we give another example. Suppose that we have obtained the measurement results shown in Table 7.5. The table contains raw data, as for instance obtained directly from a lock-in amplifier. It is admittance data according to $Y = G + jB$, and the results are given in micro-siemens (μS).

We want to visualise the data in some form of diagram (a graph or plot). However, we at once see that we have to take some serious decisions as to what to present and how to present it on the basis of *the purpose of the measurement*. Suppose, for instance, that they are from a two-electrode electrolytic cell, that we are to study electrode polarisation, and that we know that there is an effect from the series resistance of the bulk electrolyte. When this is the case, we can not present the measurement results directly—they must first be corrected. By calculations, or measurement at high frequencies, we determine the bulk (access) resistance, which physically is in series with the immittance of interest. However, our data (Table 7.5) is *admittance*. The corresponding *impedance* values must be calculated for each frequency. The series resistance must be subtracted from the real part of the impedance calculated. The result is recalculated back to admittance if the further study is to be done with admittance.

In all the plots to follow, the sample frequencies of Table 7.5 are marked. Notice how some presentations zoom in on narrow frequency ranges where more samples

Table 7.5. Measurement results to be presented

f (Hz)	G (μS)	B (μS)
0.01	23.0	0.5
0.03	23.1	0.7
0.1	23.2	0.7
0.3	23.4	1.6
1	23.8	3.7
3	25.1	7.7
10	29.8	16.4
20	36.3	23.6
50	52.5	31.8
100	68.3	32.2
200	82.0	27.0
500	92.6	17.4
1000	96.3	11.4
3000	98.7	5.6
10 000	99.5	2.4
30 000	99.8	1.1
100 000	99.9	0.5

are needed to get smooth curves, while other parts of the curve are crowded with samples. Details may be lost outside the focused frequency range with such presentations.

Plot of variables directly as given by measurement

Let us suppose that the results of Table 7.5 need no corrections. Looking at the data they seem to have a regular pattern: steadily increasing conductance G, and susceptance B going through a maximum. We start by plotting the variables directly without calculation. First we plot complex admittance as a function of frequency, with all three axes scales logarithmic (Fig. 7.33a). The curves seem to be in accordance with ordinary relaxation theory with one dispersion: conductance values have two distinct levels, and there is a transition zone centred around 80 Hz. However, a second dispersion is apparent below 0.1 Hz, discernible on the logarithmic susceptance curve only. When the admittance data is presented with linear admittance scales (Fig 7.33b), information on the second dispersion is lost.

Figure 7.34 shows the admittance data plotted directly in a Wessel diagram. The data seems to fit a circular arc, a sign that the measured data may be in accordance with a CPE model. From the diagram φ_{cpe} is estimated to be around 80° as compared to 63° in Fig. 7.33. It is therefore clear that the data are not Fricke compatible.

The frequency range defining the arc is limited, and the control of the arc form is not very accurate below 2 Hz or above 2000 Hz. The frequency range cannot be increased, because use of a logarithmic frequency scale is not possible without losing geometrical form. Nor can Wessel diagrams be changed on the linear scales of the

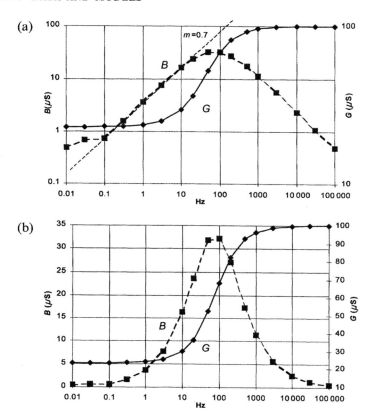

Figure 7.33. Measurement results shown directly as given in Table 7.5. (a) Triple logarithmic scales (second dispersion visible). (b) Linear admittance scales (second dispersion not visible). If B is according to eq. (10.15), the frequency exponent m at low frequencies can be found from the diagram to be 0.7. If the system were Fricke-compatible, this would have corresponded to a $\varphi_{cpe} = 63°$ for the main dispersion.

axes: the scales along the two axes must be equal. The diagram to the left does not alone give any hint of a second dispersion in the system.

Calculated parameters: Bode plot and impedance plot

A plot of the magnitude and phase shift of a variable is also called a *Bode*[21] *plot* if it is presented with logarithmic amplitude scales (e.g. in decibel), linear phase scale and logarithmic frequency scale on the x-axis. This derives from the theory of network analysis, because phase shift is a particularly important property in feedback and servo systems.

It is important to be aware of the different curves obtained with logarithmic immittance variables. With linear scales the characteristic frequency coincides with

---

[21] Hendrik W. Bode, research mathematician at Bell Telephone laboratories, author of the book *Network Analysis and Feedback Amplifier Design* (1945).

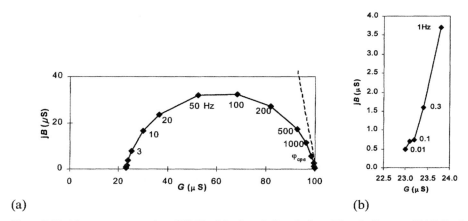

(a)   (b)

Figure 7.34. Measurement results of Table 7.5 plotted directly in a Wessel diagram (YARC). Low frequency details at the right hand side. φ_{cpe} is estimated to be around 80°. Comparing this result with that of Fig. 7.33, the system under investigation is not Fricke compatible for the main dispersion. The characteristic frequency is estimated to be $f_c = 75$ Hz.

the geometrical mean of the dispersion; this is changed with the logarithmic scales used in Bode plots (Fig. 7.35). The second dispersion is visible.

The inverse presentation with impedance **Z** is shown in Fig. 7.36. The second dispersion is discernible even with a linear scale: the reactance has a higher sensitivity in that frequency range. *The characteristic frequency is shifted* down to around 10 Hz.

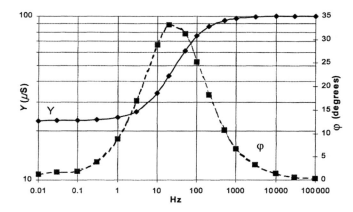

Figure 7.35. Calculated variables shown in a Bode plot: admittance with magnitude (log) and the total phase angle (linear). Data from Table 7.5. Note that the characteristic frequency f_c is a little difficult to determine directly from the Y curve because of the curveform change caused by the logarithmic admittance scale.

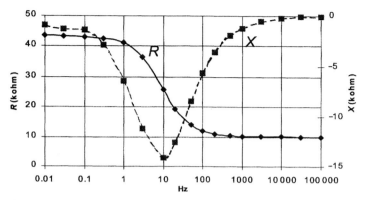

Figure 7.36. Calculated (from Table 7.5) variables: impedance shown with linear scales. Note the lower characteristic frequency ($f_c \sim 10$ Hz) compared with Fig. 7.33.

Plot of calculated capacitance and permittivity

Let us assume that we know the dimensions of the sample to be $d = 1$ mm and $A = 1$ cm^2. We can then calculate the relative permittivity according to the equations in Section 3.3. Figure 7.37 shows a plot of parallel capacitance C_p (according to $B = \omega C_p$) instead of susceptance B. Since the plot is logarithmic, the low-frequency process is easily discernible in the capacitance curve. The range of capacitance values is very large, more than seven decades. This does not necessarily reflect a true variation of a capacitor in the black-box model (cf. eq. (10.12)). Complex permittivity is shown in Fig. 7.38. ε'_r is a true copy of the C_p curve in Fig. 7.37; the dielectric loss consists of two line segments with the same slope.

Figure 7.39 shows the results in the Cole permittivity diagram. It is somewhat difficult to interpret, because both permittivity parameters are found by dividing G and B by ω. There are no circular arcs.

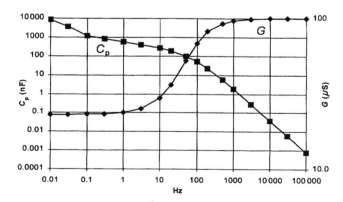

Figure 7.37. Parallel capacitance and conductance plotted with triple log-scales. Data from Table 7.5.

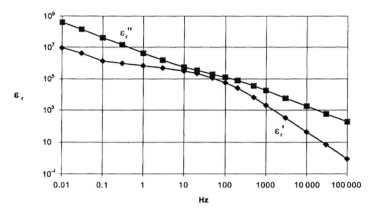

Figure 7.38. Complex relative permittivity ε_r shown with triple log-scales. Calculated dielectric variables from the data of Table 7.5.

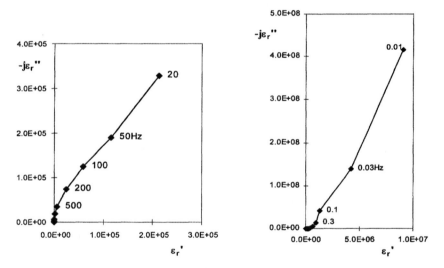

Figure 7.39. Calculated relative permittivity shown in the Wessel diagram. Data from Table 7.5. Note different ε_r' and ε_r'' scales on the low-frequency right diagram.

Plot of calculated variables: impedance plot in the Wessel diagram

In Figure 7.40 the characteristic frequency for maximum imaginary values is found at around 10 Hz, just as on the impedance plot in Fig. 7.36. The frequency range of the circular arc has also been lowered to the range > 1 Hz and < 500 Hz. Because of this, the low-frequency process is now just discernible. *The impedance locus focuses a lower frequency range than the admittance locus.*

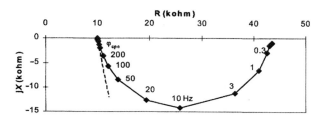

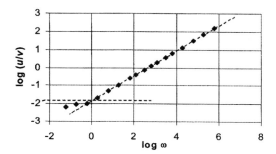

Figure 7.40. Impedance plot (Z-plot) in the Wessel plane. Data from Table 7.5. φ_{cpe} is estimated to be around 80°, as in Fig. 7.34. The characteristic frequency is estimated to be $f_c = 9$ Hz.

Figure 7.41. Ratio u/v for the YARC of Fig. 7.34 as a function of frequency with double-log scales. α (actually m) and τ_c are determined from the slope and the $\log(u/v)$-value for $\log \omega = 0$, respectively. Estimations are: $m = 0.7$, $\tau_c = 2.1$ ms, $f_c = 75$ Hz.

Check of Cole and Fricke compatibility

From the plot of the u/v ratio (Fig. 7.41) f_c is found to be 82 Hz (cf. Section 7.2.8). m is found to be 0.7, corresponding to 63°. As Fig. 7.34 shows a φ_{cpe} of about 80°, *the data is therefore not Fricke compatible.*

In conclusion, this survey has been given in some detail to show the large differences between the diagrams. The most direct and simple presentation is plotting of admittance or impedance directly *with a double logarithmic frequency scale* as a form of Bode plot. However, only plotting in the imaginary plane can reveal the existence of one or more CPE, Cole or DRT systems. The problem with Wessel diagram plots is the limited frequency range of control. *Special controls at the frequency scale ends are often necessary to ascertain the number of processes involved.* Notice also that it is possible to analyse circular arc data in the Wessel diagram as if they are Fricke and Cole compatible, even if the α-parameter should have been replaced by separate m and φ_{cpe} parameters.

CHAPTER 8

Selected Applications

Most applications of the theories of bioimmittance and bioelectricity for practical problems can be categorised into four groups:

1. Diagnostic applications
2. Therapeutic applications
3. Laboratory, *ex vivo* and *in vitro* applications
4. Perception and hazard analysis

Applications are strongly related to *applied or clinical electrophysiology*. Although there are some whole-body applications, most applications are related to specific organs—skin, heart, lung, liver, muscle, nerve, etc. Traditionally it has been a *diagnostic* field, but pacing and electroconversion, for example, have become important *therapeutic* methods. All organs develop endogenic signals reflecting physiological events: the faster the event the more evident is the signal. Signals from the peristaltic movements of intestines, for example, are very slow and need special signal processing.

Many of the diagnostic methods have achieved an extended application range by the introduction of *functional* examinations. By this is meant that signals are recorded during some form of stimulation. ECG may be recorded as a function of physical load or heart electrical pacing, EMG may be recorded during electrical stimulation or mechanical load of muscles, EEG during auditory or visual stimulation, and so on. Long-time *monitoring* is sometimes necessary, for instance searching for heart arrhythmia or epileptic spikes.

The main diagnostic task is to provide valuable information about the organ of interest, and, as described in this book, there are fundamental links between an organ's function and the corresponding electrical activity. The electrical activity is usually measured with surface electrodes, so we have the classical inverse problem: from measured surface potential back to a characterisation of the source. The source may be modelled as one main dipole (the heart vector), as several dipoles or as moving dipoles (nerve conduction). However, in the *clinical tradition* the diagnostic description is made on the basis of measured surface potential waveforms directly. Often the only computation, even in modern instrumentation, is the averaging of several events in order to increase the signal-to-noise ratio.

The *site* of an anomaly is often of special interest. Where is the infarcted region or the epileptic focus? As we have seen in Section 5.3, the space discriminative power is larger with bipolar leads than with unipolar leads, and it is larger the less the distance between the pick-up electrodes and the signal source. Distance to the signal source determines the frequency content, and therefore the information content, of the recorded signal (cf. Section 5.3.2).

In this chapter, no *general* descriptions of the different applications are given. Only aspects related to the field of bioimmittance and bioelectricity are considered. The first section is dedicated to the most important tools of our field: the electrodes.

8.1. Electrodes, Design and Properties

By electrodes we usually mean electrodes in galvanic contact with the body. This implies that *the electrode is the site of the shift from electronic to ionic conduction.* The electronic part is the metallic (carbon, etc.) part, the ionic part is the electrolyte gel or the applied or tissue liquid.

The electron–ion interface is a complicating factor: critically dependent on surface properties and therefore cleaning procedures, strongly frequency dependent and easily non-linear. To avoid such difficulties, it is tempting to try to apply fields and pick up signals by non-galvanic electrode contact with the tissue. The next subsection describes these non-galvanic contact cases.

8.1.1. Coupling without galvanic tissue contact

This book is concerned mostly with galvanic coupling between the electrodes and the tissue. The electric arc as a contact medium (Section 8.1.3) is still a galvanic contact: the current is an ionic current and not a displacement current. However, if the electrodes are isolated from the tissue by intervening air or by a thin layer of Teflon or glass, we are injecting currents into the tissue by displacement currents (cf. Section 3.4.3). There are many such situations where the sources are at a distance from the tissue of interest, and the coupling is by capacitive or inductive *fields* or electromagnetic *waves*. The question is what internal tissue fields are created by imposed external fields, and in particular the resultant fields, for example across cell membranes or inside cells. The basis for the analysis is Maxwell's equations (Section 7.2).

Examples of *electric* field coupling are a person at a distance from a high voltage power line, or capacitor-coupled short-wave diathermy currents. Static *electric* fields generate static charges on the surface of the body, and the exogenous *E*-field strength inside the body is zero. Static high-voltage fields from cathode ray tubes of PCs and TVs, for example, may transport airborne particles and microorganisms to the skin. Another example is the static field of about 200 V/m that exists outdoors as a result of atmospheric processes (in thunderstorms it may be much higher, so high that the hair on the head rises).

Pulsed *magnetic* fields are used, for example, in nerve stimulators by inducing eddy currents in the tissue. Even *static* magnetic fields in tissue are sometimes of interest, as in magnetic resonance imaging (MRI).

We are also exposed to electromagnetic *radiation*. Maxwell's equations show that an alternating *electric* field is always accompanied by an alternating *magnetic* field, and vice versa. Such an electromagnetic *wave* has a fixed relationship between electric and magnetic fields, and their vectors are perpendicular to each other. Radiation and electromagnetic waves are *far-field* effects. Far here means at several wavelengths distance from the source. The wavelength $\lambda = c/f$, where c is the velocity of light. The wavelength of 50/60 Hz power line frequencies is thus about 5000 km. At these frequencies we are therefore always in the *near-field*, and a power line electric field may be regarded as purely *electrical* if negligible ac current is flowing.

In Chapter 3 we considered a dielectric cell with an applied ac voltage, but we did not consider that at sufficiently high frequencies the wavelength of the signal would be so short that it would result in a varying E-field strength as a function of distance into the dielectric. A frequency of 300 MHz corresponds to a wavelength of $\lambda = 1$ m in air, and with measuring cell dimensions of about 10 cm the effect of standing waves and reflections must be considered. A frequency of 3 GHz corresponds to a wavelength of $\lambda = 10$ cm, and electromagnetic radiation from a source at a distance > 1 m then has the form of a planar *wave* hitting a human body.

Electric field non-galvanic coupling to tissue
Electric field coupling to tissue is a capacitive coupling. Accordingly it is more efficient the higher the signal frequency used, according to $B = \omega C$. For low frequencies, e.g. < 1 Hz and at dc, non-galvanic coupling can not be used. However, the coupling from the power lines, for example, is of interest in hazard analysis, even when the coupling is at several metres distance and the frequency is only 50/60 Hz.

Static electric fields coupled to tissue
An ideal dielectric is a material that allows internal E-fields because the material has no free charge carriers and therefore no dc conductivity. A conductor like the human body, on exposure to an external static field, does not allow internal static E-fields: the charges on the surface are rearranged so that the internal E-field strength is zero. *An externally applied static E-field to the human body therefore has no effect on internal organs, only on the skin.* However, applied *alternating* E-fields result in tissue currents and internal E-fields. At high electric fields, corona discharge will occur.

Low-frequency electric fields coupled to tissue
Humans are exposed to electric fields from power lines and TVs in their daily life. The power lines inside buildings expose us to 50/60 Hz E-fields, and in common situations some tenths of a microampere are measured if a well-insulated person is grounded via an ammeter. Such small currents are perceptible by humans under certain conditions (*electrovibration*, Section 8.14). Outdoors, a much higher current can be measured if the person is under a high-voltage power line. Calculations show that at low frequencies (e.g. power line frequencies of 50/60 Hz), the field strength coupled, for example, into a human body's interior is far smaller than the external field strength (Foster and Schwan 1989). Except for the skin, the tissues of the

body, to a first approximation, may be considered to be equipotential. However, this is not true for small-signal measurements: 1 μA 50/60 Hz through, say, an arm with a total resistance of about 500 Ω (Section 4.3) results in a voltage difference of 0.5 mV. This is certainly not a negligible signal when ECG signals of the same magnitude are to be measured. The current density distribution in such cases was studied by Guy *et al.* (1982).

The effect of an applied alternating *E*-field may be analysed as an ordinary capacitor coupled system (Chapter 3). The tissue of interest may be modelled as a part of the dielectric, perhaps with air and other conductors or insulators. The analysis of simple geometries can be done according to analytical solutions of ordinary electrostatic equations as given in Chapter 5. Real systems are often so complicated that analysis preferably is done by a finite-element (FEM) method (Section 5.6).

A simple first approximative calculation can be done with a model consisting of a conductor cylinder of diameter *d* at distance *L* from an infinite plane conductor. The capacitance per metre conductor is

$$C = \frac{2\pi\varepsilon}{\log(4L/d)} \tag{8.1}$$

With $L = 1$ m and $d = 4$ mm, *C* is about 13 pF/m (and not very dependent either on *d* or on *L*!). With an applied ac voltage *v* across the capacitor, the current through it is $i = v\omega C$ per metre, and with 60 Hz and 130 volt this gives 0.6 μA per metre wire. Such a simplified model with a person as the plane conductor is therefore in rough agreement with measured currents from power cord couplings found in actual situations. With a completely insulated person, the ac voltage of the person is the mains voltage with respect to ground reduced by the voltage division caused by the power cord/body and body/ground capacitances. If the person is grounded, the system can be regarded as a monopolar system. The grounding point is the point of high current density (active electrode). The whole body skin surface area is the low-current-density capacitively coupled neutral plate. Possible effects are to be found at the active electrode site.

Insulated electrodes
To reduce skin irritation and the use of contact paste, attempts have been made to insulate the electrode metal from the skin. With silicon chip technology it is possible to combine the fabrication of an insulating layer of SiO_2 and a high input impedance field-effect transistor, and thus obtain a low-impedance output lead not vulnerable to noise. Other possible insulating materials are tantalum oxide, barium titanate ceramics, Teflon, glass or air. Typical coupling capacitances may be in the range 50–2000 pF, with air coupling giving lower values. With such values, stray capacitances to the electrode housing can easily disturb the voltage division between the coupling capacitance and the input capacitance. This may have serious consequences for common mode signal rejection.

Insulated electrodes are used in two-electrode systems, and the bias current to the input transistor is supplied, for example, by diodes as shown in Fig. 8.1. Such

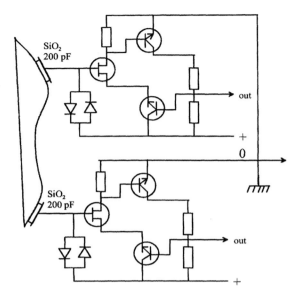

. **Figure 8.1.** Two insulated ECG electrodes. Based on Ko and Hynecek (1974), by permission.

microchips may be manufactured as low price devices for single use, but experience has shown that they are prone to serious movement artefacts and instabilities, with possible recovery time constants > 100 s after a body movement artefact (Ko and Hynecek 1974). This recovery time constant is poorly defined, because the tissue is not dc voltage defined with respect to the zero of the power supply of the two circuits. The output signal is a differential signal, but, as pointed out above, the circuits have a somewhat undefined gain so that the common mode rejection factor is poor.

A three-electrode system with the third electrode ordinarily galvanically coupled, connected to the zero of the power supply, has been found to be better, in that the CMV range is restricted. Also, two *dry* skin galvanically coupled contact electrodes and a third reference electrode for bias current was found to be a better arrangement. However, they rely on galvanic contact and are thus not true insulated electrodes.

For bioimpedance measurements, a bipolar electrode may be used covered with an insulating layer of Teflon or glass. Another example of capacitive electrode coupling is diathermy, see below.

Diathermy
Short-wave diathermy utilises a frequency around 27 MHz, corresponding to a wavelength of about 12 m. The tissue is placed in the near field of an applicator capacitor (or coil), and the local power density W_v (watt/m^3) is everywhere given by $W_v = \sigma E^2 = J^2/\sigma$. Equation (5.6) gives the corresponding temperature

rise as a function of current density (or E-field strength) under adiabatic conditions.

A degree of selectivity between tissue doses is of interest, and from these equations it may be deduced that with constant-amplitude voltage (E-field strength) the tissue of highest conductivity has the largest power dissipation and temperature rise. This is true in a homogeneous medium, but with the inhomogeneity of real tissues the opposite may also be true. Consider a capacitor of two equal slabs of materials with unequal conductivities *in series*. With dc parameters it is easy to show that the ratio between the power densities in slabs 1 and 2, when $\sigma_1 \ll \sigma_2$, is

$$\frac{W_{v1}}{W_{v2}} = \frac{\sigma_2}{\sigma_1} \gg 1$$

The current density is the same in the two slabs and is determined by the slab with lower conductivity, σ_1. Slab 1 with the *lower* conductivity also has the highest E-field strength, and will therefore have the higher power density. If the two slabs are side by side *in parallel*, the situation will be the opposite. The E-field strength is the same in both slabs, but the current densities are not. The material with *higher* conductivity will have the higher current density and the higher power density.

With *inductive* coil coupling, the circulating currents in the tissue are eddy currents, which are proportional to tissue conductivity (see later in this section). The corresponding rules for inhomogeneous materials can easily be worked out.

A frequency of 2 GHz corresponds to a wavelength of about 15 cm, and the tissue is partly in the near field and partly in the far field. At higher microwave frequencies the conditions are more and more of a far-field type, so we are dealing with electromagnetic *radiation* properly speaking (see the next subsection). The electromagnetic fields of microwave diathermy are determined not only by incident radiation but also by reflected energy from interface zones of changing impedance, e.g. at skin, fat and bone surfaces.

Examples of power absorption distributions are shown in Fig. 8.2.

Non-galvanic coupling by electromagnetic waves (far-field radiation)
Electromagnetic fields at a distance from the source greater than about 10 times the wavelength propagate as a far-field electromagnetic *wave*. Near-field components have vanished. More than 3 m away from a 1 GHz antenna the fields thus are dominated by the electromagnetic wave. Examples of electromagnetic wave coupling to tissue include microwave heating, or the coupling experienced by a person in front of a radar antenna.

Higher frequencies correspond to IR, visible and UV light. In this range electromagnetic waves behave more and more like a stream of photons—particles without charge or rest mass. Planck's law $\Delta E = h\nu$ defines the correspondence between photon energy ΔE and frequency ν. The higher the frequency, the higher the photon energy. In the UV end of the spectrum the photon energy has increased to about 5 eV, enough to excite the outer electrons of an atom. The radiation is then

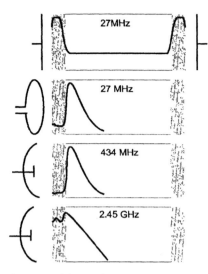

Figure 8.2. Power absorption in skin and deeper tissue at different frequencies. Upper cases: capacitive and inductive coupling, the lower ones with electromagnetic radiation. The shaded areas are poorly conducting skin tissue.

ionising, representing a rather well-defined hazard to living tissue. Even higher-energy photons, partly absorbed by tissue, are used in diagnostic radiography, but these subjects are outside the scope of this book (see e.g. Schwan and Foster (1980)).

Passage through homogeneous tissue attenuates an electromagnetic wave (Table 8.1). Fat with low conductivity has least attenuation.

In addition, reflection and scattering occur in tissue. For lower frequencies (<100 GHz) and longer wavelengths (>3 mm), the wave obeys the laws of reflection and not of scattering. *Rayleigh* scattering is the dispersion of electromagnetic radiation by *small* particles having a radius less than approximately 1/10 the wavelength of the radiation. Rayleigh scattering is strongly frequency dependent: the higher the frequency, the more powerful the scattering. When the radius of curvature of tissue is *larger* than the wavelength, geometrical scattering (planar

Table 8.1. Depth of penetration of electromagnetic waves into tissue

Frequency	Depth of penetration (cm) into tissue		
	Saline	Muscle	Fat
433 MHz	2.8	3	16
2.5 GHz	1.3	1.7	8
10 GHz	0.2	0.3	2.5

reflection mechanism) dominates. This effect is frequency independent, and the basis of lenses and the laws of reflection.

The reflection and transmission of a plane wave at a planar tissue interface depend on the frequency, the polarisation and angle of incidence of the wave, as well as the complex dielectric constant of the tissue. The coefficients of reflection R from and the transmission T through an interface of two media with intrinsic impedances Z_1 and Z_2 are

$$R = \frac{Z_2 - Z_1}{Z_2 + Z_1}$$
$$T = \frac{2Z_2}{Z_2 + Z_1}$$

(8.2)

The specific absorption rate (SAR) of non-ionising electromagnetic radiation with frequencies < 100 GHz is defined (sine waves, W/kg):

$$\text{SAR} = \frac{\omega \varepsilon_0 \varepsilon''}{2\rho} |\mathbf{E}_i|^2$$

(8.3)

This equation does not take into account possible absorption of magnetic energy related to the magnetic permeability of tissue. It also does not necessarily relate ε'' and all the absorbed power to heat and temperature rise.

Magnetic field coupling to tissue

All electric currents are accompanied by magnetic fields. With the invention of the SQUID (superconducting quantum interference device) as an extremely low magnetic field (10^{-9} tesla (T)) detector, it has become possible to measure the magnetic fields from the small endogenic currents of the body, even from the small sources in the heart and the central nervous system. This has opened up a whole new field of measurements analogous to their electric counterparts: MKG (EKG), MEG (EEG) and so on. However, these interesting subjects are outside the scope of this book, and the reader is referred to the book by Malmivuo and Plonsey (1995).

Induced (eddy) currents
A magnetic field causes an induced electromotive voltage (emv) in a conducting medium according to *Faraday's law of induction* (more generally formulated in Maxwell's equation (7.13):

$$v = -\frac{\partial \phi}{\partial t}$$

(8.4)

where ϕ is the magnetic flux (in weber or volt-second) through a closed loop, and v is the voltage between the loop ends if the loop is open. In volume conductors the loops are effectively closed and the induced potential causes a flow of electric current. Such currents induced by time-varying magnetic fields are also called *eddy currents*. The induced eddy current density will be higher for higher medium conductivity. Induced low-frequency currents will therefore follow highly conductive paths (such as blood vessels) and avoid low-conductivity areas (lung and bone tissue, myelin). With the fast switching (100 μs) of the high-gradient magnetic fields (25 mT/m) used in magnetic resonance imaging (MRI), the current density in the body approaches the threshold of perception. This is analogous to the use of skin surface coils for the stimulation of superficial nerves.

Eddy currents will set up their own magnetic fields, *opposing the external field*. The magnetic field will therefore be attenuated as function of depth (*skin effect*). The *skin depth* (depth of penetration) δ in the case of a uniform, plane electromagnetic wave propagating in a volume conductor with a magnetic permeability μ is

$$\delta = \frac{1}{\sqrt{\pi f \sigma \mu}} \tag{8.5}$$

As the wave may be reflected at the surface, the initial value corresponds to the value just inside the medium. Equation (8.5) must be used with care, because the far-field conditions of a *wave* are not fulfilled at lower frequencies.

The skin depth will be high if the conductivity is low and the frequency is low. According to Table 8.2, magnetic alternating fields are not very much attenuated by the presence of human tissue at frequencies below 1 MHz. At frequencies used in MRI, e.g. 60 MHz for 1.5 tesla systems, the attenuation and phaseshift are considerable.

Magnetic stimulation
Bickford and Fremming (1965) and Øberg (1973) demonstrated the feasibility of magnetic nerve stimulation. Barker *et al.* (1985) found that it is possible to stimulate the human motor cortex. With a suitable skin surface coil of diameter about 10 cm, a short millisecond pulse with a peak current of the order of 5000 A creates a local maximum field of a few tesla. Even if the magnetic field strength falls off with distance, the induced current in a nerve just under the coil is sufficient for excitation. The induced current is dependent on the magnetic field direction, and the excitation threshold is therefore also usually dependent on magnetic field direction. The ordinary single coil construction gives maximum field strength distributed over the whole circle corresponding to the coil windings, with only a small contribution in the coil centre. Double coils are used to focus the field strength. The inductance of the coil and the high current level and short pulse time correspond to a high voltage across the coil ends and thus a high local electric field in addition to the magnetic

Table 8.2. Skin depth as a function of frequency and material

Material	σ (S/m)	Skin depth			
		δ @60 Hz	δ @1 kHz	δ @1 MHz	δ @3 GHz
Copper	5.8×10^7	8.5 mm	2 mm	66 μm	1.2 μm
Graphite	1×10^5	210 mm	50 mm	1.6 mm	29 μm
Sea water	5	30 m	7 m	2 m	4 mm
Physiological saline	1	67 m	16 m	4.5 m	8.9 mm

field. The mechanical forces acting on the coil windings are considerable and, with repeated pulses, heat development must also be considered.

MRI (magnetic resonance imaging)

In MRI the field coupling to the body is as purely magnetic as possible. It is obtained by three sets of coils: the static main field coils (e.g. 1 T), the gradient coils (e.g. 30 mT/m with rise and fall times in the microsecond range), and the RF coils (e.g. with a separate transmitting coil of peak power 10 kW at 60 MHz and a receiving coil). However, the changing magnetic fields induce electric eddy currents, limiting the penetration and introducing phase shifts as explained.

In a superconducting coil the voltage across the coil ends is zero, and so there is no additional E-field applied to the patient. Gradient coils and RF coils, however, have a considerable voltage difference between the coil ends in order to drive the necessary current. These coils therefore produce an electric field in addition to the magnetic field.

8.1.2. The electronic conductor of the electrode

The electronic conductor of an electrode is usually a metal but it may be of carbon or a polymer. The shift from electronic to ionic conduction takes place at the interphase of the electronic and ionic conductors. The most important electrode processes and therefore electrical properties such as polarisation are associated with this interphase.

The *Ag/AgCl electrode* is one of the best electrodes in biology and medicine for dc current-carrying applications. It usually consists of silver covered by an AgCl layer, often electrolytically deposited. As the body fluids contain Cl^-, AgCl forms a non-polarisable electrode capable of passing current with less overvoltage (often called polarisation in this context) than most other types. The direct biocompatibility is questionable, however, and the use of a salt bridge often must be considered.

AgCl is a solid ionic conductor (cf. Chapter 2). With dc current flow, the AgCl layer will increase in thickness if the electrode is the anode. The polarisation impedance will go through a minimum and thereafter increase as a function of AgCl layer thickness (see below). As a cathode with dc current flow, the layer will diminish and eventually be stripped off. We are left with a pure silver surface with quite different properties, e.g. with much higher polarisation impedance and a different equilibrium potential.

Platinum is a preferred metal for direct tissue contact because it is inert and rather biocompatible. Although it can be highly polarisable by dc currents in a physiological environment, the platinum electrode is very suitable for dc potential reading applications under strict zero dc current conditions. Platinum and other noble metals and their alloys are also preferred as current-carrying electrodes in contact with living tissue, in particular for pacemaker catheter electrodes. For pacing electrodes Greatbatch (1967) found large differences between the pure noble metals and their alloys, except for Pt, Pt 90% and Ir 10%.

Both silver and platinum electrodes are often prepared by an active electrolytic process. The purpose is to lower the electrode polarisation impedance, and for the silver electrode to stabilise the electrode potential and reduce electrical noise.

The *silver* electrode is placed as the anode in a solution of, e.g., 0.9% NaCl. The optimum result is dependent on both the current density and the quantity of electricity used; often the current density is of the order of 1 mA/cm^2 and the quantity of electricity 1000 mA s/cm^2 (Geddes 1972). Optimum values are not necessarily the same for minimum impedance and maximum dc voltage stability, and they are dependent on actual electrode surface area. The best AgCl layer thickness depends on the desired mechanical durability, the minimum polarisation impedance, and also the possible quantity of electricity that may be passed in the opposite direction without stripping off the AgCl layer. Electrode material may also be sintered bulk AgCl: the surface can then be abraded and thus be used many times.

The *platinum* electrode is prepared in an electrolyte containing, e.g., 3% platinum chloride, with the platinum as the cathode. Platinum black is deposited on the surface, and here also there are optimum values for current density and quantity of electricity: a current density of about 10 mA/cm^2 and a quantity of electricity of about 30 000 mAs/cm^2 are recommended (Schwan 1963). Best results are obtained if the platinum surface is sandblasted before platinum black deposit.

The reduced polarisation impedance is due to an increased effective metal surface area (fractal surface). This is particularly important for the platinum black surface. However, the surface may be fragile, and a protein layer formed with tissue contact may easily smooth the micro-rough surface and increase polarisation impedance. Platinum black electrodes are best stored in distilled water and short-circuited (Schwan 1963).

Pure noble metals are too soft for many applications, and alloys with more than one noble metal are often used to improve the mechanical properties. Iridium alloys may also have interesting catalytic properties.

Many other metals are used for practical reasons, e.g. *titanium* for its particular high biocompatibility. *Stainless steel* is used for instance in needle electrodes because of its strength, non-corrosive properties and low price. However, stainless steel may be unsuitable for low-noise, small signal measurements. *Tin* and *lead* alloys are used for their low-noise properties and low melting points which mean that they can easily be formed or moulded. Thin *nickel* plates are used because they can be made flexible, though not as flexible as carbon rubber plates. Nickel may give allergic skin reactions. In dc therapy and skin iontophoresis, the pharmaceutical or bactericidal properties of a metal may be of interest, for example those of *silver, iron, aluminium* or *zinc*.

Carbon is of special interest because it is x-ray translucent but is still an electronic conductor. Combinations of carbon and rubber-like materials are used because they are soft and can be adapted to the anatomy. Usually they are of black colour and are made as multi-use, large, flexible plates (e.g. as neutral electrodes in monopolar systems) or as small skin electrodes for nerve stimulation.

Some new materials based on conductive *polymers* are electronic conductors. They may be flexible, but special consideration must be given to the ionic contact medium.

8.1.3. The contact electrolyte

The contact medium between the electronic conductor of the electrode and the tissue is by definition an ionic conductor. Isolating materials such as glass that depend purely on displacement currents are a special group, and are treated in Section 8.1.1.
The purpose of a contact medium is

- To control the metal–electrolyte interface.
- To form a high conductance salt bridge from the metal to the skin or tissue.
- To ensure small junction potentials.
- To enable the metal–electrolyte interphase to be kept at a distance from the tissue.
- To fill out spaces between an electrode plate and the tissue.
- To moisten a poorly conducting skin with electrolytes.

The contact medium represents an electrolytic volume dc resistance in series with the polarisation and tissue impedances. Contact media of special interest are

1. Tissue fluids
2. Tap water
3. Saline or salt bridge electrolytes
4. Gel or paste with wet electrolytes
5. Hydrogels (solid gels) with electrolytic conductance, with or without adhesive properties
6. Ionic polymers
7. Electric arcs

The mechanical or viscous properties of the contact medium are important, and often the electrolyte is thickened with a gel substance or contained in a sponge or soft clothing. Electrodes are often delivered as pregelled devices for single use, and must then be stored safely to prevent drying out. The medium may also contain preservatives to increase storage life, or quartz particles for purposes of abrasion on the skin.

The conductivities σ of some often used contact creams and pastes are as follows. Redux creme (Hewlett Packard) 10.6 S/m, Electrode creme (Grass) 3.3 S/m, Beckman–Offner paste 17 S/m, NASA Flight paste 7.7 S/m, NASA electrode creme 1.2 S/m. In comparison 0.9% NaCl (by weight) physiological saline solution has a conductivity of 1.4 S/m, and muscle tissue has less than half of this but is very anisotropic. Most gels are therefore strong electrolytes. NASA Flight paste, for instance, contains 9% NaCl, 3% KCl and 3% $CaCl^2$, in total 15% (by weight) of electrolytes. Thick EEG paste may contain as much as 45% of KCl.

Generally, the ionic mobility and therefore the conductivity in a high-viscosity paste is lower than in a liquid. In a *hydrogel* this is very much the case. Hydrogels are "solid gels" with natural or synthetic hydrocolloids (McAdams and Jossinet 1991b). They do not wet the skin. In contrast to wet gels, the effective electrode area when applied to the skin is fairly constant and well defined. Hydrogels as skin contact media have been found to give smaller parallel dc conductance and higher capacitance coupling than wet contact media (McAdams and Jossinet 1991). This implies more undesirable properties the lower the frequency of interest, but more desired properties in high-frequency applications such as for electrosurgery plate electrodes.

Wet electrolytes of high concentration ($> 1\%$) penetrate the skin actively, with a time constant often quoted to be of the order of 10 min (Tregear 1966; Almasi and Schmitt 1970; McAdams and Jossinet 1991b). However, actually the process is not exponential (as diffusion processes are not), and may go on for hours and days (Grimnes 1983a) (cf. Fig. 4.24). The penetration is stronger the higher the electrolyte concentration, but also with more accompanying skin irritation. NaCl is better tolerated by human skin at high concentration than are most other electrolytes.

Skin impedance is usually much higher than electrode polarisation impedance (Grimnes 1983a). As they physically are in series (Fig. 8.3), electrode polarisation impedance can usually be neglected in skin applications.

In some skin applications the electrode polarisation impedance may still be a source of error. With hydrogel contact electrolyte, the series resistance of the contact medium may be a disturbance at higher frequencies. When the stratum corneum is highly penetrated by electrolytes, the skin impedance may be so low that the electrode polarisation impedance becomes important.

Skin impedance is of great concern for both recording and current-carrying applications. The frequency band and the resolution in both space and time may depend on the skin impedance. High electrode impedance and relatively low amplifier input impedance has been a problem in ECG right from the time of Einthoven, who used large electrolyte containers with the limbs deeply immersed. A classical study of ECG applications is that by Geddes and Baker (1966).

Barnett (1937), Rosendal (1940), Almasi and Schmitt (1970) studied skin impedance, Grimnes (1983a) studied the site dependence. Rosell et al. (1988) measured the impedance of skin coated with gel, but otherwise unprepared, and

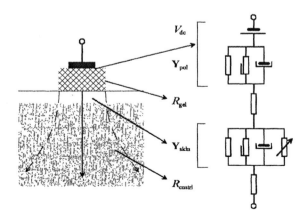

Figure 8.3. The total electrode/skin/deeper-tissue equivalent circuit. R_{cnstri} is related to the current constrictional zone of the electrode (cf. Fig. 5.2).

found that for 1 cm^2, the impedance at 1 Hz varied from 10 kΩ to 1 MΩ and at 1 MHz was always close to 120 Ω. They measured at sites typically used for ECG, impedance plethysmography, impedance cardiography and electrical impedance tomography and stressed the importance of considering these values for designers of biopotential amplifiers. McAdams and Jossinet (1991b) re-examined the problem with respect to the increased demands posed by the renewed interest in high-resolution ECG recording.

Skin abrasion and tattooing

With a dry environment, the outer surface layers of the stratum corneum have the lowest water content and therefore the highest impedance. Even rather light abrasion is surprisingly effective in removing the surface layers and providing direct access to the deeper part of the stratum corneum with a higher water content.

Other methods have also been tried to avoid the high impedance of the skin. Vitreous carbon buttons have been implanted in the skin, and tattooing techniques have been used for depositing colloidal carbon in skin (Hoenig *et al.* 1978). Black dot areas of about 8 mm in diameter showed reduced impedance at 1 Hz. The effect was reported to be equally active after more than 2 years.

Electric arc

A special ionic contact medium, to skin or any tissue surface, is gas *plasma*. Plasma is an ionised, and therefore conductive, gas. The ionisation process is induced by high electric field strength and the process is enhanced in inert gas atmospheres such as argon.

In its simplest form, an arc may be drawn from a needle electrode to the skin surface. The arc will seek highly conductive spots such as the sweat duct orifices. The arc may be dc or ac driven, and in air about 2000 V per millimetre of distance from the skin is necessary for ignition with a steady-state current in the micro-ampere range.

A single spark is the usual contact medium with electrostatic discharges in every-day situations (Section 8.14).

In electrosurgery, argon is used as a gas medium for radiofrequency coagulation of tissue surfaces. The argon gas flow guides the arc and facilitates the spread of coagulation over a larger area. At the same time, oxygen reactions are impeded, and thus also the carbonisation of tissue.

8.1.4. Dc voltage and noise generation

Non-polarised (equilibrium, zero current) conditions
Metal–liquid potential
The voltage measured is the difference between the half-cell potentials of two electrodes (Chapter 2). The half-cell potential is dependent on the electrolyte and metals involved.

Two equal *AgCl electrode* plates immersed in a homogeneous chloride electrolyte will in practice generate < 1 mV dc, or under well-controlled conditions just some few microvolts.

The *noble metals* in saline solution are highly polarisable, and the dc voltage will be unstable and poorly defined unless special precautions such as the use of electrometer input amplifiers are taken.

Different electrode surfaces easily generate hundreds of millivolts of equilibrium voltage. If an AgCl surface is stripped of its coating so that a pure silver surface appears, the dc voltage will change and the generated noise will increase. In monopolar systems a large indifferent electrode is often made with a different metal surface (e.g. nickel) from the monopolar electrodes (e.g. AgCl), and large dc voltages will be generated.

Noise
Noise by definition means an ac voltage superimposed on the equilibrium dc potential. The noise can take the form of pulse noise, white noise or $1/f$-noise. The frequency range is therefore important, generated noise is defined in rms values, (cf. Section 6.2.4) e.g. $\mu V/\sqrt{Hz}$, but is often also given as peak-to-peak (p-p) values.

1. There is less noise the larger the electrode area (EA), because of the averaging effect.
2. There is more noise the more polarisable the electrode (poorly defined half-cell potential).
3. There is more noise the more diluted the contact electrolyte.

With a usual ECG signal bandwidth of 0.1–100 Hz, an AgCl electrode in 0.9% saline may generate a signal of the order of 10 μV p-p. A pure silver plate may easily generate 10 times this amplitude. Sudden spikes of millisecond duration and hundreds of microvolt amplitude may also be generated. An important noise source is a non-uniform electrode surface, with local current exchanges between impurity centres in local corrosion processes.

Stirring is another source of noise as a result of mechanical disturbance of the concentration gradients in the double layer. To minimise this effect it is important to stabilise the solution with respect to the metal surface. A protection cup (Fig. 8.10) or a salt bridge separation to the tissue is often used in order to minimise the metal–electrolyte interphase movements.

Liquid–liquid potential (salt bridge function)
It is useful to consider a contact electrolyte as a salt bridge, emphasising the physical separation between the metal–electrolyte interphase and the electrolyte–tissue contact zone. Just as a metal–electrolyte junction generates a dc potential, so does the junction of two different liquids. This potential is called a *liquid junction potential* (Φ_{lj}):

$$\Phi_{lj} = \frac{\mu^+ - \mu^-}{\mu^+ + \mu^-} \frac{RT}{nF} \ln \frac{c_1}{c_2} \qquad (8.6)$$

where μ^+ and μ^- are the mobilities of cations and anions, respectively.

For instance, for a junction of different concentrations of NaCl solution with $c_1 = 10c_2$ the dilute side is 12.2 mV negative with respect to the other side. The salt bridge usually represents a low-resistance bridge to the tissue, and makes an effective contact with only a small liquid junction potential. The solution may be in the form of a liquid, a paste, a gel or a hydrogel. The ionic mobilities are less the higher the viscosity of the medium, and the liquid junction potential thus changes according to eq. (8.6).

The liquid junction dc potential generated by a salt bridge is minimised if the mobilities of the ions used are as similar as possible (cf. eq. 8.6). This is the case for K^+ and Cl^-; a salt bridge therefore often contains a strong KCl electrolyte. It may then be necessary to impede the strong electrolyte from reaching the tissue or measuring cell by introducing a rather tight plug or filter. The influence of the cations Na^+, K^+ and Ca^{2+} is different on living cells and tissue, so the choice between NaCl, KCl or $CaCl^2$ may be very important.

In electrophysiology requiring dc-stability, AgCl electrodes are widely used. They may be used with a salt bridge filled with saturated KCl. Even if the salt solution is immobilised with agar gel at the tissue side, potassium is known to influence excitable cells, and it may be preferable to use NaCl 0.9% instead. A liquid junction dc potential must then be accounted for.

Adsorbed species at the electrode surface may change the half-cell potentials.

Dc measurements of skin potentials are of particular interest. Often palmar skin is at about -20 mV with respect to an invasive electrode or a non-palmar skin surface electrode. Such a voltage is very dependent on the contact electrolyte used, depending on the liquid junction potential between the salt bridge and the natural electrolytes (sweat) of the skin. It is found that the measured dc voltage is very dependent not only on the concentration of the salt but also on the cations (e.g. Na^+, K^+ or Ca^{2+}) used.

Reference electrodes

The most cited reference electrode is the *platinum/hydrogen* electrode, and electrode dc potentials are often given relative to such an electrode. However, it is not often used in practical work. In the field of bioimpedance, by far the most important non-polarisable electrode for stable dc potential measurement is the AgCl electrode. This is because all tissue liquids contain some Cl^- ions, and because the electrode can be made very small, e.g. with just a small chlorided silver wire. The half-cell potential relative to a standard hydrogen electrode at 25°C in an aqueous solution is $+0.222$ V. Under ideal conditions such an electrode can be reproducible to ± 20 μV. Thus two equal electrodes in the same solution should have zero potential difference to within ± 40 μV. This precision assumes virtually no dc current flow.

The *hydrogen/platinum* reference electrode is an important device for absolute calibration, although it is impractical in most applications. The platinum electrode metal is submerged in a protonic electrolyte solution, and the surface is saturated with continuously supplied hydrogen gas. The reaction at the platinum surface is a hydrogen redox reaction: $H_2 \rightleftharpoons 2H^+(aq) + 2e$, of course with no direct chemical participation of the noble metal. Remember that the standard electrode potential is under the condition pH 0, hydrogen ion activity 1 mol/L, at the hydrogen reference electrode. Thus the values found in tables must be recalculated for other concentrations. Because of this reaction it is a hydrogen electrode, but it is also a platinum electrode because platinum is the electron source or sink and perhaps a catalyst for the reaction. A solution of iron ions and the redox reaction $Fe^{3+} + e^- \rightleftharpoons Fe^{2+}$ is another example of a platinum electrode exchanging electrons with a solution.

Polarising (current-carrying) conditions
Metal–liquid voltage
During current flow, an overvoltage is created as well in the double layer, in the diffusion layer and the resistive voltage drop in the bulk electrolyte. Many applications are in the non-linear region of current–voltage ratios. With dc currents the overvoltage may increase gradually and attain a value of many volts with polarisable electrodes; several effects of electrolysis at the electrode surface may accompany it.

In *iontophoresis* or *anti-hyperhidrosis* treatment, electrode applications are with continuous dc current. During such dc conditions the volume of the electrolyte compartment is of interest with respect to the buffering capacity for the products of electrolysis.

Pacemaker pulses are usually short ac pulses with a zero dc component, and the overvoltage is mainly determined by the double layer capacitance.

Defibrillator shocks are usually monophasic (Fig. 8.23) with a large dc component. Geddes *et al.* (1975a) measured the current–voltage characteristics with a 5 ms duration heavily damped sinusoidal defibrillator pulse. Standard defibrillator electrodes 3.5 inches in diameter (60 cm^2) were used with current pulses up to 80 amperes. The electrodes were face-to-face at 1 cm distance with the space filled with an 8.4 ohm-cm electrode paste. It was found that the impedance of both electrodes (defined as the ratio of *peak* voltage to *peak* current) at such current levels was only a fraction of one ohm. With the usual thoracic tissue impedance of about 50 ohm, little energy is therefore lost in electrode polarisation processes. The 0.01 Hz impedance of the same electrode pair at small linear ac current levels was found to be about 2 kohm.

8.1.5. Polarisation immittance

By electrode polarisation immittance we mean the immittance of the electronic/ionic interphase as described in Section 2.5. In this interphase many of the processes occur physically *in parallel* (Fig. 2.17), and there is good reason to study them by using *admittance*. As the electrode is *in series* with the biomaterial of interest, there is also good reason to focus on electrode *impedance* when studying polarisation immittance as a source of error in tissue characterisation. Some authors use the term *electrode polarisation impedance* exclusively to refer to the impedance of the constant phase element (CPE) according to eq. (7.21), but with, for instance, a pregelled electrode it is more reasonable to let the gel series resistance be a part of the electrode and the polarisation immittance concept.

The metal/electrolyte interface area is called the *electrode area* (EA). EA may be the plane area or may include a surface roughness or fractal factor. The interface area of the contact medium and the tissue is called the *effective electrode area* (EEA). With skin surface electrodes often EEA > EA (Fig. 8.10). With electrolyte-filled glass micro-electrodes often EEA < EA (Fig. 8.14). *On tissue*, the contact area is very important:

1. Large EA implies low electrode polarisation impedance.
2. Small EEA may imply geometrically dependent current constriction effect (cf. Fig. 5.2) and a larger influence from the tissue or electrolyte volume near the electrode.
3. Large EEA for *recording* electrodes implies an averaging effect and loss of spatial resolution.
4. Large EEA for *stimulating* electrodes implies higher excitable tissue volume and a summation effect in the nerve/muscle system.

The immittance per electrode area (EA) is dependent on double-layer capacitance, possible redox reactions and sorption processes at the interface, as well as diffusion processes in the electrolytic solution (cf. Fig. 2.17). In series with the interface impedance there is the bulk volume resistance of the contact medium obeying Ohm's law $J = \sigma E$. By determining and then subtracting this series resistance, the polarisation immittance properly speaking is revealed.

Platinum black electrode

The admittance is shown in Fig. 8.4 for black platinum in a physiological saline solution, one of the lowest polarisation impedance systems known. Schwan's data

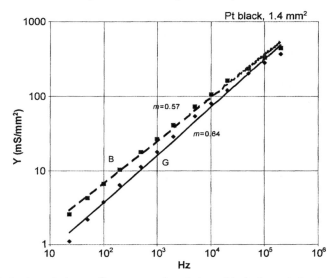

Figure 8.4. Polarisation admittance for a monopolar platinum black electrode in physiological saline solution. Constrictional series resistance subtracted. Data from Schwan (1963), by permission.

are for a 1.4 mm^2 platinum disk, but here given as admittance parallel values. If inversely proportional to EA, the values correspond to an impedance $|Z|$ less than 1 Ω for 1 cm^2 at 1000 Hz. The exponential factor for the whole frequency range for the susceptance B is $m = 0.57$. If Fricke's law is valid, this corresponds to a phase angle $\varphi = m \times 90° = 51.3°$ and a loss tangent $\tan \delta = 0.43$. The conductance G exponent factor is $m = -0.64$ corresponding to a phase angle $\varphi = -57.6°$. As is easily seen in Fig. 8.4, the ratio B/G and therefore also $|\varphi|$, diminishes with increasing frequency. Instead of whole range exponents, local exponents for a limited frequency range may be examined, and a useful accordance with Fricke's law is found even if φ is a function of frequency (Schwan 1963).

Figure 8.5 shows the calculated capacitance values of the platinum black electrode. It is not a pure double layer capacitance (Figs 2.8 and 2.9) as it is too frequency dependent. It must be due to redox or sorption processes at the platinum fractal surface. Under optimum conditions it is possible to obtain capacitance values of the order of 500 μF/mm^2 at 20 Hz (Schwan 1963). A polished and etched platinum surface has a capacitance of the order of 0.5 μF/mm^2 at 20 Hz, 1/30 of the value for the platinum black surface shown in Fig. 8.5.

AgCl wet gel electrode
Figure 8.4 showed that platinum black is a preferred electrode metal element with respect to low polarisation impedance. However, it may easily be polarised by a dc current, and for such cases an Ag/AgCl electrode is usually preferred. Typical immittance plots for one pregelled Ag/AgCl ECG electrode obtained from two electrodes in front-to-front contact (no tissue between) is shown in Fig. 8.6. The data were obtained with a constant amplitude current of 1 μA/cm^2 rms, corresponding to a single electrode maximum (1 mHz) voltage of 2.5 mV rms.

A possible interpretation of these data is that at high frequencies the series resistance of the electrolyte dominates (frequency-independent value of about

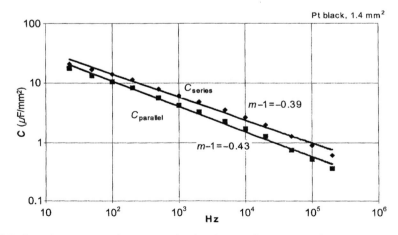

Figure 8.5. Capacitance values of the polarisation immittance of the platinum black electrode of Fig. 8.4.

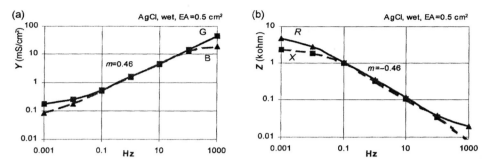

Figure 8.6. Measured impedance and calculated admittance of one commercial AgCl pregelled wet electrolyte ECG electrode.

10 ohm, right hand diagram of Fig. 8.6), and the low value indicates a strong, high-conductivity electrolyte. At low frequencies the parallel admittance of the polarisation process at the AgCl surface dominates. Accordingly the *high-frequency* results must be analysed from the *impedance* plot. The low-frequency admittance plot (left) indicates a dc leakage parallel to the polarisation process of ~0.1 mS. In the medium frequency range from about 0.1 Hz to 100 Hz (3 decades), the phase angle is constant and equal to about 43°, so the polarisation immittance is dominated by a CPE. The frequency exponent m of the admittance in the same range is found to be 0.46, which multiplied by 90° is also about 43°. The CPE is therefore of a Fricke type (cf. Section 7.2.3).

Figure 8.7 gives the capacitance values of the Ag/AgCl electrode. The values at 100 Hz are roughly only 1/20 of the values for the platinum black surface.

A possible electrical equivalent circuit is shown in Fig. 8.8. Such equivalent circuits are often given in the literature in the most simplified way with ideal components as shown for 10 Hz in Fig. 8.8b. Important characteristics of an

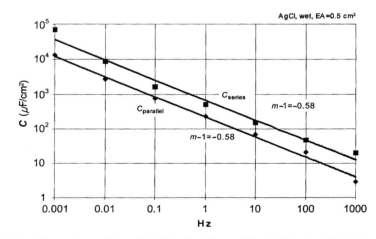

Figure 8.7. Capacitance values of the polarisation immittance of the AgCl wet gel electrode of Fig. 8.6.

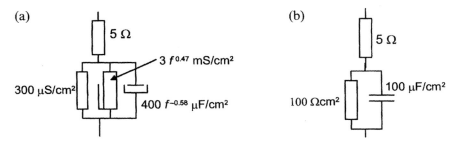

Figure 8.8. Equivalent circuits for the electrode polarisation impedance found with a particular AgCl/ wet-gel electrode. (a) With frequency dependent CPE components ($m = 0.47$). (b) Very simplified version with ideal components, valid around 10 Hz (ECG).

electrode are lost by such a simplification. For two-electrode tissue measurements, the immittance of the equivalent circuit of Fig. 8.8 is a source of error physically in series with the tissue, and must either be negligible or be subtracted as impedance from the measured impedance.

Hydrogel/aluminium electrode
Figure 8.9 shows the same data for hydrogel and aluminium electrode metal. Series resistance flattens out at about 300 ohm at higher frequencies. This corresponds to a conductivity of the hydrogel equal to only $\sigma = 6$ mS/m (gel thickness 1 mm). Accordingly, the polarisation impedance down to about 0.1 Hz of this hydrogel electrode is purely resistive and is dominated by the frequency-independent resistance of the gel.

The conductivity of the hydrogel contact medium is much smaller than for the wet gel (Fig. 8.6 showing a series resistance of only about 10 ohm). With such a high series resistance, the visible constant phase frequency range of the CPE is reduced to a small zone around 0.1 Hz. The polarisation admittance $|Y|$ of the metal–electrolyte interphase is about 20 μS/cm^2 at 0.01 Hz, and consequently much lower

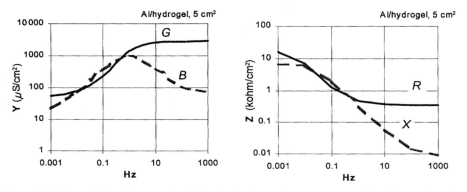

Figure 8.9. Measured impedance and calculated admittance for one commercial hydrogel/aluminium ECG electrode.

than for the AgCl/wet-electrolyte case with about 600 μS/cm^2 at 0.01 Hz. At the lowest frequencies B is almost proportional to frequency, indicating a relatively static capacitance ($B = \omega C$). Because of the large series resistance, it is difficult to assess accordance with Fricke's law directly. However, it is possible to subtract the presumed static series resistance from the measured R, and then continue the analysis from there. Notice that the admittance waveform at the left hand side of Fig. 8.9 corresponds to a single dispersion.

The series capacitances of some other metal/electrolyte interfaces are given by Geddes (1972). They are always lower than those of the AgCl or platinum black electrodes in 0.9% NaCl. For stainless steel it was given as 40 μF/cm^2 at 20 Hz ($\sim 500\,\mu$s/cm^2).

The series resistance in these models may be regarded as an access resistance to the electrode interface, dependent on electrolyte conductivity and geometry but not linked to the polarisation processes. As such it belongs to the treatment of contact media, and should be subtracted from measured electrode impedance when analysing polarisation immittance. The series resistance may be divided into a segmental and a constrictional part. The constrictional part has been given for some geometries permitting analytical solutions, e.g. for spheres, ellipsoids and disks, in Section 5.2. The series resistance may often be regarded as a salt bridge with a high-conductivity electrolyte solution. The resistance is therefore usually low, but dependent on bridge length and cross-sectional area. The salt bridge resistance may be a problem at high frequencies. Electrodes for plethysmography, for instance, are used around 50 kHz, and the polarisation impedance is then so low that the series resistance may be the dominant factor. This is particularly true with hydrogel contact media as shown in Fig. 8.9. In four-electrode systems the resistance of the salt bridge between the potential reading electrodes and the electrolyte of the measuring cell may also attain problematically high values.

Difference between electrolyte and tissue as contact media
A contact electrolyte is a homogeneous material. The current distribution may be uniform with the metal closing the cross-sectional area of a measuring cell tube. Otherwise there will be a current spread from the electrode with a constrictional effect near the electrode, perhaps with an edge effect with a surface plate electrode. Even if the current density may always be unevenly distributed, there is one fundamental difference if the metal is in direct contact with tissue. At low frequencies the cells will have a *shadowing* effect because the poorly conducting cell membranes force the current to go around the cells in the interstitial liquid. The current density may vary locally and have much higher values than the average. The effective polarisation impedance values may be higher than those found with a homogeneous electrolyte.

8.1.6. Electrode design

Skin surface electrodes
The most frequently used skin surface electrode is the recessed type (Fig. 8.10a). The metal plate is at a certain distance from the tissue surface. The electrolyte is guarded

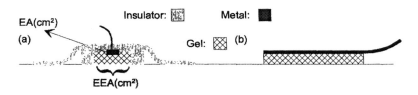

Figure 8.10. Skin surface electrode designs. (a) Recessed metal with gel in rigid cup. (b) Electrode with hydrogel contact electrolyte.

in a more or less rigid container for mechanical support and to minimise evaporation. The electrolyte solution is often supported or contained in a sponge. In this way the metal/solution interphase is clearly defined, and the total electrode impedance can easily be measured by connecting two electrodes face to face. Streaming potentials (electrokinesis) can be annoying if the solution is moved with respect to the metal during movement of the patient, but with the recessed electrode the interphase is stabilised and motion artefacts are minimised.

The area of the skin wetted or in contact with the electrolyte solution is called the *effective electrode area* (EEA) of the electrode. EEA is the dominating factor determining electrode/skin impedance. EEA may be much larger than the metal area in contact with the solution (EA), which determines the polarisation impedance. The electrode is fixed with a tape ring outside the EEA. Electrolyte penetrating the tape area increases EEA but reduces the tape sticking area. Pressure on the electrode does not squeeze electrolyte out onto the skin surface because of the rigid container construction.

Figure 8.10b shows an electrode with solid contact gel. The gel is sticky and serves both as contact electrolyte and electrode fixation. For this electrode EEA = EA. The electrolyte conductivity is rather low because the solution is a gel with low ionic mobility (cf. Section 8.1.5). Electrolyte series resistance may be the dominating factor of electrode/skin impedance at high frequencies. This may introduce problems in use, e.g., for impedance plethysmography around 50 kHz. With this electrode type the skin is not wetted. With such constructions it is possible to make an electrode with a very large EEA > 100 cm^2 and to obtain efficient skin contact over the whole contact area by using a rather thick gel. For neutral electrode plates in high frequency electrosurgery, the conductivity of the gel may deliberately be chosen to be low, so that the electrode functions mostly by capacitive coupling.

Band electrodes are used as large-surface, low-impedance electrodes with the additional attractive feature of reducing the current constrictional effect found with smaller disk electrodes. This is important in two-electrode applications such as impedance plethysmography and estimation of body composition. In Fig. 8.11 the usual configuration is shown with the current-carrying electrodes C and C' as the outer electrodes. Consider the effect of swapping the current-carrying and recording electrodes.

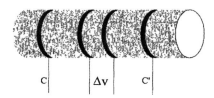

Figure 8.11. Band or catheter ring electrodes.

Multiple-point electrodes

The problem with a plate electrode in some applications is the covering effect on the tissue. With a multiple-point electrode, for instance, humidity can escape from the surface. Depending on tip sharpness and electrode pressure, the points may penetrate the superficial layers.

Concentric ring electrodes

Barnett (1937) used a three-electrode system with a $6\,\mathrm{cm}^2$ ring electrode as measuring electrode on the upper arm (Fig. 8.12). Barnett used monel metal covered with flannel soaked in normal saline. The current-carrying counterelectrode was positioned symmetrically on the other side of the arm. The reference electrode was concentric to the measuring ring electrode. This is an efficient geometry for eliminating much of the deep tissue series impedance. Barnett measured in the frequency range 2–42 kHz. With a four-electrode technique he also measured the deep tissue contribution.

Yamamoto *et al.* (1986) used a two-electrode quasi bipolar system, with the large electrode having about 10 times the area of the small one. The concentric

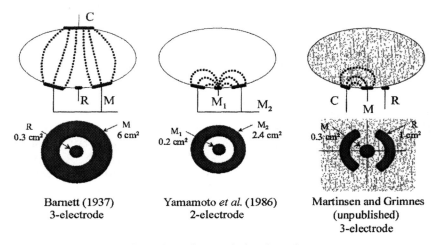

Figure 8.12. Concentric ring electrodes.

electrodes were spring-mounted in a tube in order to secure a constant electrode mechanical pressure on the tissue surface. With a third current-carrying electrode, they have also been used as a part of a three-electrode system (Martinsen *et al.* 1996).

An interesting version is a three-electrode concentric ring system with the outer ring split into two parts, one used as reference electrode (R) and the other as current-carrying (C) electrode. In this way the reference electrode picks up a potential nearer to that of the measuring electrode, so that the deeper tissue series impedance contribution is reduced.

Invasive needle electrodes (Fig. 8.13)
Such electrodes are used for EMG and in neurology. They may be of stainless steel, with outer diameter around 1 mm or less. As seen, the shaft may be used as a reference electrode. Because of the small EA and according to eqs (5.1) and (5.9) the total electrode impedance may be high and dominated by electrode polarisation impedance.

Microelectrodes for intracellular recordings
Monopolar recordings are made with a small transmembrane electrode in contact with the cytoplasm and a larger extracellular electrode in the interstitial liquid. The tip of the electrode must be small enough to penetrate the membrane without too much damage or too much plasma/electrolyte leakage. Squid giant axons have diameters up to 1000 μm, and in such studies the diameter of the penetrating part may be of the order of 50 μm. For ordinary cell sizes, tips down to 1 μm diameter are used.

The electrode itself may be the end of a metal needle or the end of a fluid-filled glass capillary (micropipette). The electrical characteristics of these two types are

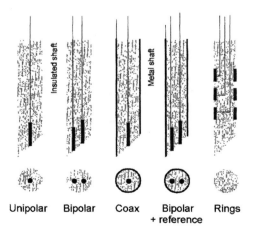

Figure 8.13. Invasive needle electrodes.

very different, and they have their characteristic application areas. The metal electrode has poor dc properties, and is best for recording fast action potentials. The micropipette electrode is better suited for recording dc or slowly varying dc potentials.

The metal needle can be made of tungsten, stainless steel, platinum alloys, etc., and it is possible to fabricate tip diameters down to 1 μm. It is necessary to isolate the shaft except at the tip; a lacquer or glass coating can do this. A description of how to make these electrodes can be found in Geddes (1972). The metal surface exposed to the electrolyte is very small (of the order of 10 μm^2), and accordingly the polarisation impedance may be several megaohm, increasing at lower frequencies and making the electrodes less suited for dc measurements. The frequency response depends on the polarisation impedance and the distributed stray capacitance. The stray capacitance depends on the length of the needle shaft in contact with fluids, and on the thickness of the isolating layer.

The glass micropipette electrode (Fig. 8.14) is filled with an electrolyte, usually 3M KCl. Such a high concentration is used to reduce the electrode resistance, but the infusion of this electrolyte into the cell must be minimised. A liquid junction potential of some millivolts will be generated by the concentration gradient at the tip.

The electrode is fragile. During muscle studies the glass capillary end may easily break when the cells are excited. Micropipette electrodes are made with orifice diameters down to 1 μm. The length and diameter of the small tip determine to a large extent the electrical properties of the electrode, and its resistance may easily reach 100 MΩ. The electrode surface is often AgCl, and the electrode area (EA) can easily be made so large that electrode polarisation impedance error is negligible. It is a purely resistive electrode, however, with distributed stray capacitance due to the glass dielectric in contact with the inner electrolyte and external tissue liquids.

The ordinary mode of use of a micropipette electrode is monopolar. Double pipette electrodes are sometimes used with only one penetration of the cell membrane. One pipette is current carrying and the other is recording in a three-electrode setup.

Cellular patch clamp
Instead of introducing microelectrodes into the cell, a part of the cell membrane (a patch) may be attached to the electrode (glass pipette) by suction. The seal between the glass end and the membrane may be so effective that the local membrane potential is recorded. The cell membrane may be ruptured with voltage pulse or additional suction, and contact to the cell interior so obtained (cf. Section 8.12.1).

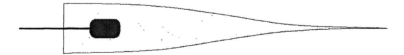

Figure 8.14. Glass micropipette electrode.

In vitro measuring cells

Some material constants with a homogeneous cylinder of biomaterial are: $\sigma = GL/A$, $\varepsilon = CL/A$, and resistivity $\rho = RA/L$. The ratio A/L (dimension: metre), or L/A (dimension $1/m$) is the *cell constant* of a measuring cell. In general the cell constant can be determined for other geometries than the cylinder, so that it is easy to derive the unknown conductivity, permittivity or resistivity from measured results. It is of course important that the measuring cell is completely filled with the sample biomaterial. If sample size varies, it is easier to use a *guard ring* (Fig. 8.15) surrounding the measuring electrode and kept at the same voltage. The effect of stray fields or current constriction is then reduced.

If the measurement is to be performed under *ex vivo* conditions, the tissue must be kept alive by controlling the temperature and the extracellular fluids. With dead samples, control of temperature and extracellular liquid may be necessary. For "dry" dead samples, control of the *ambient air humidity* may also be necessary.

One way of reducing the effect of electrode polarisation is to increase the impedance of the sample by increasing its length. Several methods are possible to compensate further for the polarisation impedance:

1. Four-electrode system (tetrapolar) (cf. Section 5.4.2).
2. Measure electrode polarisation impedance separately and subtract.
3. Substitute the unknown with a known sample for calibration.
4. Vary the measured sample length, e.g. in suspensions.

Methods (2) and (3) are based upon the assumption that the metal/liquid interphase and thus the polarisation impedance is invariable. This is not always the case. Measuring on "dry" samples, for instance, implies poor control of the contact electrolyte. Also, a sample may contain local regions of reduced conductivity near the electrode surface. The currents are then canalised with uneven current density at the metal surface (shadowing effect). Electrode polarisation impedance, in particular at low frequencies, is then dependent on the degree of shadowing.

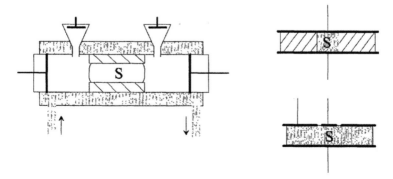

Figure 8.15. Measuring cells for tissue samples (S). Left: four-electrode liquid-filled system, temperature-controlled. Upper right: Sample confined by insulating material. Lower right: guard ring (kept equipotential with centre electrode).

VHF/UHF electrodes

The measurement of biomaterials is easier at very high frequencies because electrode polarisation impedance has more or less disappeared as a source of error. However, there are new problems associated, in particular, with the effect of stray capacitance, but also with the self-inductance of the leads. Special designs are therefore necessary for frequencies > 100 MHz.

The wavelength at 3 GHz is 10 cm, and in the GHz range the dimensions of the sample are comparable with the wavelength. Measuring principles may be based upon use of transmission lines, investigation of reflection and standing wave patterns, or determination of resonance. The arrangement of the measuring cell can be with the sample in the end of a coaxial arrangement, or with the sample in a closed system in the form of a waveguide. See Schwan (1963) for further details.

8.2. ECG

The heart is a large muscle, contracting in a rhythmic way. Because it is so important an organ and such a well-defined signal source, all sorts of recording electrodes are in use, from invasive intracardial catheter electrodes for local His-bundle recording, to multiple sterile electrodes placed directly on the heart surface during open heart surgery (epicardial mapping).

The classical ECG examination is with skin surface electrodes on the four limbs. It is somewhat unusual that pick-up electrodes are placed so far away from the organ of interest. The main reason is standardisation and reproducibility. In our language, the limbs are salt bridges to the thorax. By these, the coupling to the thorax is completely defined, and the position of the electrodes on each limb is not at all critical because each limb is equipotential. The right leg (RL) electrode is used as reference electrode. "Reference electrode" may here mean two things: either the reference electrode for unipolar electrodes (not usually the case for the RL electrode), or the reference electrode for the electronic circuitry (usually the case for the RL electrode; in older equipment also grounded to the house appliance ground wire, cf. Sections 6.2.8 and 8.14.5). The three other electrodes are used for bipolar voltage recording, which makes three possible potential differences; in the field of ECG they are called *leads* I, II and III (Fig. 8.16). The signal amplitude is of the order of 1 mV, and the frequency content is 0.05–100 Hz with dc voltages filtered out.

In the days of Einthoven,[22] no amplification tube had yet been invented, and the signal from the electrodes had to drive a galvanometer directly. It was important to find a lead (Section 5.3) with maximum signal and therefore the *unipolar augmented* leads were developed. An artificial reference voltage was obtained by summing the voltage from two of the limb electrodes; the third was the unipolar electrode. In all, therefore, there were three unipolar augmented leads: aV_1, aV_2 and aV_3.

---

[22] Willem Einthoven (1860–1927), Dutch physician. Nobel prize laureate in medicine 1924 (on ECG).

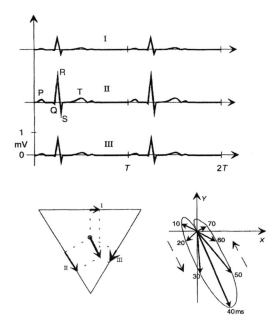

Figure 8.16. I, II and III leads with ECG waveforms. Lower part: Einthoven's triangle with the heart dipole vector; two-dimensional vectorcardiography in the body transversal plane, with the heart dipole vector locus given for each 10 ms in the diastole.

To obtain the most discriminating leads, the electrodes must be as near to the heart as possible. Eight electrodes are positioned around the chest as unipolar electrodes; as reference electrode either the right arm is used (chest-right arm, *unipolar CR leads*), or the sum of left arm, right arm and left leg voltages (*unipolar V-leads*).

The original bipolar I, II and III leads are derived from three connections to the thorax volume conductor containing the beating heart. It was discovered early that if the voltage differences (scalars) of the three leads were treated as vectors on a flat surface with directions as the sides of a triangle (the Einthoven triangle), one vector potential was the sum of the two other vectors. The scalar potentials and the triangle define a *heart vector*. It must be considered as the electrical *dipole equivalent* of the heart electrical activity. The Einthoven triangle defines the vector transfer function **H** (cf. Section 5.3.3) so that the dot product **H** · **m** is the scalar potential measured. The heart vector **m** is a model that in a very simple form gives approximately the same current and potential distribution in a volume conductor as the real heart does in its thorax volume. However, the problem is not only to what degree one dipole mimics the heart electrical activity, but also to what degree the total model can copy the anatomy of the thorax. The anatomy of the thorax is a hindrance to faithful assessment of the electrical activity of the organ of interest (the inverse problem:

from measured surface potential to characterising the electrical properties of the source itself).

Vectorcardiography is based upon the heart vector in a three-dimensional Cartesian diagram. Seven strategically positioned skin surface electrodes are used to pick up the signals. The heart vector is projected into the three planes. Three loci of the vector tips in the three planes are the basis for the clinician's description.

In principle this is the same data used for interpreting the curves from leads I, II and III. However, it violates a very long tradition in electrophysiology. For nearly 100 years it has been based upon interpreting curves obtained with standardised electrode positions. From long experience, an information bank has been assembled giving the relationship between waveforms and diagnosis. The reasons for these relationships may be unknown, and not of great concern as long as the empirical procedure furnishes precise diagnostic results. Doctors the world over are trained according to this tradition. The interpretation of EEG waveforms is another striking example of such a practice.

During open heart surgery a total of 11 sterile electrodes are used for epicardial mapping. They have direct contact with the surface of the heart, and the spread of signal on the heart surface is examined by sampling at millisecond intervals. Infarcted regions are revealed by abnormal potential spread. The signal amplitude is large, and the discriminative power is also large.

Another invasive ECG recording technique is with intracardial catheter electrodes, e.g. for the study of *His*[23] *bundle* transmission. Both unipolar and bipolar leads are used, and the catheter is always advanced into the right atrium via the venous vessels. From there it is further advanced through the tricuspid valve into the ventricle. Being so near to the source, small position changes have a large influence on the recorded waveform. This is acceptable because usually the time intervals are of greatest interest. Since the electrodes are very near or on the myocardium, the recorded signals have several millivolt amplitudes and a frequency content up to and above 500 Hz.

8.3. Impedance Plethysmography

Plethysmography is the measurement of volume. Dynamic plethysmography is usually associated with volume changes due to the beating of the heart but may also be related, for example, to respiration or peristaltic movements of the alimentary canal. During the heart systole with increased blood flow, the volume of a limb, for example, increases owing to the increased blood volume (*swelling*). Measurements may be based upon mechanical dimensional change (strain-gauge plethysmography), light absorption (photo-plethysmography), x-ray absorption or immittance change. Areas of application are rather diversified: heart stroke volume (cardiac output), respiration volume, fluid volume in pleural cavities, oedema, urinary

---

[23] Wilhelm His Jr (1863–1934), Swiss physician.

bladder volume, uterine contractions, detection of vein thrombosis. Estimation of volume from immittance measurement is based upon two effects:

1. A geometry-dependent effect illustrated by the ratio A/L in the equation $G = \sigma A/L$. The resulting effect will be dependent on the constraints on the measured tissue volume: if the volume increase results in a swelling of length L, conductance will fall. If the volume increase results in a swelling of diameter/circumference/cross-sectional area, the conductance will increase. The volume increase may also occur outside the measured tissue volume; then measured conductance will not necessarily change with the geometrical volume increase.
2. The blood will usually have a higher conductivity than the surrounding tissue.

For the further analysis of these effects it is useful to set up some simple *models*. If swelling is longitudinal, the increase in volume Δv is modelled as a resistor in series, and a resistance model is preferable. If the tissue length L is considered constant (as is supposed in many cases), the cross-sectional area swelling is modelled as a conductance increase in a parallel model. In the simple models to be presented the length L of the measured volume is considered constant, and therefore a conductance model is used. In all models the biomaterials are considered incompressible. As we shall see, all these simple models show a direct proportionality between the unknown volume change Δv and the measured conductance change ΔG. The geometry of the models is shown in Fig. 8.17.

Static cylinder models with biomaterial of constant length L, uniform current density and conductivity

One-compartment model. Cylinder surrounded by air
Basically plethysmography is about measurement of an absolute volume. For the single cylinder the absolute volume v is easily found from $G = \sigma A/L$:

$$v = G\rho L^2 \tag{8.7}$$

In many applications the absolute volume may remain unknown, the emphasis instead being on the relative volume change $\Delta v/v$. The relative conductance change $\Delta G/G$ is also of special interest, because this ratio is related to the signal-to-noise

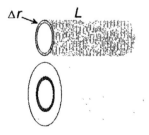

Figure 8.17. Cylinder models of constant length L, small volume increment $\Delta v = L\pi\,\Delta r^2$. Upper: one-compartment model. Lower: two-compartment model.

ratio, which should be as high as possible. With constant L a swelling occurs with an increase in cross-sectional area, and therefore in conductance:

$$\frac{\Delta G}{G} = \frac{\Delta v}{v} \quad \text{or} \quad \frac{\Delta G}{\Delta v} = \frac{\sigma}{L^2} \quad \text{or} \quad \Delta v = \rho L^2 \Delta G \tag{8.8}$$

From the equation to the left it is clear that relative volume changes can be found without knowing the dimensions of the cylinder. In order to have a high sensitivity (large ΔG) for a given volume change Δv, the length L should be as short as possible.

Two-compartment model. Cylinder (e.g. blood) surrounded by another layer of biomaterial (e.g. tissue)

$$\frac{\Delta G}{G} = \frac{\Delta v}{\Delta v + v_A + v_t} \tag{8.9}$$

$\Delta v + v_A$ are the volume of the inner cylinder, v_t is the volume of the surrounding tissue. Equation (8.9) shows that the sensitivity falls with larger tissue volume v_t. Thus high-sensitivity plethysmography poses the same problem as EIT: to selectively measure immittance in a well-defined small volume. In plethysmography the measurement should be confined as much as possible to the volume where the volume change occurs.

The effect of different conductivities
In the two-compartment, constant-length, parallel cylinder model analysed above, different conductivity in the inner cylinder will change ΔG and therefore sensitivity. With a conductivity σ_t of tissue and σ_b of blood, the relative conductance change is

$$\frac{\Delta G}{G} = \frac{\Delta v}{\Delta v + v_A + (\sigma_t/\sigma_b)v_t} \tag{8.10}$$

Therefore changes in the σ_t/σ_b ratio change the calibration factor between sought volume Δv and measured ΔG. When blood volume is the variable of interest, blood has a higher conductivity than all other tissue (except some special cases of body liquids such as concentrated urine). From eq. (8.10) it is clear that if $\sigma_b > \sigma_t$ the contribution from the tissue volume is less and the sensitivity higher. σ_b is dependent on both the haematocrit and the velocity-dependent orientation of the erythrocytes.

  Shankar *et al.* (1985) used an *in vitro* flow circulation system to simulate the physiological conditions. They pointed out that the blood resistivity change peaks with velocity and hence earlier than the blood volume change, which peaks with pressure. Although they found the blood resistivity change to be strong enough to change the morphology of the impedance pulse obtained from an artery, this time difference implied that the resistivity change resulted in a contribution of less than 5.5% of the height of the impedance pulse.

Dynamic cylinder model

Two-compartment model. Inner cylinder (e.g. blood) with a locally increased diameter representing a volume increase (bolus) passing the measured volume, surrounded by another layer of biomaterial (e.g. tissue), Fig. 8.18.

During the heart systole the inner cylinder is filled with blood (inflow). Later, during the diastole the blood is transported further (outflow) but is also returned via the venous system. It is to be expected that the waveforms of measured ΔG depend on whether the measured volume is in the limbs or head (periphery) or in the chest (central). The sensitivity with a tetrapolar electrode system will in principle be dependent on the bolus length with respect to the measured length (see Fig. 8.18). For the contribution of each volume element, see below under the heading geometrical considerations. For more realistic models for the thorax, see Section 8.3.1.

Geometrical considerations

As pointed out (eq. 8.9), sensitivity—and therefore the calibration factor—is dependent on the tissue volume in parallel with the zone of increased volume.

With disk-formed surface electrodes the constrictional resistance increase from the proximal zone of the electrodes may reduce sensitivity considerably. A prerequisite for two-electrode methods is therefore large band electrodes with minimal current constriction. In general a four-electrode (tetrapolar) system is preferable, it may then be somewhat easier to confine the measured tissue volume to the zone of volume increase.

To analyse the situation with a *tetrapolar* electrode system in contact with, e.g., a human body, we must leave our simplified models and turn to lead field theory (Section 5.3.3). The total resistance measured is the ratio of recorded voltage to injected current. The recorded voltage is the sum of potential difference contributions from each small volume dv in the measured volume. In each small volume the resistance contribution is the vector dot product of the space vectors $\mathbf{J}_{reci}$ (the local current density due to a unit reciprocal current applied to the recording electrodes (dimension: $1/m^2$)) and $\mathbf{J}_{cc}$ (the local current density due to a unit current applied to the true current-carrying electrodes (dimension: $1/m^2$)):

$$R = \int_v \rho\, \mathbf{J}_{reci} \cdot \mathbf{J}_{cc}\, dv \qquad (8.11)$$

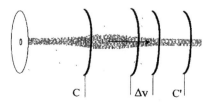

Figure 8.18. Tetrapolar electrode system and the effect of a bolus of blood passing the measured volume.

The sensitivity of a resistance measurement to a certain volume can be determined by studying the dot product in that volume. It is important to realise that in some volume positions where the vectors are perpendicular to each other, the measured resistance is insensitive to local resistivity changes. In other positions the angle is $> 90°$, and in those positions the sensitivity is reversed.

In a tetrapolar system it is always possible to swap the recording and current-carrying electrode pairs.

It is also possible to have the electrode system situated *in the volume of interest,* e.g. as needles or catheters. Such volume calculation, say of cardiac output, is used in some implantable heart pacemaker designs.

Tissue immittance values, signal frequency and electrode polarisation impedance
As pointed out, the parallel conductance model is preferable when tissue length is considered constant. Of course, admittance may also be used, with possible additional information from the quadrature channel. However, even if a frequency is chosen for maximal quadrature current, the current will usually be less than 1/10 of the in-phase signal, and will therefore have poorer signal-to-noise ratio. If blood volume is the variable of interest, blood conductance is not very frequency dependent at frequencies < 1 MHz. The admittance of the surrounding tissue will decrease with lower frequencies, and it is to be expected that this will reduce the tissue shunt effect and increase sensitivity. The lower frequency limit is also determined by hazard considerations. Current levels used are often a few milli-amperes, and in order to be below the threshold of perception the frequency has to be higher than about 10 kHz.

At a frequency around 50 kHz, measured resistance is from about 500 Ω (finger) to $< 20 \Omega$ (chest) (cf. Section 4.3.6). Hydrogel as contact electrolyte may then be impractical because of its low conductivity. As already mentioned, band electrodes are also important in two-electrode systems in order to reduce the constrictional resistance. Even with an optimal electrode system, the relative conductance signal $\Delta G/G$ is usually around 0.1% and $< 0.5\%$.

Swanson and Webster (1983) analysed the errors in four-electrode impedance plethysmography due to inadequate instrumentation, improper electrode application and physiological changes. Their paper gives a detailed list of instrumentation requirements that ensures an error of less than 5% from each error source in most applications.

8.3.1. Impedance cardiography (ICG)

Impedance plethysmography of the thorax gives somewhat different waveforms from those of limb plethysmography. A common tetrapolar system uses two band electrodes around the neck, one band electrode corresponding to the apex of the heart, and a fourth further in the caudal direction. Another system uses two measuring systems, each with two excitation and two recording electrodes, eight electrodes in total. Four electrodes are connected around the neck, the others at the lower thorax. However, Qu *et al.* (1986) used impedance cardiography for monitoring cardiac output during stress tests and found that a spot electrode array increased the signal-to-noise ratios significantly compared to a band electrode array both at rest and during four levels of exercise. Motion artefacts were found to be at a minimum when the electrodes were placed in the sagittal plane of the body. The suggested placement was current-injecting electrodes on the back-side of the body on the upper neck and lower thorax and potential pick-up electrodes on the front side on the lower neck and over the sternum at the fourth rib.

The amplitude of the impedance change ΔZ as a function of the heartbeat is about 0.5% of the basic Z value. The ΔZ waveform is similar to the aortal blood pressure curve. The *first time derivative* of $\Delta Z \, dZ/dt$ is called the impedance cardiographic curve (ICG). By adding information about patient age, sex and weight it is possible to estimate the heart stroke volume and cardiac output.

The contribution of the flowing blood is of course complicated. Blood streams from the right side of the heart to the lungs, where the air volume increases the sensitivity to the increased blood volume. At the same time the left side of the heart pumps blood into the aorta. On both sides the heart itself is emptied of blood during the systole.

The signal must be corrected for the rather large variations due to respiration. Open-heart pre- and post-operative values are not comparable, because the surgery changes the tissue properties and immittivity distribution.

8.4. EEG, ENG/ERG/EOG

The electrical activity of 10^{11} brain cells is recorded on the skin of the skull with a standardised electrode network of 21 electrodes. The leads may be bipolar or unipolar. The signal amplitude is only of the order of 50 μV, and the frequency content 1–50 Hz, so dc voltages are filtered out. The low-frequency content is a clear sign that the skull has a detrimental effect on the signal transmission. Even so, the number of electrodes indicates that the information content is sufficient to roughly localise a source. It is believed that the brain centres are less synchronised the higher the activity, resulting in smaller amplitude and more high-frequency content of the EEG. The waves have largest amplitude and lowest frequency content during sleep. The electrodes used are rather small, often made of tin/lead with collodium as contact/fixation medium.

The brain can be stimulated, for example, by hearing a sound or by looking at changing patterns. The EEG signal can be time-averaged on the basis of synchronisation pulses from the stimulator. Electrical activity and the brain's electrical response can be extracted from noise by this method. In this way the hearing of small children and babies can be examined.

Electrocorticography with electrodes placed directly on the cortex during surgery permits direct recording of high-amplitude, high-frequency EEG.

8.4.1. ENG/ERG/EOG

EOG (electro-oculography) is an electrophysiological method in which dc potentials are utilised, and therefore AgCl electrodes are used. The dc potential is dependent on the position of the eye, and is of particular interest, for example, when the eyelids are closed (REM sleep). As a dc recording method, EOG tends to be prone to drift, which makes the spatial localization of the point of gaze problematic. It is also sensitive to facial muscle activity and electrical interference. The signals are due to a potential between the cornea and the fundus of the eye with a functioning retina, and are not from the ocular muscles (Geddes and Baker 1989).

ENG (electronystagmography) is also the recording of corneo-retinal potentials, usually used to confirm the presence of nystagmus (special eye movements). The electrodes are placed to the side (laterally), above, and below each eye. A reference electrode is attached to the forehead. A special caloric stimulation test is performed, with cold and/or hot water introduced into the canal of one ear. The electrodes record the duration and velocity of eye movements that occur when the ear is temperature stimulated.

ERG (electroretinography) records the ac potentials from the retina. The electrode system is unipolar, with a gold foil or AgCl recording electrode embedded in a special saline-filled contact lens in contact with the cornea. The eye may be considered as a fluid-filled sphere in contact with the retina as a thin, sheet-like bioelectric source. The ERG signal caused by a light flash is a very rapid wave with an initial rise time of less than 0.1 ms (early receptor potential) and an amplitude around 1 mV, followed by a late receptor potential lasting many milliseconds.

8.5. Electrogastrography (EGG)

The typical electrogastrographic (EGG) signal due to stomach activity is recorded with a bipolar lead using a pair of standard ECG electrodes on the skin some 4 cm apart. The signal is typically of about 100 μV amplitude, and periodic with a period of about 20 s (0.05 Hz fundamental). The best position for the EGG electrodes is along the projection of the stomach axis on the abdomen.

Internal electrodes are also used, but are in general not considered to provide more information than external EGG signals. Of course the internal electrodes are nearer to the source, implying higher-amplitude signals with more high-frequency content. Because of the very low-frequency spectrum, external noise from slowly varying skin potentials tends to be a greater problem than with internal recordings.

8.6. EMG and Neurography

EMG

To record signals from muscles (electromyography, EMG), both skin surface electrodes and invasive needles are used. The distance to the muscle is often short, and the muscle group large, so signals have high amplitude and high-frequency content. But the signal is usually composed of many muscle groups, and the signal looks rather chaotic. The muscle activity is related to the rms value of the signal. If the muscle activity is low and controlled, single motor units become discernible with needle electrodes. EMG is often recorded in connection with active neurostimulation, involving both the muscle and the nervous system.

The frequency EMG spectrum covers 50–5000 Hz, and the amplitude may be several millivolts with skin surface electrodes.

ENeG

The sum of activities from a nerve bundle (electroneurography) may be picked up by skin surface electrodes, e.g. on the arm where the distance from the electrodes is not too many millimetres. Action potentials from single nerve fibres must be *measured* with invasive needle electrodes, which may be bipolar or unipolar (Fig. 8.13). To find the right position, the signal is monitored during insertion, often guided by sounds from a loudspeaker.

Stimulation may be done with electrodes of the same design: either transcutaneously right above the bundle of interest, or by needles. Muscles in the hand, for example, are stimulated by stimulating efferent nerves in the arm. EMG electrodes can pick up the result of this stimulation, for instance for nerve velocity determination. It can also be picked up by neurographic electrodes, but the signal is much smaller and must therefore be averaged with multiple stimuli. However, the method is more of interest because there is more information in the response waveform.

8.7. Electrotherapy

Transcutaneous stimulation (TENS) for pain relief
TENS (transcutaneous electrical nerve stimulation) is electrical stimulation through surface electrodes. The advantage of not using syringe injections is obvious and the electrical pulses stimulate the body's own mechanisms for obtaining pain relief. There are three theories of how the pain relief is achieved.

1. Gating theory. Pain perception is controlled by a gate mechanism in the synapses, particularly in CNS of the spine. This gate is controlled by separate nerve fibres, as a result of stimulation and with pulses of high frequency (50–200 Hz), these fibres are stimulated and pain relief is obtained.

2. Endorphins. The body uses natural forms of morphine called endorphins for pain relief. The secretion of endorphins is obtained with low-frequency (2–4 Hz) stimulation. These low frequencies correspond to the rhythmic movement of an *acupuncture* needle, in classical acupuncture it is necessary also to stimulate motor nerve fibres. The effect of endorphins is probably in the higher centres of the CNS.

3. Vasodilatation. This effect is usually linked to pain in cold extremities. Increased blood flow may increase the temperature from the range 22–24°C to 31–34°C, also in the extremities not stimulated. The effect must therefore be elicited in higher centres of the CNS.

The afferent pain nerves have a higher threshold and rheobase than sensory and motor nerves. Thus it is possible to stimulate sensory and motor nerves without eliciting pain. Very short pulses of duration 10–400 μs are used, with a constant-amplitude current up to about 50 mA and a treatment duration of 15 min or more. The skin electrodes may be bipolar or monopolar. Their position is in the region of the pain: an electrode pair may, for example, be positioned on the skin on the back of the patient, or implanted with thin leads out through the skin. The electrode pair may also be positioned outside the pain area, e.g. at regions of high afferent nerve fibre densities in the hand.

Electroacupuncture
The secretion of endorphins is obtained with low-frequency (2–4 Hz) stimulation, corresponding to the rhythmic movement of an acupuncture needle. Instead of or in addition to the mechanical movement, the needle is used as monopolar electrode, pulsed by a low frequency in the same frequency range (1–4 Hz). This is not a TENS method strictly speaking, as the electrode is invasive and the current is not transcutaneous.

8.7.1. Electrotherapy with dc

Applied dc through tissue for *long* duration, e.g. > 10 s, is a method almost 200 years old, and is traditionally called *galvanisation.* Today the dc effect is often ignored, even if it is quite clear that dc through tissue has some very special effects. Generally the physical/chemical effects of a dc through tissue include:

- Electrolysis (local depletion or accumulation of ions)
- Electrophoresis (e.g. protein and cell migration)
- Iontophoresis (ion migration)
- Electro-osmosis (volume transport)
- Temperature rise

Some of these effects are particular to long-duration (> 10 s) dc or very low-frequency ac, in particular electrolysis. Other effects are in common with ac.

Short-term effects of dc through the skin are limited to the sweat ducts. The current density is much smaller in the stratum corneum, but long-term currents may have an effect. Electrolytic effects influence the skin proximal to the electrode. Possible effects in deeper layers are erythema (skin reddening) and hyperaemia (increased blood perfusion) due to the stimulation of vasomotor nerves. If the dc is applied transcutaneously, there is always a chance of unpleasant prickling, reddening and wound formation in the skin under the electrode.

Iontophoresis

Iontophoresis in the skin is described in Section 8.13.4. If a drug is in ionic form, the migration velocity and direction are determined by the polarity of the dc. The ions may be transported from the electrode, and thus have an effect both in the tissue during passage and when assembled under the other electrode. Anaesthetic agents may be introduced into the skin by iontophoresis, for minor surgery or for the treatment of chronic pain. Antibiotics and metallic silver have been introduced iontophoretically, as well as zinc for ischaemic ulcers. The mechanism of iontophoresis is of course accompanied by a possible electrophoretic action (see below), but is not necessarily as pH-dependent as the latter. The advantage of iontophoretic instead of local syringe injection is not always obvious.

Electrotonus

Tonus is the natural and continuous slight contraction of a muscle. Electrotonus is the altered electrical state of nerve or muscle cells due to the passage of dc. Subthreshold dc currents through nerves and muscles may make the tissue more (excitatory effect) or less (inhibitory effect) excitable. Making the outer nerve cell membrane less positive lowers the threshold and has an excitatory effect (at the cathode, *catelectrotonus*); the anode will have an inhibitory effect (*anelectrotonus*). This is used in muscle therapy with diadynamic currents (see below).

Wound healing

Many controlled studies have shown that a small, microampere, current as a long-term treatment leads to accelerated healing. There are two classes of cases: accelerated healing of bone fractures, and of skin surface wounds. Ischaemic dermal ulcers are treated with dc, and the healing rate is approximately doubled. It has been found that a monopolar cathodic application during the first days, followed by an anodic application, gives the best results. In a skin wound it is believed that positive charge carriers (ions and proteins) are transported to the liquid wound zone by endogenic migration.

An increased rate of bone formation has also been found when small currents are applied to each side of a bone fracture (bipolar electrode system). This is of particular interest in cases of bone fractures that will not mend in the natural way: so called *non-union*. Nordenstrøm (1983) described the use of dc treatment by application of needles into tumours for the treatment of cancer.

Dc ablation

At higher current densities the acid under the anode leads to coagulation and the alkali under the cathode to liquification of the tissue. Warts can thus be treated; and with a needle cathode in the hair follicle, the local epidermis is destroyed and the hair removed.

Dc shock pulses

Dc shock pulses have also been used for destroying calculi in the urinary tract— *electrolithotrity*.

Hydrogen production

The application of a dc current to the inside of the eyeball by a needle electrode is used to produce bubbles of hydrogen in the aqueous humour, *electroparacentesis*.

8.7.2. Electrotherapy of muscles

Electrotherapy is a broad term and should, for instance, include pacing, defibrillation and electroconvulsion. Here we will keep to the traditional meaning, which is more limited to methods stimulating *muscles*, either directly or via the nerves. Usually such stimulation uses square wave pulses and also provides data of diagnostic value.

Figure 8.19 shows the minimum stimulus current to an efferent nerve fibre as a function of pulse duration for obtaining a certain muscle response. The coupling (synapses) between the nerve axon end plates and the muscle cells is an important part of this signal transmission line. It is not possible to reduce the current below a certain minimum level, the *rheobase* value. The pulse length with current amplitude 2 × rheobase value is called the *chronaxie*.

Pflüger's law relates the muscle effect to the leading or trailing edge of the pulse, and to anodal or cathodal polarity:

Leading edge cathodal	strongest effect
Leading edge anodal	
Trailing edge anodal	↓
Trailing edge cathodal	weakest effect

If a linear triangular current pulse is used instead of a square wave pulse, an *accommodation* effect will appear, particularly at pulse duration > 1 ms. A slowly increasing dc does not excite a nerve to the same extent as a dc step change, and the accommodation implies that the threshold current amplitude will be larger with triangular than with square waveforms. The current–time curves can be recorded for diagnostic purposes: the curves are quite different for degenerated muscles.

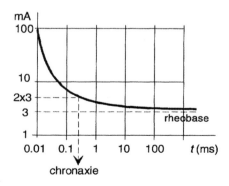

Figure 8.19. Minimum stimulus current to an efferent nerve fibre bundle for obtaining a certain muscle response, as a function of pulse duration.

The pulses may be of unidirectional current (interrupted dc, *monophasic*), which implies that the current has a dc component. High-voltage pulsed galvanic stimulation is also used, with pulse currents up to some amperes but pulse duration only a few microseconds. If dc effects are to be avoided (e.g. to reduce electrolytic effects or electrode metal corrosion), the current used is *biphasic*. *Faradic* currents are biphasic currents of the type generated by an induction coil. If the pulses are slowly increased in amplitude, then reduced, and after a pause increased again, we have a *ramp* or *surged* current. As many effects are current controlled, it is often better to use a constant-amplitude current mode than a constant-amplitude voltage mode for the stimulator output.

Pulse waveform treatment of innervated muscles (faradisation)
Short (0.5–5 ms) triangular pulses are used for tetanic muscular contractions, interrupted or with varying amplitude. Intervals between pulses are 10–25 ms. For muscle pain relief, rectangular pulses of length 2 ms and interval 7 ms are used. Transcutaneous nerve stimulation (TENS) for pain relief is different, being based on stimulating afferent nerve fibres with much shorter pulses (e.g. 0.2 ms).

Pulse waveform treatment of denervated muscles
As there is no innervation, the stimulation is directly to the muscle. The paralysed part of the muscle mass can be stimulated selectively, because such muscles have smaller accommodation at long pulse duration. Very long (1 s) triangular waveforms are used, with even longer intervals between the pulses.

Diadynamic currents for the treatment of pain and increase of blood perfusion
This is the combination of two currents: a pulse current superimposed on dc. Each current is separately adjusted. Often the dc level is first increased slowly so that no perception occurs (*electrotonus*), and then with a constant dc flowing the pulse amplitude is increased until a weak vibration is felt. The pulse waveform may be a 50 Hz power line half-rectified (50 Hz) or fully rectified (100 Hz) current.

Interferential currents
Two electrode pairs are used to set up two different current paths crossing each other in the target tissue volume. Each pair is supplied by a separate oscillator, adjusted to, e.g., 5000 and 5100 Hz. The idea is that the target volume is treated with the frequency difference, 100 Hz. The advantage is the possible selective choice of a limited treated volume deep in the tissue, together with lower electrode polarisation and skin impedance, plus less sensation in the skin and the tissue outside the treated volume.

If the current level is low enough for linear conditions, the resultant current density in the tissue is the linear summation of the two current densities (cf. Section 6.2.2). According to the superposition theorem in network theory and Fourier analysis, the new waveform $f(t)$ does not contain any new frequencies. The linear summation of two currents at two different frequencies remains a current with a frequency spectrum consisting of just the two frequencies—no current is created at any new frequency.

The current density in the treated volume must be high enough to create non-linear effects. The process can then be described mathematically by a multiplication (cf. Section 6.2.2). If $\omega_1 \approx \omega_2 \approx \omega$, $f(t)$ is a signal of double the frequency and half the amplitude, together with a signal of the low beat frequency $\omega_1 - \omega_2$, also of half the amplitude. The double-frequency signal is not of interest, but the low beat frequency $\omega_1 - \omega_2$ is now present in the tissue.

A third electrode pair can be added, with additional flexibility of frequency and amplitude selection. In any case it must be taken into account that muscle impedance may be strongly anisotropic, with a possible 1:10 ratio between different directions.

8.8. Body Composition

Measured body impedance is higher at lower frequencies because of the capacitive properties of the cell and other membranes of the body. By measuring at selected frequencies it is possible to estimate the extracellular/intracellular body fluid balance. Because fat has lower admittivity than muscle tissue, it is also possible to estimate the ratio of muscle to fat mass. The intention is often to determine total body water, extracellular/intracellular fluid balance, muscle mass and fat mass. Application areas are as diverse as sports medicine, nutritional assessment and fluid balance in renal dialysis and transplantation.

One of the first to introduce the method was Thomasset (1965), using a two-electrode method. With just two electrodes it is important to use large-area band electrodes in order to reduce the contribution from the current constrictional zones near the electrodes. With a tetrapolar electrode system it is easier to select the preferred volume to be measured. The small circumference of the lower arm, wrist and fingers causes those body segments to dominate measured impedance in a so-called whole body measurement (cf. Fig. 4.30). With measuring electrodes, say, on one hand and one foot, the chest will contribute very little. By using more than four electrodes it is possible to measure body segments such as arms, legs and chest selectively. Cole multifrequency modelling was early used by Cornish *et al.* (1993). Lozano *et al.* (1995) also used segmental multifrequency impedance measurements to estimate the values of intracellular and extracellular impedance by adjusting the parameters of the Cole equation for each segment measured. Their results showed that there is a sharp disequilibrium between the intracellular and extracellular compartments in the very first dialysis period and they stressed the importance of continuously monitoring segmental impedance during dialysis.

One method uses eight electrodes, two electrodes at each hand and foot. The body impedance is then modelled in five segments (Fig. 8.20). It is easy to see that all segmental impedances can be determined by letting two limbs be current carrying and using a third limb for zero-current potential reading (five electrode lead, pentapolar). The leads are then varied in succession. The system will be highly sensitive for the detection of asymmetrical limb bioimpedance. Cornish *et al.* (1999) provided a set of standard electrode sites for bioimpedance measurements.

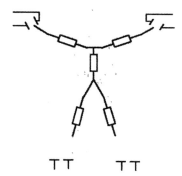

Figure 8.20. Human body divided into five impedance segments, octopolar electrode system.

Body *position* is important because it influences the distribution of both blood and the fluids in the stomach/intestine tissue. Direct body-segment to body-segment skin contact must be avoided in order to obtain stable readings. The feet should therefore be kept at a distance from each other, and the arms should be held out from the chest. Scharfetter *et al.* (1998) also analysed the artefacts produced by stray capacitance during whole-body or segmental bioimpedance spectroscopy, and proposed a model for simulating the influence of stray capacitance on the measured data.

The impedance of the chest segment is the most difficult one to determine accurately, both because it is much lower than the impedance of the limbs and because it varies with respiration and heartbeats.

Several indexes have been introduced in order to increase the accuracy of the estimation of, e.g., total body water (TBW). Sex, age and anthropometric results such as total body weight and height are parameters used. An often used index is H^2/R_{segm}, where H is the body height and R_{segm} is the resistance of a given segment. This is therefore actually a *conductance* index. Calibration can then be done by determining the k-constants in the following equations:

$$TBW = k_1 \frac{H^2}{R_{segm}} + k_2 \qquad (8.12)$$

$$TBW = k_1 \frac{H^2}{R_{segm}} + k_2 W + k_3 \qquad (8.13)$$

Such equations are not directly derived from biophysical laws but have been empirically selected because they give the best correlation. The correlation coefficients according to eq. (8.12) can be better than 0.95, and it can be slightly improved by also taking into account the body weight W (eq. 8.13). R_{segm} is determined at a single frequency, usually around 50 kHz. Actually it is of course an impedance in the sense $|Z_{segm}|$ or Z'_{segm}. The calibration can be done with results from more cumbersome methods such as those using deuterium, underwater weighing or dual energy x-ray absorption. The method was evaluated by US National Institutes of Health (NIH 1994).

The measuring frequencies are selected in the kilohertz range, and the current is usually around 0.5 mA. Higher current levels are difficult to use because the threshold of perception is reached at the lowest frequencies (cf. Section 8.14).

8.9. Cardiac Pacing

Pacing of the heart may be done transcutaneously, but this is accompanied by pain. The usual method is with two epicardial electrodes and leads out through the chest to an external pacemaker, or with an implanted pacemaker.

Implanted pacemakers are of many models. Let us consider a demand pacemaker, with special recording ring electrodes on the catheter for the demand function. If QRS activity is registered, pacing is inhibited. The pacemaker housing may be of metal (titan) and function as a large neutral electrode. Pacing is done with a small catheter tip electrode, either unipolar with the neutral electrode or bipolar with a catheter ring proximal to the tip electrode.

As can be seen from Fig. 8.21, the chronaxie is less than about 500 μs, so there is an energy waste in choosing the pulse duration much larger than 100 μs.

A pacemaker may be externally programmed by magnetic pulses, but because of this a pacemaker is to a certain degree vulnerable to external interference. Typical limits are: static magnetic field <1 gauss, 40 kV arcing >30 cm distance (car ignition system), radar 9 GHz E-field <1.2 kV/m.

Typical pacemaker data are: pulse amplitude 5 mA, impedance monopolar electrode system 1 kΩ, load voltage 5 V, and lithium battery 6.4 V with capacity 1800 mAh. The stimulus electrodes are ac coupled in the pacemaker's output stage, so that no dc can pass and unduly polarise the electrodes. The electrodes are made of noble metals so as to be biocompatible, and consequently they are highly polarisable (Section 8.1). The monopolar pacing electrode system impedance is not very

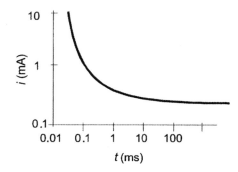

Figure 8.21. Current–time curves for heart pacing with a square wave pulse delivered during diastole with intracardial catheter electrode.

dependent on faradaic impedance because the admittance of the double-layer capacitance is large at the frequencies used.

Intracardial electrodes together with the use of electrosurgery are treated in Section 8.11.

8.10. Defibrillation and Electroshock

8.10.1. Defibrillation

Defibrillator shocks are the largest electric shocks used in clinical medicine. Current up to 50 A is applied for some milliseconds through the thorax, driven by ~ 5 kV. The electrode system is usually bipolar with two equal electrodes of surface ~ 50 cm^2 (in adults; defibrillation of children is rare). Electrodes are positioned so that as much as possible of the current is passing the heart region. With a more unipolar system with one electrode under the shoulder, the current path is more optimal, and this is used if the defibrillation is planned (electroconversion).

Earlier, conductive paste was used on the skin; today contact pads are used because they are quick to apply. They also make it possible to avoid the usual contact paste, which is easily smeared out on the skin surface and causes stray currents (either short-circuiting the shock energy or representing a hazard for the personnel involved). The current density is so high that reddening of the skin often occurs, especially at the electrode edge (cf. Section 5.2.2). The large electrode and the large current cause an extremely low-ohmic system: 50 ohms is the standardised resistance of the complete system with two electrodes and the tissue in between. The resistance falls for each shock given; this is attributed to tissue damage.

We must assume that it is the local current density in the heart that is the determining parameter for a successful conversion. As this is unknown, it is usual practice to characterise the shock in energy (watt-second = joule). This refers to the capacitor used to store the energy (Fig. 8.22). Stored energy is $CV^2/2$, so a shock dose is simply chosen by choosing the charging voltage. Maximum stored energy is usually 400 Ws. Not all the energy will be dissipated in the patient system, a part will be

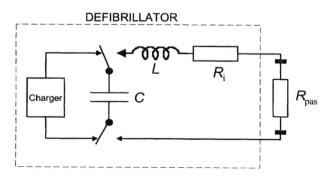

Figure 8.22. Defibrillator circuit. Typical values are $C = 20$ μF, $L = 100$ mH, $R_i = 15$ Ω, $R_{pas} = 50$ Ω.

dissipated in the internal resistance R_i of the coil used to shape the waveform of the discharge current pulse. External shock is given transcutaneously, so the voltage must be high enough to break down the skin even at the lowest dose. For internal, direct epicardial application (internal shock), sterile electrode cups are used directly on the heart without any paste or pad. The necessary dose is usually less than 50 Ws.

There is a range of accepted current duration. Figure 8.23 shows some current discharge current waveforms. Note that some models use biphasic waveforms, some use truly monophasic.

Defibrillators are also made as implanted types, using intracardial catheter electrodes. In order to reduce energy consumption, new waveforms have been used: the exponential truncated waveform. It may be monophasic or biphasic. The idea of the biphasic waveform is that the second pulse will cancel the net charge caused by the first pulse and thereby reduce the chance of refibrillation.

Tissue impedance measurements with defibrillator electrodes are used in some external and some internal defibrillator models. Measuring current and voltage during a shock gives a high current level, minimum value, non-linear region of the peak voltage to peak current ratio. Between shocks, the small signal, linear impedance is also monitored. The measured impedance value is used to customise both waveform and energy level for each shock given.

8.10.2. Electroshock (brain electroconvulsion)

Electroshock therapy is a somatic method used in psychiatry for the treatment of depressions. The traditional current waveform is a quarter-period power line 50 Hz sine wave, starting at the waveform maximum. Pulse duration is therefore 5 ms, followed by a pause of 15 ms. Automatic amplitude increase, or pulse grouping, is used. It is now often replaced by another waveform, a train of pulses of 1 ms duration with a total energy around 20 joule. It is believed that with this waveform the memory-loss problems are less. The corresponding voltage and current are

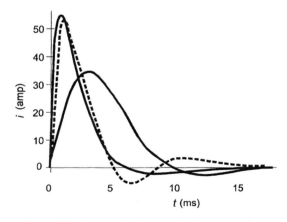

Figure 8.23. Some typical current discharge waveforms.

several hundred volt and milliamp. Large bipolar electrodes are used on the temples. The positioning is usually bilateral, but ipsilateral positioning is also used.

Electroconvulsive therapy (ECT) is a much discussed procedure, partly because it has been perceived as a brutal medical treatment. It is performed under anaesthesia and, because of the heavy muscle contractions, muscle relaxants are given. The shock elicits a seizure not very different from a grand mal epileptic attack; the seizure is to last longer than 25 s. The effect is presumably due to the enormous synchronised activity of the whole central nervous system. The treatment is usually repeated several times within a few weeks' span. The treatment is often followed by a loss of memory for recent events, and the therapeutic effect is not permanent.

8.11. Electrosurgery

High-frequency (also called radiofrequency, RF) current in the frequency range 0.3–30 MHz is used to cut or coagulate tissue. The method must not be confused with electrocautery. In *electrosurgery* the current passes through the tissue, with heat development in the tissue and with cold electrodes (diathermy). With *electrocautery* the current is passed through a wire and not through tissue, and the wire is accordingly heated. Bipolar forceps are used for microsurgery and represent a dipole current source in the tissue (cf. Section 5.2.3). A unipolar circuit (called monopolar in the field of surgery) is used in general surgery. The neutral electrode is a large flexible plate covered with sticky hydrogel for direct fixation to the skin. The neutral plate is often split in two, and a small current is passed between the two plates via the skin and tissue. Impedance is measured, and if this impedance is outside pre-set or memory set limits, the apparatus will warn as a sign of poor and dangerous plate contact.

The active electrode may be handheld, or endoscopic: long and thin types are either flexible or rigid.

Figure 8.24 illustrates the monopolar circuit. The monopolar cut electrode is often a small area blade, the coagulation in the form of a sphere (cf. Section 5.2).

The waveform used is more or less pure sinusoidal in *cut mode*, but is highly pulsed with a crest factor of 10 or more for the *spread coagulation mode*. In spread coagulation, tissue contact is not critical: the current is passed to the tissue mostly by *fulguration* (electric arcing). The electromagnetic noise generated may be severe over a wide frequency spectrum, and this causes trouble for medical instrumentation connected to the same patient.

Electrosurgery is based upon the *heating effect* of the current, and this is proportional to the square of the *current density* (and the electric field) and tissue *conductivity*. The power volume density W_v falls extremely rapidly with distance from the electrode, as shown by the equation for a voltage-driven half-sphere electrode at the surface of a half-infinite homogeneous medium (eq. 5.5). With constant-amplitude current, the power volume density is

$$W_v = \frac{i^2}{4\pi^2 \sigma r^4} \qquad \text{half sphere} \qquad (8.14)$$

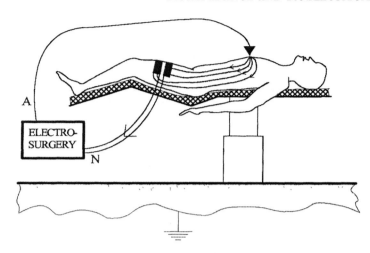

Figure 8.24. Monopolar electrosurgery.

Tissue destruction therefore occurs only in the immediate vicinity of the electrode. Power dissipation is linked with conductance, not admittance, because the reactive part simply stores the energy and sends it back later in the ac cycle. Heat is also linked with the rms values of voltage and current—ordinary instruments reading average values cannot be used. The temperature rise ΔT is given by eq. (5.6). Because heating is current *density* dependent, the heat effect is larger the smaller the cross-sectional area of an electrode, or at a tissue constriction zone. This is an important reason for the many hazard reports accompanying the use of electrosurgery in hospitals. Another is that the whole patient is electroactive in the normal mode of use of electrosurgery. The RF potentials of many body segments may easily attain some tenths of a volt rms in normal mode operation, and insulation of these body segments is critical.

High frequencies are chosen to avoid nerve and muscle stimulation, cf. the sensitivity curve of Fig. 8.34. The output is neither constant-amplitude voltage nor constant-amplitude current. The optimal output characteristic is linked with the very variable load resistance: tissue resistance increases when tissue is coagulated, fat has higher resistance than muscles and blood, and the contact geometry is very dependent on the electrode chosen and the way it is held by the surgeon. If a constant-amplitude current is chosen, power will be proportional to load resistance, and tissue will quickly be carbonised in high-resistance situations. If constant-amplitude voltage were chosen, power would be inversely proportional to load resistance, and when tissue layers around the electrode coagulated, current would stop flowing. Modern instrumentation therefore measures both output voltage and current, and regulates e.g. for an isowatt characteristic.

Typical power levels in unipolar electrosurgery are about 80 W (500 ohm, 200 V, 400 mA rms), in bipolar work 15 W (100 ohm, 40 V, 400 mA rms). The frequency content of the sine wave is the repetition frequency, usually around 500 kHz. In pulsed mode the frequency content is very broad, but most of the energy will be in the frequency band 0.5–5 MHz. In pulsed mode the peak voltage can reach 5000 V or more, so insulation in the very humid surroundings is a problem.

The arc formed, particularly in coagulation mode, is a source of both noise and rectification. Rectification is highly undesirable, because LF signal components may be generated that excite nerves and muscles. In order to avoid circulating rectified currents, the output circuit of a monopolar equipment always contains a safety blocking capacitor (Fig. 8.25). LF voltage is generated, but it does not lead to LF current flow. Even so, there may be local LF current loops in multiple arc situations (Slager *et al.* 1993). The resulting nerve stimulation is a problem in certain surgical procedures.

Argon gas is sometimes used as an arc guiding medium. The argon gas is made to flow out of the electrode mainly for two purposes: to facilitate and lead the formation of an arc between the electrode and tissue surface, and to impede oxygen reaching the coagulation zone. In this mode of operation, no physical contact is made between the metal electrode and the tissue: the surgeon points the pen towards the tissue and coagulation is started as though by a laser beam (with which it is often confused). The gas jet also blows away liquids on the tissue surface, thus facilitating easy surface coagulation.

Implants
The use of monopolar electrosurgery on patients with metallic implants or cardiac pacemakers may pose problems. Metallic implants are usually considered not to be a

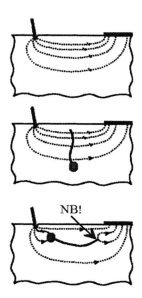

Figure 8.25. Monopolar electrosurgery and an implant, e.g. a pacemaker with intracardial catheter electrode. Importance of catheter direction with respect to current density direction.

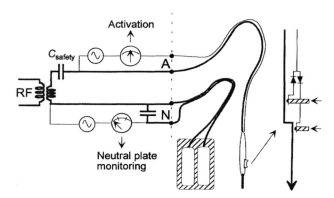

Figure 8.26. Typical electrosurgery output circuit. Note double plate neutral electrode for monitoring of skin contact. Safety blocking capacitor to prevent rectified LF currents in tissue.

problem if the form is round rather than pointed (Etter *et al.* 1947). The pacemaker electrode tip is a small-area electrode, and relatively small currents may coagulate endocardial tissue. The pacemaker catheter positioning should therefore not be parallel with the electrosurgery current density lines. This is illustrated in Fig. 8.26 for a heart pacemaker implant.

Ablation
With catheters it is possible to destroy (without removing) tissue with RF currents in a minimally invasive procedure. In cardiology this is called *ablation*. Both dc and RF current have been tried for this purpose. The choice of bipolar or monopolar catheter technique is important (Anfinsen *et al.* 1998).

8.12. Cell Suspensions

8.12.1. Electroporation and electrofusion

It is possible to open a cell membrane, e.g. for gene transfer, by applying a sufficiently strong and short electric field pulse: this is *electroporation*. *Electrofusion* is the connection of two separate cell membranes into one by a similar pulse. It is believed that both processes are based on the same field-induced restructuring of the bilayer lipid membranes (BLMs), a process which may be reversible or irreversible. It is not yet clear to what extent the process is different from the cell membrane breakdown mechanism found in an excitable cell.

It is known that an ordinary cell membrane can not withstand a prolonged dc potential difference ΔV_m more than about 150–300 mV without irreversible damage. For short pulses in the μs range, it has been found that at a threshold voltage Δv_m of about 1 V the cell membrane becomes leaky and rather large macromolecules pass in and out of the cell (lysis). The following expression for the electric field in the membrane E_m is valid if the membrane thickness d is much less than cell radius r, and if the conductivity of the membrane material is much less than both the internal and external (σ_o) electrolyte conductivity:

$$E_m = 1.5 \, (r/d) \, E \, (1 - e^{-Kt}) \cos \theta \qquad (8.15)$$

E is the electric field in the external homogeneous medium, $K = \sigma_o/3rC_m$, and θ is the angle between the E-field and the cell radius r (cell centre is origin).

For electroporation a threshold voltage of about 1 V across the cell membrane has been found. Such a cell membrane potential difference corresponds to the order of 2–20 kV/cm in the suspension according to cell size, type, etc. It may still be a reversible electroporation as long as it is caused by a single pulse of a short duration, e.g. of the order of 20 μs. If a train of such pulses is applied, the cell is killed because of the excessive material exchange. It is believed that a large part of the material exchange (lysis) is an *after-field* effect lasting up to 0.1 s or more. If the electroporation is reversible, the pores or cracks then reseal. Electrofusion is certainly an irreversible after-field effect.

The primary field effect shows threshold behaviour, about the same value for poration and fusion: the electric field effect in the cell membrane lipid bilayer is a molecular rearrangement with both hydrophobic or hydrophilic pore formation. Hydrophilic pores are considered to be water filled, with pore walls that may comprise embedded lipids. The threshold field strength has been found to be inversely proportional to the cell diameter. At the time of pulse application, cell fusion may occur if two cells are in contact with each other, DNA uptake may occur if DNA is adsorbed to the cell surface. Cells may be brought into contact with each other by means of the pearl chain effect (see below). The electroporative cell transformation probability due to DNA entrance is low, typically 10^{-5}. Field values above threshold are believed to increase the pores in number and size, until a critical value is reached where complete membrane rupture occurs (irreversible non-thermal breakdown). The difference between the threshold level and the critical level is not large, so overdoses easily kill the cells. It is interesting to speculate whether electroporation is a mechanism in defibrillator shock treatment. The field strength used is lower (of the order of 500 V/cm), but the pulse duration is longer, of the order of some milliseconds.

The breakdown process (formation of pores) is not immediate. Initial current after pulse application is the charging current due to the membrane capacitance. Membrane capacitance is composed of both the bilayer dielectric and the electric double layers at the membrane surfaces. First these layers are polarised, that is the charge distribution is changed. After some microseconds the permeability for both ions and electrically silent non-charged particles starts to increase, owing to the increase in the size of small pores that exist even at zero applied field. The cell membrane is a very dynamic system in which pores are continuously created and resealed. The number and size distribution of the pores can be analysed according to stochastic models. The mean diameter is of the order of 1 nm, but the number of pores per cell is surprisingly low, of the order of one pore per cell. A typical system is a suspension of many cells, and all measured effects are statistical mean values.

The usual source for the electric field pulse is the discharge of a charged capacitor, e.g. a 25 μF capacitor charged to 1500 V. The charge voltage and the distance between the capacitor plates determine the E-field strength, and the capacitance together with the system resistance determines the time constant of the discharge current waveform. The circuitry is very similar to the defibrillator circuit shown in Fig. 8.22, except that the inductor extending the time constant into the millisecond range is not necessarily used. The pulse is accordingly a single exponentially decaying dc pulse, and the time constant is dependent on the liquid conductivity. With more complicated circuitry it is possible to make a square-wave high-voltage pulse generator. Because it is dc, there may be appreciable electrolysis and change in pH near the electrodes. To keep the necessary voltage low, the distance between the electrode plates is small.

It is possible to use a radiofrequency (RF) pulse instead of dc. The RF causes mechanical vibrations in addition to the electrical effects, and this may increase the poration or fusion yield. As the effect is so dependent on the cell diameter, it may be difficult to fuse or porate cells of different sizes with dc pulses; the threshold level for the smallest cell will kill the largest.

8.12.2. Cell sorting and characterisation by electrorotation and dielectrophoresis

The principles behind electrorotation, dielectrophoresis and other electrokinetic effects are described in Section 2.4.6. The direction and rate of movement of bioparticles and cells due to these mechanisms depend on the dielectric properties of the cell. These dielectric properties may to some extent reflect the type of cell or the condition of the cell and there is consequently a significant potential for the utilisation of these techniques for cell sorting or characterisation.

Electrorotation has been used to differentiate between viable and non-viable biofilms of bacteria. Because of the small size of bacteria, determination of their dielectric properties by means of electrorotation is impractical. By forming bacterial biofilms on polystyrene beads, however, Zhou *et al.* (1995) were able to investigate the effect of biocides on the biofilms.

Masuda *et al.* (1987) introduced the use of travelling-wave configurations for the manipulation of particles. The frequency used was originally relatively low, so that electrophoresis rather than dielectrophoresis was predominant. The technique was later improved by, among others, Fuhr *et al.* (1991) and Talary *et al.* (1996), who used higher frequencies at which dielectrophoresis dominates. Talary *et al.* (1996) used travelling-wave dielectrophoresis to separate viable from non-viable yeast cells and the same group have used the technique to separate erythrocytes from white blood cells (Burt *et al.* 1998).

Hydrodynamic forces in combination with stationary electric fields have also been used for the separation of particles. Particles in a fluid flowing over the electrodes will to different extents be trapped to the electrodes by gravitational or dielectrophoretic forces. Separation is achieved by calibration of, for example, the conductivity of the suspending medium or the frequency of the applied field. This approach has been used for separation between viable and non-viable yeast cells

(Markx *et al.* 1994), different types of bacteria (Markx *et al.* 1996), and leukaemia and breast cancer cells from blood (Becker *et al.* 1994, 1995). Dielectrophoresis has also been used successfully for other types of bioparticles like DNA (Washizu and Kurosawa 1990), proteins (Washizu *et al.* 1994) and viruses (Schnelle *et al.* 1996).

Another interesting approach to particle separation is called field-flow fractionation, and this technique can be used in combination with dielectrophoresis (Davis and Giddings 1986). Particles are injected into a carrier flow and another force, e.g. by means of dielectrophoresis, is applied perpendicular to the flow. Dielectric and other properties of the particle will then influence the particle's distance from the chamber wall and hence its position in the parabolic velocity profile of the flow. Particles with different properties will consequently be released from the chamber at different rates and separation hence achieved. Washizu *et al.* (1994) used this technique for separating different sizes of plasmid DNA.

8.12.3. Cell-surface attachment and micromotion detection

Many types of mammalian cells are dependent on attachment to a surface in order to grow and multiply. Exceptions are the various cells of the blood and cancer cells which may spread aggressively (metastases). To study cell attachment, a microelectrode is convenient as shown in Section 5.2, the half-cell impedance is more dominated by electrode polarisation impedance the smaller the electrode surface. Figure 8.27 shows the setup used by Giaever's group (Giaever and Keese 1993).

A monopolar electrode system with two gold electrodes is used. A controlled current of 1 μA, 4 kHz is applied to a microelectrode <0.1 mm^2, and the corresponding voltage is measured by a lock-in amplifier. With cell attachment and

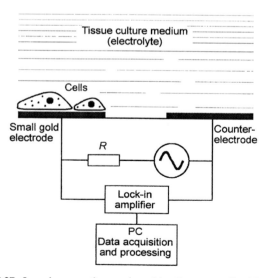

Figure 8.27. Impedance motion sensing with cells on a small gold electrode.

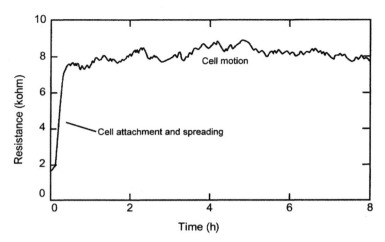

Figure 8.28. Changes in electrical resistance reflecting attachment and motion of cells on a small gold electrode.

spreading, both the in-phase and quadrature voltage increase as the result of cell surface coverage. It is possible to follow cell motion on the surface, and the motion sensitivity is in the nanometre range. The method is very sensitive to subtle changes in the cells, e.g. due to the addition of toxins, drugs and other chemical compounds. It is also possible to study the effect of high-voltage shocks and electroporation.

Figure 8.28 shows an example of cell attachment and motion as measured with the Electric Cell-substrate Impedance Sensing (ECIS) instrument, which is a commercially available version of the system (Applied BioPhysics Inc., Troy, NY, USA).

8.12.4. Coulter counter

The principle of the Coulter counter is based on letting cells in suspension pass a capillary. If a cell has different electrical properties from the liquid, the impedance of the pore will change at each cell passage. Cell counting is possible, and it is also possible to obtain information about each cell's size, form or electrical properties. Figure 8.29 shows the basic setup of the two-electrode conductance measuring cell. Typical dimensions (diameter, length) for a capillary are 50 μm and 60 μm (erythrocytes) or 100 μm and 75 μm (leukocytes).

8.13. Skin and Keratinised Tissues

8.13.1. Electrical assessment of stratum corneum hydration

Stratum corneum hydration is essential for proper function and appearance of the skin. The moisture content can be measured *in vitro* by means of gravimetry or

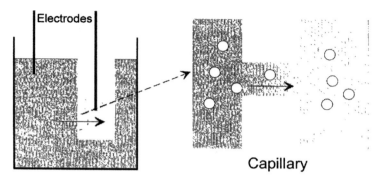

Figure 8.29. The measuring capillary of a Coulter counter.

electron microscopy, or by magnetic resonance techniques *in vivo*. The resolution of the latter technique, however, is currently not sufficiently high to enable isolated measurements on the stratum corneum. Compared to these techniques, assessment of stratum corneum hydration by means of electrical measurements would represent a tremendous reduction in instrumental cost and complexity.

A prerequisite for using electrical measurements in this way is of course a detailed knowledge of how the different parts of the skin influence the electrical impedance. Furthermore, the current and potential distribution in the skin will also be determined by the electrode geometry, which must be taken into account. As explained in Section 4.3.5, the impedivities of the stratum corneum and the viable skin converge as the measuring frequency is increased. Measurements at high frequencies will hence normally be influenced largely by the deeper layers of the skin. The frequency must therefore be kept low to achieve isolated measurements on the stratum corneum. A frequency scan, i.e. impedance spectroscopy, can not be utilised in stratum corneum hydration measurements owing to the problems of interacting dispersion mechanisms explained in Section 7.2.12. Contrary to certain opinions (Salter 1998), the mere fact that the current distribution in the different skin layers will differ between different measuring frequencies is enough for discarding the multiple frequency approach on stratum corneum *in vivo* (Martinsen *et al.* 1999). Further complications are introduced by the dispersions of the electrode impedance and deeper skin layers, and also by the Maxwell–Wagner type of dispersion that is due to the interface between the dry stratum corneum and the viable epidermis.

Since the sweat ducts largely contribute to the dc conductance of the skin, the proper choice of electrical parameter for assessment of stratum corneum hydration is consequently low-frequency ac conductance or susceptance.

Examples of available commercial instruments include the Corneometer, the DermaLab, the Nova DPM (Dermal Phase Meter), the SensoDerm and the Skicon.

The Corneometer applies a sawtooth voltage with a fundamental frequency of approximately 100 kHz to a set of closely separated electrodes in a two-electrode system. The centre-to-centre distance between these electrodes is about 0.1 mm. The electrodes are separated from the skin surface by a thin layer of glass. The results from the Corneometer are expressed in arbitrary units. Using closely separated electrodes may focus the measurements on the stratum corneum at higher frequencies also, but both the electrode size and separation must be small compared to the stratum corneum thickness in order to achieve this. The risk of only measuring excess moisture on the skin surface excludes the practical use of an electrode configuration in which the electrodes are separated by only a few micrometres. This technique will hence only be suitable on skin sites with a rather thick stratum corneum.

The DermaLab is based on a 100 kHz susceptance measurement using a probe with three concentric electrodes with a maximal diameter of 12.5 mm. Little is published about the instrument and a more detailed assessment of the method is therefore impossible.

The Nova DPM uses a range of frequencies up to 1 MHz, and the measured result is described as a cumulative value of reactance measurements and is presented in so-called DPM-units. The actual method is unfortunately not published, which impedes professional evaluation also of this instrument.

The SensoDerm is based on a low-frequency (88 Hz) susceptance measurement with a three-electrode system. This ensures isolated measurements on the stratum corneum without influence from the sweat ducts. The results are expressed in $\mu S/cm^2$ or in per cent water content by weight.

The Skicon measures conductance at 3.5 MHz with a two-electrode system. The high frequency justifies the use of conductance instead of susceptance, but also leads to significant influence from viable skin. The Skicon presents the measured data in μmho (μS).

8.13.2. Electrical assessment of skin irritation, dermatitis and fibrosis

Irritant contact dermatitis is a localised, superficial, nonimmunological inflammation of the skin resulting from the contact with an external factor. The dermatitis may be acute, e.g. if the influence from the external source is strong and of short duration, or of a more chronic kind if the influence is weaker but prolonged. The difference between irritant and allergic contact dermatitis is subtle, and depends mainly on whether the immune system is activated or not. Established signs of irritation are oedema, erythema and heat, and any electrical parameter sensitive to these physiological changes could serve as a possible parameter for the assessment of skin irritation. As for other diagnostic bioimpedance measurements, the parameter should be immune to other, irrelevant changes in the skin. To eliminate the large variations in interpersonal electrical impedance baselines, normalisation by means of indexes are often used rather than absolute impedance values.

A depth-selective skin electrical impedance spectrometer (formerly called SCIM) developed by S. Ollmar at the Karolinska Institute is an example of a commercial instrument intended for quantification and classification of skin irritation. It measures impedance at 31 logarithmically distributed frequencies from 1 kHz to 1 MHz, and the measurement depth can to some extent be controlled by electronically changing the virtual separation between two concentric surface electrodes (Ollmar 1998).

Ollmar and Nicander (1995), Nicander *et al.* (1996), Nicander (1998) used the following indexes:

Magnitude index (MIX) = $|\mathbf{Z}|_{20\ kHz}/|\mathbf{Z}|_{500\ kHz}$
Phase index (PIX) = $\varphi_{20\ kHz} - \varphi_{500\ kHz}$
Real part index (RIX) = $R_{20\ kHz}/|\mathbf{Z}|_{500\ kHz}$
Imaginary part index (IMIX) = $X_{20\ kHz}/|\mathbf{Z}|_{500\ kHz}$

where $\mathbf{Z}$, R, X and φ have their usual meaning. The authors found significant changes in these indexes after treatment with sodium lauryl sulphate, nonanoic acid and benzalkonium chloride, and the measured changes correlated well with the results from subsequent histological examinations. The stratum corneum is soaked with saline before the measurements in order to provide good contact between the electrode system and the skin surface and to focus the measurements on the viable skin, although the barrier function of intact stratum corneum will still give a considerable contribution. The choice of frequencies for the indexes hence seems reasonable. The group is extending its impedance spectroscopy technology to further applications. Examples are the detection of other conditions and diseases in the skin or oral mucosa (Emtestam and Nyrén 1997; Lindholm-Sethson *et al.* 1998; Norlén *et al.* 1999), the early detection of transplanted organ complications (Ollmar 1997; Halldorsson and Ollmar 1998) and assessment of skin cancer (Emtestam *et al.* 1998).

Subcutaneous fibrosis is a common side effect of radiotherapy given, for example, to women with breast cancer. Nuutinen *et al.* (1998) measured the relative permittivity of the skin at 300 MHz with an open-ended coaxial probe, and found that the permittivity values were higher in fibrotic skin sites than in normal skin. Based on *in vitro* experiments with protein–water solutions indicating that the slope of the dielectric constant vs the electromagnetic frequency is a measure of the protein concentration, Lahtinen *et al.* (1999) demonstrated that skin fibrosis can also be measured with the slope technique. Both Nuutinen *et al.* (1998) and Lahtinen *et al.* (1999) found a significant correlation between the permittivity parameters and the clinical score of subcutaneous fibrosis obtained by palpation. Finally, radiation-induced changes in the dielectric properties were also found in subcutaneous fat by modelling the skin as a three-layer dielectric structure (Alanen *et al.* 1998).

Because of the obvious advantages of skin impedance measurements as a non-invasive *in vivo* diagnostic method, we will probably (and hopefully) in the future see a large number of suggested approaches for the detection and monitoring of different diseases that in some way are manifest in the physiological data of the skin.

8.13.3. Electrodermal response (EDR)

The sweat activity on palmar and plantar skin sites is very sensitive to psychological stimuli or conditions. One will usually not be able to perceive these changes in sweat activity as a feeling of changes in skin hydration, except, for example, in stressing situations such as speaking to a large audience. The changes are easily detected by means of electrical measurements, however, and since the sweat ducts are predominantly resistive, a low-frequency conductance measurement is appropriate (Grimnes 1982). EDR measurements have for many years been based upon dc voltage or current, and accordingly the method has been termed galvanic skin response (GSR).

The measured activity can be characterised as exosomatic or endosomatic. The *exosomatic* measurements are usually conducted as resistance or conductance measurements at dc or low-frequency ac. Resistance and conductance will of course be reciprocal when using dc excitation, but when ac excitation is used it is important to remember that resistance generally is a part of a series equivalent of a resistor and a capacitor, while conductance is a part of a parallel equivalent of these components. In this case it is obvious that resistance and conductance are no longer reciprocal, as discussed in Section 7.2.2, and conductance should be preferred to resistance since ionic conduction and polarisation basically appear in parallel in biological tissue.

The *endosomatic* measurements are carried out as dc voltage measurements. The mechanisms behind the changes in skin potential during sympathetic activity are still not understood, but processes like sodium reabsorption across the duct walls and streaming potentials in the sweat ducts should be taken into account.

The so-called "lie detector" is perhaps the most well-known instrument in which the electrical detection of this activity is utilised. There are, however, several other applications for such measurements, mainly within two categories: neurological diseases or psychophysiological measurements. Examples of the first category are neuropathies (from diabetes), nerve lesions, depressions and anxiety. The later category may include emotional disorders and lie detection. Qiao *et al.* (1987) developed a method to measure skin potential, skin electrical admittance, skin blood flow and skin temperature simultaneously at the same site of human palmar skin. This was done in order to be able to investigate a broader spectrum of responses to the activity of the efferent sympathetic nerve endings in palmar blood vessels and sweat glands.

Both *evoked* responses (e.g. to light, sound, questions, taking a deep breath) and *spontaneous* activity may be of interest. Measurement of spontaneous EDR is used in areas such as sleep research and the detection of the depth of anaesthesia and in sudden infant death syndrome research.

Figure 8.30 shows detection of EDR by means of 88 Hz conductance (G) and susceptance (B) measurements (Martinsen *et al.* 1997a). The measurements were conducted on palmar sites of the right and left arm. In Fig. 8.30, both hands show clear conductance waves, but no substantial susceptance waves. No time delay can be seen between the onset of the conductance waves in the two hands, but the almost undetectable susceptance waves appear a few seconds after the changes in conductance. This indicates that these changes have different causes. The rapid conductance change is presumably a sweat duct effect and the slower change in susceptance is most probably due to a resultant increased hydration of the stratum

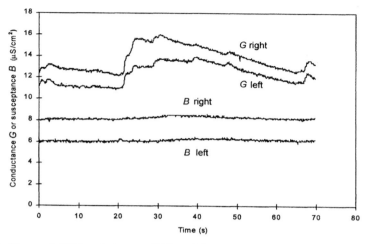

Figure 8.30. Measured 88 Hz admittance EDR activity on palmar skin sites. A deep breath at approximately 20 s on the time scale triggered the response. From Martinsen *et al.* (1997a), by permission.

corneum itself. There are no susceptance waves that could indicate any significant capacitance in the sweat ducts.

Venables and Christie (1980) give detailed suggestions on the analysis of EDR conductance waves based on the calculation of amplitude, latency, rise time and recovery time. They also give extensive statistical data for these parameters in different age groups. As already mentioned several times in this book, the use of absolute values for the electrical properties of tissue is hazardous owing to their liability to measurement error and their dependence on electrode size, gel composition and ambient environment. The use of indexes or other relative parameters would presumably prove beneficial also in EDR measurements. Mørkrid and Qiao (1988) analysed the use of different parameters from the Cole admittance equation in EDR measurements and proposed a method for calculating these parameters from measurements at only two frequencies.

8.13.4. Iontophoretic treatment of hyperhidrosis

Hyperhidrosis is a state of extreme sweat secretion in palmar, plantar or axillary skin sites. The disorder can be treated with drugs (e.g. anticholinergics), tap water iontophoresis or surgical sympathectomy. Tap water iontophoresis has been in use at least since the beginning of this century, and represents a simple, effective, but somewhat painful cure. In its simplest form, the setup used comprises two water-filled metal tubs and a dc supply. In the case of palmar hyperhidrosis, the hands are placed in the two tubs and a dc current is driven from one hand to the other through the upper body. This treatment has now been discontinued in Norway because the current is driven through the heart region. The Drionic is an example of an alternative device for the treatment of palmar hyperhidrosis where this problem is solved. When one hand is placed on the Drionic, half the hand is connected to one electrode through a wet sponge, and the other half to the other electrode through another wet sponge. The current is hence driven locally in the hand, and only one hand is treated at a time.

How tap water iontophoresis can impede excess sweating is still not fully understood. One theory suggests abnormal keratinisation in the epidermis as a result of the current being shunted through the sweat ducts, leading to a plugging of the sweat orifices. This plugging can not be found on micrography of the skin, however, and a more plausible theory is presumably that the current leads to a reversible destruction of the sweat glands.

8.13.5. Iontophoresis and transdermal drug delivery

The transport of charged substances through the skin was shown early by the famous experiment of Munk (1873). He applied an aqueous solution of strychnine in HCl under two electrodes attached to the skin of a rabbit. Without current flow nothing happened to the rabbit; with application of a dc current for 45 min, the rabbit died.

Abramson and Gorin (1939, 1940) found that timothy pollen could be transported into the skin by electrophoresis. They studied the transport of dyes into human skin by electrophoresis. Without the application of electricity, no particular skin marks were seen after the dye had been in contact with the skin for some minutes. With an applied dc current, and after the superfluous dye had been wiped off, small dots were seen corresponding to the pores of the skin. Positively charged methylene blue was transported into the skin under an anode, and negatively charged eosin under a cathode. Some pores were coloured with only one of the types.

Iontophoresis of pilocarpine is the classical method for obtaining sweat for the cystic fibrosis test (Gibson and Cooke 1959). The penetration of pilocarpine in the skin enhances sweat production. The test is usually performed on children with both electrodes placed on the underarm (for safety reasons the current should not pass the thorax). A 0.5% solution of pilocarpine is placed under the positive electrode, and the dc current is slowly increased to a maximum of about 1.5 mA. The iontophoresis time is about 5 minutes.

A skin surface *negative* electrode attracts water from deeper layers, a positive electrode repels water. This is an *electro-osmotic* effect and not iontophoresis (Abramson and Gorin 1939; Grimnes 1983b).

The conductivity of human skin is very unevenly distributed. The current pathways have been found to be the pores of the skin, particularly the sweat ducts, and only to a small extent through the hair follicles (Abramson and Gorin 1940; Grimnes 1984).

Transdermal drug delivery through iontophoresis has received widespread attention. A long-term delivery with transdermal dc voltage of < 5 V is used (Pliquett and Weaver 1996). High-voltage pulses up to 200 V decaying in about 1 ms have also been used on human skin for enhancement of transport by electroporation (Pliquett and Weaver 1996). The effect was found to be due to the creation of aqueous pathways in the stratum corneum.

8.14. Threshold of Perception, Hazards

8.14.1. Threshold of perception

The perception of a current through human skin is dependent on frequency, current density, effective electrode area (EEA), and skin site/condition. Current duration also is a factor, in particular in the case of dc determining the quantity of electricity and thereby the electrolytic effects according to Faraday's law (Section 2.4.1).

Dc

If a dc source coupled to two skin surface electrodes is suddenly switched on, a transient sensation may be felt in the skin. The same thing happens when the dc current is switched off. This proves that many nerve endings are only sensitive to *changes* in a stimulus, and not to a constant stimulus. At the moment a dc is switched on, it is not only a dc but it also contains an ac component. Dc must therefore be applied with a slow increase from zero up to the desired level if the threshold of dc perception is to be examined.

Dc causes ion migration (iontophoresis) and cell/charged particle migration (electrophoresis). These charge carriers are depleted or accumulated at the electrodes or when passing ion-selective membranes in the tissue. In particular, almost every organ in the body is encapsulated in a *macro-membrane* of epithelial tissue. There are, for example, three membranes (meninges) around the brain and central nervous system (pia mater, arachnoidea, dura mater). There are membranes around the abdomen (peritoneum), the fetus, the heart (pericardium) and the lungs (pleura), inside the blood vessels (endothelium), and around the nerves (myelin, neurolemma in the hand). At some tissue interfaces and at the electrodes, the chemical composition may gradually change.

A sensation will start either under one of the electrodes (anode or cathode) or in the tissue between. The chemical reaction at an electrode is dependent on the electrode material and electrolyte, but also on the current level (cf. Sections 2.4 and 8.1). A sensation around threshold current level develops slowly and may be difficult to discern from other sensations, e.g. the mechanical pressure or the cooling effect of the electrode. Once the sensation is clear, and the current is reduced slowly to avoid ac excitation, the sensation remains for some time. This proves that the current does not trigger nerve ends directly, but that the sensation is of a chemical, electrolytic nature as described by the law of Faraday. The after-current sensation period is dependent on the perfusion of the organ eliciting the sensation.

On palmar skin, with a surface electrode of varying area A, the current I_{th} or current density J_{th} at the threshold of perception was found to obey the following equation for a sensation within 3 min after current onset (Grimnes *et al.* 1998):

$$J_{th} = J_0 A^{-0.83} \qquad \text{or} \qquad I_{th} = I_0 A^{0.17} \tag{8.16}$$

The perception was found to be localised under the monopolar electrode, never in the tissue distal to the electrode. Surprisingly, according to eq. (8.16), the threshold as a function of electrode area A is more dependent on *current* than current density. There may be more than one reason for this:

1. A spatial summation effect in the nervous system. The current *density* is reduced when the same current is spread by a larger electrode, but at the same time a larger number of nerve endings are excited. Consequently the current threshold is not greatly altered when electrode area is changed.
2. The dc current is not evenly distributed under a plate electrode (eq. 5.10), the current density being higher at the edge. The conductance of a surface sphere or plate electrode is proportional to radius or circumference, not to area (eqs 5.1 and 5.9).
3. The dc current is probably concentrated to the sweat ducts and the nerve endings there (Grimnes 1984).

A practical use of dc perception is the old test of the condition of a battery by placing the poles on the tongue. This test is actually also done clinically: *electrogustometry* is the testing of the sense of taste by applying a dc to the tongue.

Sine waves

The lowest level ($< 1~\mu A$) of 50/60 Hz perception is caused by *electrovibration* (Grimnes 1983d). It is perceived when the current-carrying conductor slides on dry skin. Dry skin is a poor conductor, so that potential differences of several tenths of volts may exist across the dielectric, which is the stratum corneum of the epidermis. With dry skin only a small microampere current flows. The electric field sets up an electrostatic compression force in the dielectric, pressing the stratum corneum to the metal plate. In the stratum corneum there are no nerve endings, and consequently there is no perception. However, if the skin is made to slide along the metal, the frictional force will be modulated by the electrostatic force and be felt as a lateral mechanical vibration synchronous with double the frequency of the ac voltage. Even if the voltage across the dielectric is > 20 V at threshold, the corresponding (mainly capacitive) current may be $< 1~\mu A$. If the skin is at rest, or if the skin is wet, no sensation is felt.

The second level (1 mA (Dalziel 1954)) is due to the *direct electrical excitation* of nerve endings, which must be a function of the local current (density). The electric current threshold of perception with firm hand grip contact and contact area of several square centimetres is around 1 mA. Threshold current has a surprisingly small dependence on contact area. The reason for this is mentioned above in the discussion of dc perception. With a small contact area around 1 mm^2, the threshold of perception is around 0.1 mA.

Interpersonal variation and the dependence on age and sex are small. Skin condition is not important as long as there are no wounds. Skin site may be important. On the fingertips the density of nerve ending is large, but the stratum corneum is thick and the current will be rather uniformly distributed. Other sites

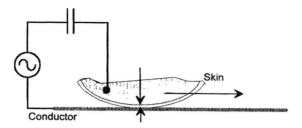

Figure 8.31. Electrovibrational perception mechanism.

may have much thinner skin and lower density of nerve endings but possess conductive sweat ducts that canalise the current.

Frequency dependence
For sine waves the maximum sensitivity of the nervous system is roughly in the range 10–1000 Hz (cf. also Fig. 8.34). At lower frequencies each cycle is discernible, and during each cycle there may be enough charge to give electrolytic effects. At frequencies > 1000 Hz the sensitivity is strongly reduced, and at > 100 kHz no perception remains because the levels are so high that electrical stimulation is overshadowed by the heating effect of the current. This is the frequency range for electrosurgery.

A single pulse or a repetitive square wave may give both dc and ac effects. In both cases the duration of the pulse or square wave is an important variable (cf. rheobase and chronaxie).

The exponentially decaying discharge waveform corresponds to the case of electrostatic discharge (see below).

Electrostatic discharge pulse
The perception of an electrostatic discharge is an annoyance, and in some situations a hazard. It is particularly troublesome indoors during the winter with low relative humidity (RH). Low RH reduces the conductivity of most dielectrics, e.g. the stratum corneum (cf. Section 4.3.5), and also the conductivity of clothing, construction materials, trees, concrete, etc. A person may be charged up to more than 30 kV under such circumstances, and with a body capacitance to the room (ground) of about 300 pF, the electrical energy is of the order of 0.1 joule. A smaller discharge, near the threshold of perception, typically has a time constant of a few μs and peak current around 100 mA (Fig. 8.32) and was obtained by discharging a capacitor of 100 pF charged to 1.4 kV. The point electrode is brought nearer to the skin until an arc is formed, heard and perceived in the skin.

The maximum current is determined by the voltage drop in the arc (probably less than 100 V) and the resistance in the skin. The arc probably has a very small cross-sectional area, so most of the resistance is in the proximal zone in the stratum corneum. The current density is probably far out in the non-linear breakdown region of the skin, but because of the short pulse duration the impedance is

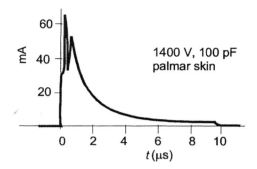

Figure 8.32. Capacitor discharge current through palmar skin. Monopolar electrode: 1.3 mm diameter pin of steel, sharpened at the tip. Indifferent electrode on the underarm. From Gholizadeh (1998).

presumably also determined by the capacitive properties of stratum corneum. A rough calculation based on eq. (5.9) with $\sigma = 0.1$ S/m and calculated resistance from measured current maximum of 20 kΩ gives the arc contact diameter with the skin as $2a = 200\ \mu$m.

The charge transferred around threshold level is of the order of 0.2 C, corresponding to a stored energy is about 10 μJ. The formation of an arc in the air between the conductor and the skin is possible when the voltage difference is larger than about 400 V. The arc discharge at threshold can be heard as a click and felt as a prick in the skin.

8.14.2. Electrical hazards

Electromagnetic field effects
Coupling without galvanic tissue contact is covered in Section 8.1.1. However, electromagnetic hazards are outside the scope of this book. There is a vast amount of experimental data on this subject, and the interested reader is recommended to the CRC handbook of Polk and Postow (1986).

Direct current
The risk of sudden death is related to stimulation of the cells of three vital organs of the body: the heart, the lungs and the brainstem. Involuntary movements may lead indirectly to sudden death (loss of balance, falling). Heating and electrochemical effects may also be fatal by inducing injuries that develop during hours and days after the injury. In electrical injuries the question often arises whether the current is evenly distributed in the tissue or follows certain high-conductance paths. Current marks and tissue destruction often reveal an uneven current distribution (cf. Ugland 1967).

The current path is important, and organs without current flow are only indirectly affected: to be directly dangerous for the healthy heart, the current must pass the heart region.

Cell, nerve and muscle excitation

Heating effects are certainly related to current density in volume conductors, but this is not necessarily so for nerve and muscle excitation. Excitation under a plate electrode on the skin is more highly correlated to current than to current density (cf. Section 8.14.1). The stimulus summation in the nervous system may reduce the current density dependence if the same current is spread out over a larger volume of the same organ. Therefore, and for practical reasons, safe and hazardous levels are more often quoted as current, energy or quantity of current in the external circuit, and not as current density in the tissue concerned.

Macro/microshock

A *macroshock* situation occurs when current is spread to tissue far from the organ of interest, usually the heart. The current is then spread out more or less uniformly, and rather large currents are needed in the external circuit in order to attain dangerous levels (usually considered as > 50 mA at 50/60 Hz) (Fig. 8.33).

The heart and the brainstem are particularly sensitive for small areas of high current density. Small area direct contact occurs, for example, with pacemaker electrodes, catheter electrodes and current-carrying fluid-filled cardiac catheters. Small contact area implies a monopolar system with possible high local current densities at *low current levels* in the external circuit. This is called a *microshock* situation. The internationally accepted 50/60 Hz safety current limit for a part in direct contact with the heart is therefore 10 μA in normal mode, and 50 μA under single fault condition (e.g. if the patient is in contact with mains voltage owing to insulation defects). The difference between macro- and microshock safety current levels is therefore more than 3 decades.

The heart is most vulnerable to electric shock in the repolarisation interval, that is in the T-wave of the ECG waveform. Therefore the probability of current passage

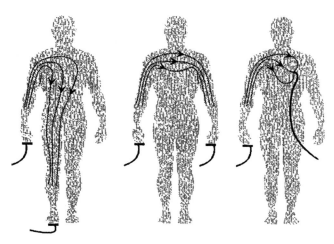

Figure 8.33. Macroshock and microshock (right) situations.

during the ~100 ms duration of the T-wave is important. If the current lasts for more than one heart cycle, the T-wave is certainly affected. For short current durations, <1 s, the risk of heart stoppage is determined by the chance of coincidence with the T-wave.

Let-go current
Let-go current threshold (15 mA at 50/60 Hz) is the current level at which the current density in muscles and nerves is so large that the external current controls the muscles. As the grip muscles are stronger than the opening muscles of the hand, a grip around the current-carrying conductor can not be released by the person experiencing the shock. Let-go current levels are therefore the most important data for safety analysis. The result in Fig. 8.34 shows that 1% of the population have a let-go threshold as low as 9 mA at power-line frequencies.

    Fatal levels are reached at current levels > 50 mA at 50/60 Hz if the current path is through vital organs: heart, lung or brainstem (cf. the electric chair, Section 8.14.3).

Heating effects
Joulean heating is dependent on the in-phase components of potential difference and current density. The resulting temperature rise is dependent on the power density, the specific heat of the tissue, and the cooling effect of blood perfusion (cf. eq. (5.6)).

    The tissue damage is very dependent on exposure time. Cells can tolerate long-time exposure of 43°C; above about 45°C, the duration becomes more and more critical. In high-voltage accidents the heating effect may be very important, and victims are treated as thermal burn patients. In particular, special attention is paid to the fluid balance, because electrical burn patients tend to experience renal failure more readily than those with thermal burns of equal severity. As electric current disposes thermal energy directly into the tissue, the electrical burn is often deeper than a thermal burn caused by thermal energy penetrating from the surface. The

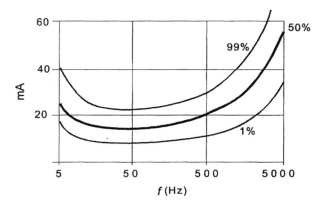

Figure 8.34. Let-go current as a function of frequency.

general experience is therefore that an electrical burn is more severe than it may appear in the first hours after the injury.

Electrolytic effects
Electrolytic effects are related to dc, applied or rectified by non-linear effects at the electrodes or in the tissue. Also, with very low-frequency ac (e.g. < 10 Hz), each half-period may last so long as to cause considerable irreversible electrolytic effects. With large quantities of electricity ($Q = It$) passed, the *electrolytic* effects may be systemic and dangerous (as from lightning and high voltage accidents). The risk of skin chemical burns is greater under the cathode (alkali formation) than under the anode (acid formation); the natural skin pH is on the acidic side (pH < 5.5).
Nerve damage is often reported in high-voltage accidents.

Current-limiting body resistance
The most important current-limiting resistance of the human body is that of dry skin. This may be impaired by high-field electrical breakdown, skin moisturising or a skin wound. Skin breakdown may occur below 10 V ac at 50/60 Hz owing to electro-osmotic breakdown (Grimnes 1983b).
Without the protective action of the skin, the *internal body resistance* may be divided into a *constrictional zone resistance* with increased current density near an electrode, and *segmental resistances* of each body segment with rather uniform current density. With small area electrode contact the constrictional zone resistance will dominate (cf. Fig. 5.2). The segmental resistance may be estimated from $R_{sr} = L/\sigma A$. With constant σ the segmental resistance depends on the ratio L/A, and accordingly varies according to body or limb size (cf. Section 4.3.6).

Table of threshold values
From the threshold current levels, the corresponding voltages are found by estimating the minimum current-limiting resistance. These worst-case minima are found by assuming no protective action from the skin, only from the volume resistance of the living parts of the body. The levels are summarised in Table 8.3.

Table 8.3. Summarising 50/60 Hz threshold levels, perception and hazard

Current threshold	Voltage threshold (very approximate)	Organs affected	Type	Comments
0.3 μA	20 V	Skin	Perception threshold	Electrovibration, mechanical
10 μA	20 mV	Heart	Microshock hazard	Myocardial excitation
1 mA	10 V	Skin	Perception threshold	Nerve excitation
15 mA	50 V	Muscles	Let-go	Loss of muscle control
50 mA	250 V	Heart, lung, brainstem	Macroshock hazard	Life threatening

Table 8.4. Threshold/maximum electric charge values (microcoulomb) for single monophasic pulses through the chest

Pulse type	Charge (μC)
TENS threshold	3
TENS, max	7
Safe level	20
Hazard threshold	75
Heart pacing	100

The question of current canalising effects in tissue is unsolved. It is well known that the myelin sheath around the nerves serves electrical insulation and current canalisation to the Ranvier nodes. But the extent of current canalisation in many parts of the body is largely unknown. The blood has a high conductivity, but what are the electrical properties of the endothelium of the blood vessels? Do electrosurgery currents in the liver follow the bile duct? The current through skin is canalised through the sweat ducts. The high acid concentration in the stomach must produce high electrical conductivity.

Single pulses
Trigger and safety levels have been examined by Zoll and Linenthal (1964) for external pacing of the heart. For TENS (transcutaneous electrical nerve stimulation) single monophasic pulses of duration <1 ms, the FDA have set up the threshold/maximum values shown in Table 8.4 for the electric charge through the thorax.

8.14.3. Lightning and electrocution

Lightning
An average lightning stroke may have a rise time of 3 μs and duration of 30 μs, energy dissipated 10^5 J/m, length 3 km, peak current 50 000 A, power 10^{13} W. After the main stroke there are continuing currents of typically 100 A and 200 ms duration.

The mechanical hazards are due to the pressure rise in the lightning channel. The energy per metre is equivalent to about 22 g of TNT per metre. A direct hit from the main stroke is usually lethal, but often the current path is via a tree, the ground (current path from foot to foot, current through tissue determined by the *step voltage*), or from a part of a house or other building. The current path is of vital importance: for humans a current from foot to foot does not pass vital organs, for a cow it may do so.

It is believed that there are around 500 deaths caused by lightning per year in the United States.

Electrocution

The current path in an electric chair is from a scalp electrode to a calf electrode. The current is therefore passing the brain and the brain stem, the lung and the heart. It is believed that the person becomes unconscious immediately after current onset, but it is well known that death is not immediate. The electric chair was used for the first time in 1890, and the first jolt was with 1400 V 60 Hz applied for 17 seconds, which proved insufficient. At present a voltage of about 2000 V applied for 30 seconds is common, followed by a lower voltage for, say, a minute. The initial 60 Hz ac current is about 5 A, and the total circuit resistance is therefore around 400 ohm. The power is around 10 kW and the temperature rise in the body, particularly in the regions of highest current densities in the head, neck and leg region, must be substantial. Because of the cranium, the current distribution in the head may be very non-uniform. Temperature rise is proportional to time and the *square* of current density according to eq. (5.6), so there are probably local high-temperature zones in the head.

The scalp electrode is a concave metal device with a diameter about 7 cm and an area of about 30 cm^2. A sponge soaked with saline is used as contact medium.

8.14.4. Electric fences

The electric fence is used to control animals and livestock. There are two types of controllers: one type delivers a continuous controlled ac current of about 5 mA. The other delivers a capacitative discharge, like the working principle of a defibrillator. The repetitive frequency is around 1 Hz, and the capacitor is charged to a dc voltage up to 10 kV. The large voltage ensures that the shock will pass the animal's hair-covered skin. The shock is similar to an electrostatic discharge, although the electric fence shock energy is higher and the duration longer. The capacitor is charged to an energy typically in the range 0.25–10 joule.

8.14.5. Electrical safety of electromedical equipment

Special safety precautions are taken for electromedical equipment. Both patient and operator safety is considered (as well as damage to property). *Electromedical equipment* is equipment that is situated in the patient environment and is in physical contact with the patient, or that can deliver energy (electrical, mechanical or radiation) to the patient from a distance. Equipment for *in vitro* diagnosis is also important for patient safety with respect to correct diagnosis, but as long as it is not in the patient environment the safety aspects are different.

The basis for the national or international standards (IEC, UL, VDE, MDD (the European Medical Device Directive), etc.) is reduction of the risk of hazardous currents reaching the patient or the operator under normal conditions. Even under a *single fault condition*, patient safety must not be impaired.

The part of the equipment in physical contact with the patient is called the *applied part*. It may ground the patient (type B applied part), or keep the patient floating with respect to ground (BF, body floating; CF, cardiac floating) by means of

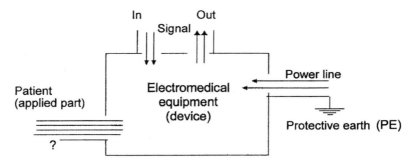

Figure 8.35. Basic parts of a grounded (class I) electromedical device.

galvanic separation circuitry (magnetic or optical coupling, battery-operated equipment). In most situations, higher safety is obtained by keeping the patient floating. If the patient by accident comes into contact with a live conductor, the whole patient will be live, but *little current* will flow.

Figure 8.35 shows the most important parts of an electromedical device. The power-line and earth connection are shown to the right. The signal connections to the outside world are shown in the upper part. Important safety aspects are linked with these signal input and output parts: they may be connected to recorders, printers, dataloggers, data networks, coaxial video cables, synchronisation devices, etc. These devices may be remotely situated and outside the electrical control of the treatment room. With a floating applied part, hazardous currents from the outside do not reach the patient: the galvanic separation protects both ways.

The device may be grounded for safety reasons (safety class I, as shown in Fig. 8.35, maximum resistance in the *protective earth* (PE) conductor between power plug and chassis 0.2 Ω), or double insulated (safety class II).

Leakage currents are currents at power-line frequency (50 or 60 Hz); they may be due to capacitive currents even with perfect insulation, and are thus difficult to avoid completely. *Patient leakage currents* are the leakage currents flowing to the patient via the applied part. Patient *auxiliary currents* are the functional currents flowing *between* leads of the applied part, e.g. for bioimpedance measurement. They are not leakage currents and therefore are usually not at power line frequency. *Earth leakage current* is the current through the ground wire in the power line cord (not applicable for double-insulated devices). According to IEC, earth leakage should be < 500 µA during normal conditions for all types B, BF or CF (Table 8.5). *Enclosure leakage current* is a possible current from a conductive, accessible part of the device to earth. Grounded small devices have zero enclosure current during normal conditions, but if the ground wire is broken, the enclosure leakage current is equal to the earth leakage current found under normal conditions.

The current limits according to IEC-60601 (1988) are shown in Table 8.5.

The insulation level is also specified. It is defined in kilovolt, creepage distances and air clearances in millimetres. Important additional specifications are related to maximum exposed surface temperature, protection against water penetration (drop/

Table 8.5. Allowable values of continuous leakage and patient auxiliary currents (μA) according to IEC-60601

Currents	Type B		Type BF		Type CF		
	N.C.	Single fault	N.C.	Single fault	N.C.	Single fault	
Earth leakage	500	1000	500	1000	500	1000	Higher values, e.g., for stationary equipment
Patient leakage	100	500	100	500	**10**	50	
Patient leakage				5000		**50**	Mains on applied part
Patient leakage		5000					Mains on signal part
Patient auxiliary	100	500	100	500	10	50	ac
Patient auxiliary	10	50	10	50	10	50	dc

N.C. = normal conditions.

splashproof), cleaning–disinfection–sterilisation procedures, and technical and users' documentation.

A non-medical device such as a PC may be situated within a patient environment (instrument B in Fig. 8.36), but in itself it must not have an applied part. An electromedical device (instrument A) must be inserted between B and the patient. The connection between A and B is via the signal input/output of device A. If the instrument B has higher earth leakage current than 500 μA, an insulation power line transformer or extra ground must be provided. The reason for this is that a person can transfer the enclosure leakage current by touching the enclosure of B and the patient simultaneously. During the single fault condition of a broken ground wire to B, the earth leakage current of B may then be transferred to the patient. This would not happen with the interconnected signal ground wires as shown in Fig. 8.36, but could happen if A and B were in the same rack with one common power cord.

Electromedical equipment may have more than one applied part (Fig. 8.37). The producer must basically declare the intended use of his equipment. If an applied part

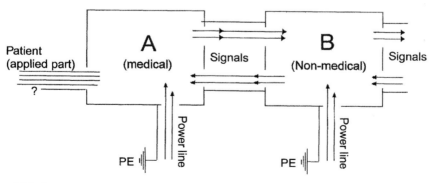

Figure 8.36. A non-electromedical device (B) within the patient environment. According to IEC-60601-1-1.

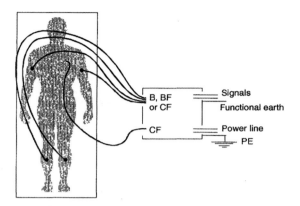

Figure 8.37. Electromedical equipment with two applied parts.

is intended to be used in direct connection with the heart, it must be of type CF. The same instrument may have another applied part intended to be used with skin surface electrodes or sensors. That applied part may be of type B, BF or CF. A plug in the instrument may be marked with type B, but a box with a galvanic separation may be inserted in the cable so that the applied part is converted from type B to BF or CF.

CHAPTER 9

History of Bioimpedance and Bioelectricity

We may imagine that the first sensory experience with electricity was of electrostatic discharges created by rubbing. Magnetic stones were known very early, and the Arabs are believed to have used such a stone floating as a compass around AD 700. *Leonardo da Vinci* experimented with lodestone and iron, and knew that the forces penetrated a wooden wall. *William Gilbert*[24] was the first scientist to devote a whole book exclusively to electromagnetism: *De Magnete* from the year 1600, written in Latin. The book is actually considered the first great scientific work published in England. Gilbert was the first to use the word "electricity"; to distinguish between static electricity and magnetism; and to consider the earth as a giant magnet. Bioelectricity was not mentioned.

A device generating static electricity was first made by *Otto von Guericke* in 1663, using a rotating sphere of sulphur. A more efficient machine with a rotating glass sphere was invented by *Francis Hauksbee* in 1704. He also experimented with evacuated glass bottles and observed the light generated in high electric fields. By 1740 electrostatic machines with a rotating glass disk had become popular and were in widespread use in Europe. In 1745 a new cheap and convenient source of static electricity was invented: the low-loss, high-voltage capacitor in the form of the Leyden jar. This was a glass bottle with a metal foil on the outside and a conductor on the inside. Not surprisingly, the ability to store electricity in a jar that can be "filled" and "emptied" led people to think of electricity as a fluid. The Leyden jar spread very rapidly in Europe and America.

Benjamin Franklin[25] had an electrostatic machine in Philadelphia, and in the winter of 1746–47 he began to investigate electrical phenomena. He suggested an experiment to prove the identity of lightning, but this was first carried out in France. He is believed to have tried the dangerous experiment of flying a kite in a thunderstorm. In 1763 professor Richman in St. Petersburg was killed by such an experiment. He is believed to be the first victim of experimenting with electricity. Franklin and his associates concluded early that the corona discharge of "Electrical Fire" or "St. Elmo's fire" was a discharge that equalised bodies with an excess and deficiency of "electrical fire". Franklin introduced the concept of positive and

---

[24] William Gilbert (1544–1603), British physician/physicist/natural philosopher at the court of queen Elizabeth I.

[25] Benjamin Franklin (1706–1790), American printer and publisher, author, inventor and scientist, diplomat and a religious protestant. Made several stays in London and Paris.

negative electricity. He suggested that buildings could be protected from lightning by erecting pointed iron rods. The term *Franklin currents* means currents of electrostatic origin and is named after him. His papers were collected in 1751 in the book *Experiments and Observations on Electricity*, soon translated into French (1752), German (1758) and Italian (1774).

The French abbot *Jean-Antoine Nollet*[26] was interested in bioelectric phenomena and made use of electrostatic machines and Leyden jars for electrotherapy. His book *Lettre sur l'électricité* was published in Paris in 1753, and in it he referred to Franklin's work. It is said that under the French king Louis XV (king from 1715 to 1774) the whole court "se fait électricer".

All these experiments were carried out with static electricity. The history of continuously flowing electricity *started* with *bio*electricity, and in particular with *Luigi Galvani* (1737–1798) at the university of Bologna. On 6 November 1780 he discovered that while an assistant was touching the sciatic nerve of a frog with a metal scalpel, the frog's muscle moved when he drew electric arcs on a nearby electrostatic machine (Galvani's first experiment, performed in his home). Galvani's frogs were placed on iron gratings, and he used bronze hooks to move them. He then discovered that the muscle twisted at the mere touch of the hook to the spinal cord. This is known as Galvani's second experiment. His explanation based on "animal electricity" was challenged by Volta, leading to the famous Galvani–Volta controversy. Galvani was a physician (an obstetrician) and a natural philosopher, and he also examined the organs of electric fishes. Many expressions reveal this historical origin, we speak of *galvanism, galvanic current* (= dc) and galvanic separation; in modern terms galvanic means related to dc current. A *galvanostat* is a dc constant-current source, and a *galvanometer* is a dc current meter.

It was *Alessandro Volta* (1745–1825), professor in physics at Como (later in Pavia and then Padua), who found the correct explanation of Galvani's second experiment: Galvani actually experimented with dc created by different metals in contact with the same electrolyte: the animal's own body fluids. He used the frog muscle both as a part of a battery, and as the first ammeter! The concept of animal electricity was abandoned, but reappeared later under the term animal magnetism, referring to hypnosis (mesmerism). Volta invented a new source of continuous electricity, the Volta battery.[27]

The work of *Michael Faraday* was important both in electrochemistry and for the discovery of magnetic induction. He was also interested in bioelectricity. On his European round tour he passed through Genoa in 1813, where he studied the electrical discharge from the torpedo fish. In 1820 *Hans Christian Ørsted* published

---

[26] L'Abbé Nollet was a member of l'Académie Royale des Sciences in Paris, the Royal Society in London and the Institute of Bologna, professor at the College of Navarra, and "Maitre de Physique" for the Dauphin. His favourite public experiment was to discharge a Leyden jar through many series-coupled persons. On one occasion he did it in front of King Louis XV with a chain of 180 Royal Guards, on another through a row of Carthusian monks *more than a kilometre long*! At the discharge, the white-robed monks reportedly leapt simultaneously into the air. He also discovered the osmotic pressure across semipermeable membranes.

[27] In 1801 Volta demonstrated his battery in Paris before Napoleon, who made Volta a count and senator of the kingdom of Lombardy.

his discovery of the relationship between flowing electricity and magnetism. It was presented in Latin, but within the same year it was translated into French, Italian, German, English and Danish. In 1831 Faraday invented the induction coil, which became very important for bioelectric research and practical use. *Faraday stimulation* means stimulation with high voltage/current pulses, faradic current. The induction coil was further developed by *Nikola Tesla*, who discharged a capacitor through a coil with just a few windings, air-coupled to a secondary coil of several hundred windings. Tesla currents were therefore high-voltage damped oscillating pulses in the lower MHz range.

Parallel with the discovery of new sources of electricity, the *detection* of small bioelectric currents became possible. Soon after Ørsted's discovery in 1820, the first *galvanometers* appeared. The problem was twofold: to increase sensitivity and to make the instruments follow the rapid changes of muscle and nerve currents. *Carlo Mateucci* measured muscle current impulses in 1838, and in 1843 *Du Bois-Raymond* measured the current impulse from a frog nerve. He also studied fishes that are capable of generating electrical currents. He created the field of scientific electrophysiology and his book *Untersuchungen über die Tierischer Elektrizität* in 1848 is a very early work.

Richard Caton registered currents from the brain (early form of EEG) in 1875. The problem of registering the activity of the heart was more difficult because the galvanometers of the time were not sufficiently rapid. *August Waller* recorded the human ECG in 1887 with the capillary electrometer (a voltage reading device), but the QRS-complex was highly distorted because of too slow a response. *Willem Einthoven* presented a sensitive and quick quartz *string* galvanometer in 1903, and with this device he registered more accurate ECG curves.

Hermann von Helmholtz measured the conduction velocity of a nerve cell axon around 1850. He formulated the very basic theorems of superposition and reciprocity, and also some very important laws of acoustics. *Hermann Müller* in Kønigsberg/Zürich during the 1870s found the capacitive properties of tissue and the anisotropy of muscle conductance, based also on ac measurements. Based upon Faraday's work, *James Clerk Maxwell* set up his famous equations in 1864. He more specifically calculated the resistance of a homogeneous suspension of uniform spheres (also coated, two-phases spheres) as a function of the volume concentration of the spheres. This is the basic mathematical model for cell suspensions and tissues still used today. In 1891, the first electrotherapeutic congress was held in Frankfurt am Main. Around 1900 it was well known that large high-frequency currents of more than one ampere could pass the human body with only heat sensation (*Arsène d'Arsonval* 1893), but that small low-frequency currents excited the nerves without heating effects.

The study of bioelectricity before 1900 can be divided in five periods according to the sources of electricity available:

• Continuous static electricity from 1663—*von Guericke's* electrostatic device (high voltage)
• Static electricity discharge from 1745—the *Leyden* jar (high voltage)

- Continuous dc current from 1800—*Volta's* electrochemical battery (low voltage, high current)
- High voltage/current pulses from 1831—*Faraday's* induction coil
- Continuous ac current from 1867—*Werner von Siemens'* rotating dynamo (high power)

These were the giants of the last centuries. We call them giants because their influence was of a very general and deep nature, and the whole of society was somewhat aware of their achievements. The nineteenth century was important for a broad understanding of some of the chemical and physical processes behind bioelectricity. In the twentieth century the specialisation has gradually increased. The achievements are more specialised with a narrower impact, and not so well known to the "public at large". On the other hand, the *technology*, in the basic sense of the word, the know-how of how to construct devices and have them produced and marketed, belongs entirely to the twentieth century. These products and their importance for new medical procedures can be seen by everyone. The bringing of medical instrumentation to the marketplace started with the x-ray machine and ECG at the beginning of the twentieth century; electrosurgery and diathermy equipment appeared in the 1930s, the EEG in the 1940s, the pacemaker and the defibrillator in the 1960s.

Rudolf Hoeber discovered the frequency dependence of the conductivity of blood, and postulated the existence of cell membranes (1911). *Philippson* in 1921 measured tissue impedance as a function of frequency, and found that the capacitance varied approximately as the inverse square root of the frequency. He called this a polarisation capacitance similar to that found for the metal/electrolyte interphase. In the late 1920s *Gildemeister* found the constant phase character of tissue, and *Herman Rein* found electro-osmotic effects.

Kenneth S. Cole (1928a,b)

It was the Cole brothers who paved the way for an analytical, mathematical treatment of tissue immittance and permittivity. K. S. Cole worked for a period in Debye's laboratory. In the theoretical part (1928a) he *calculated* the impedance of a suspension of spheres, where each sphere was coated with a layer having capacitive properties. He found expressions for the impedance at dc and infinite frequency (r_0 and r_∞, both purely resistive). He introduced a constant phase element (CPE, defined in the paper by the phase angle $\phi_3 = \text{arccotan}(m)$, and $m = r_3/x_3$, accordingly using m completely differently from Fricke: ideal resistor has $m = \infty$ and $\phi_3 = 0°$), and found as the impedance locus for such a system a circular arc with the centre below the real axis in the Wessel diagram. *A plot of impedance (X and R) in the Wessel diagram with the purpose of searching for circular arcs, may accordingly be called a Cole-plot.*

Cole discussed the three-component electrical equivalent circuit with two resistors (one ideal, lumped, physically realisable electronic component; one frequency dependent not realisable) and a capacitor (frequency dependent) in two different configurations. He discussed his model first as a descriptive model, but later discussed Philippson's explanatory interpretation (extra/intracellular liquids and cell membranes).

In the experimental paper (1928b) he presented the measuring cell and tube oscillator, and the results obtained with a suspension of small eggs. The results were in accordance with the theory outlined in (1928a).

Peter Debye (1929)

In Debye's classic book *Polar Molecules* he regarded molecules as spheres in a continuous medium having macroscopic viscosity. The model was particularly based upon gases and dilute solutions of polar liquids. From the model he deduced the equation

$$\varepsilon^* - \varepsilon_\infty = \frac{\varepsilon_0 - \varepsilon_\infty}{1 + j\omega\tau_0} \qquad \text{The Debye equation} \qquad (9.1)$$

where ε^* is the complex permittivity. Debye's work was not centred on biological materials; it was hardly known at the time that many large organic molecules are strongly polar.

Hugo Fricke (1932)

Fricke[28] showed that the electrode polarisation capacitance often varies as f^{-m} (in this book written[29] f^{m-1}), and that there is a basic empirical relationship between the exponent m and the phase angle of the electrode polarisation impedance (*Fricke's law*): $\varphi = (1 - m)90°$ (in this book $\varphi = (m)90°$). He found that the frequency exponent m usually is frequency dependent; accordingly, Fricke's law does not necessarily imply a constant phase element (CPE). However, for certain electrodes m is frequency independent over an extended frequency range. Such a constant phase element we may call a Fricke CPE. Fricke did not use circular analysis in the Wessel plane. He laid the basis for the Maxwell–Wagner dispersion model.

The ideal capacitor has (Fricke's symbols) $m = 0$ and $\varphi = 90°$, the ideal resistor $m = 1$ and $\varphi = 0°$. His model was a purely descriptive model.

Kenneth S. Cole (1932)

Cole repeated the presentation from 1928, but now with a quasi-four-element equivalent circuit with two static resistors; his z_3 is a constant phase element (CPE). *The model implies that the two resistors are not a part of the polarisation process.* This is explicitly stated in Cole (1934). He did not discuss a micro-anatomical or relaxation theory explanatory model. He pointed out that different equivalent circuits may equally well mimic measured data, all are possible descrip-

---

[28] Hugo Fricke (1892–1972), Danish physicist. PhD in Copenhagen under Niels Bohr. Emigrated in the early 1920s to the United States and lived for several years at Cold Spring Harbor.

[29] The use of phase angles, loss angles and the parameters m and α has been very confusing since the days of Cole and Fricke. This book is based on the following philosophy: The use of the phase angle coefficient α follows the tradition of the Cole brothers with α always positive. The frequency dependence of a capacitance is then $C = C_1 f^{m-1}$, $0 \leqslant m \leqslant 1$, according to eqs (7.25) and (7.28).

tive models. He did point out the similarity between data from tissue/cell suspensions and polarisation on metal–electrolyte interphases.

Kenneth S. Cole (1940)

This is an important and original paper. Here the famous Cole equation was presented:

$$z = z_\infty + \frac{r_0 - r_\infty}{1 + (j\omega\tau)^\alpha} \tag{9.2}$$

For the first time there was a mathematical expression for the *impedance* dispersion corresponding to the circular arc found experimentally. The equation introduced a new parameter: the somewhat enigmatic constant α. Cole interpreted α as a measure of *molecular interactions* with no interactions $\alpha = 1$ (ideal resistor and capacitor). Comparison was made with the impedance of a semiconductor diode junction (selenium barrier layer photocell).

In fact the Cole–Cole (1941) permittivity equation (see below) was briefly introduced in this paper ("in manuscript").

Kenneth S. Cole and Robert H. Cole (1941)

Here the famous Cole–Cole equation was presented. The emphasis turned from impedance to permittivity. The two brothers[30] did not link the paper to biological data (with the exception of two references): it is a general paper about dielectrics. It is the first paper by Kenneth S. Cole in which the concepts of dielectrics, Debye theory and relaxation theory and dispersion are used. The Cole–Cole *equation* was presented:

$$\varepsilon^* - \varepsilon_\infty = \frac{\varepsilon_0 - \varepsilon_\infty}{1 + (j\omega\tau_0)^{1-\alpha}} \tag{9.3}$$

Here * means that the permittivity is complex, τ_0 is introduced as a generalised time constant. As a permittivity equation they used $1 - \alpha$, not α, as exponent. The equation was derived from the Debye equation simply by analogy, based upon the overwhelming amount of experimental data for all sorts of dielectrics giving impedance loci of *arcs* of circles with their centres below the real axis (and not complete Debye half-circles) in the Wessel diagram.

This is an empirical, purely descriptive equation, with no direct explanatory power. The problem with the Cole–Cole equation has always been its small explanatory power, which has led to endless debates about its interpretation. However, the original paper introduced the concept of a distribution of relaxation times and linked this with α. This was then the beginning of an explanatory model.

---

[30] Cole brothers: Kenneth S. Cole (1900–1984) and Robert H. Cole (1914–1990), American physicists. Kenneth S. Cole worked with biological systems and cell suspensions, both passive impedance and non-linear action potentials studies. One of the founders of the US Biophysical Society. Robert Cole worked with the dielectric properties of non-biological materials. They had one common publication in 1941, in which they introduced the famous Cole–Cole dielectric equation. Kenneth Cole introduced the corresponding impedance equation one year earlier.

The paper also presented an equivalent electrical circuit for the Cole–Cole equation. The permittivity was modelled as two ideal, lumped capacitors and one frequency-dependent impedance (not physically realisable) modelled as a constant phase element. They stressed that this impedance was *"merely one way of expressing the experimental facts, and that it and its real and imaginary parts have no conventional meaning"*. The constant phase impedance was purely a descriptive model, although they discussed several possible explanations, for example that the relaxation processes in the dielectric were characterised by a distribution of relaxation times.

From this paper a Cole–Cole plot should be defined as a plot of the complex *dielectric constant* in the Wessel diagram to search for one or more circular arcs.

In the late 1930s *Cole* and *Curtis* extended their investigations to the non-linear effects of excitable membranes. After the war, *Hodgkin* and *Huxley* revealed some of the main mechanisms of nerve transmission, for which they won the Nobel prize in 1963.

Herman Paul Schwan

H. P. Schwan (born 1915) is one of the founders of biomedical engineering as a new discipline.[31] Before the war, at the laboratory of Rajewski at the Kaiser Wilhelm Institut für Biophysik, he had already started with some of the most important topics in the field: the low-frequency blood and blood serum conductivity, the counting of blood cells, the selective heating and body tissue properties in the UHF frequency range, electromagnetic hazards and safety standards for microwaves, tissue relaxation and electrode polarisation. He also worked with the acoustic and ultrasonic properties of tissue. In 1950 he revealed for the first time the frequency dependence of muscle tissue capacitance, and interpreted it as a relaxation phenomenon. He introduced the concept of dispersion and was first to describe the α dispersion in muscle tissue (Schwan 1954). Two of the most cited articles in the field of biomedical engineering appeared in 1957 and 1963: "Electrical properties of tissue and cell suspensions", and the more methodologically focused article "Determination of biological impedances". In the 1957 article Schwan introduced the α, β and γ classification (cf. Section 3.8). He pioneered low-frequency precision measurements (Schwan *et al.* 1962), four-electrode techniques and gigahertz measurements. Later he turned also to dielectrophoresis, electro-rotation, and non-linear phenomena of interfacial polarisation (Schwan's law of linearity; McAdams and Jossinet 1994). He is much appreciated for a number of frequently cited review articles, lately also about the history of our field (Schwan 1992, 1993). Some of his students and close collaborators at the University of Pennsylvania are and have been Edwin Carstensen, Kenneth Foster, David Geselowitz, Dov Jaron, Mariam Moussavi, Banu Onaral and Shiro Takashima.

---

[31] "Working to establish a new discipline. Herman P Schwan and the roots of biomedical engineering." In Nebecker F. (1993) *Sparks of Genius*. IEEE Press.

CHAPTER 10

Appendix

10.1. Vectors and Scalars, Complex Numbers

Suppose we have a black box with an internal electric network and one port = two external terminals. Suppose that with sinusoidal excitation all voltage differences and all currents inside are also sinusoidal. Suppose that inside there are only resistors. Then all current maxima and voltage maxima occur simultaneously—there are no phase shifts.

Suppose now that inside the box there is a circuit with resistors and capacitors. Then the voltage across a capacitor is also sinusoidal, but the voltage maximum occurs *after* the current maximum. There is a time lag, a phase shift. The phase shift is measured in degrees, and one complete period is 360°. In such circuits we must keep track not only of the magnitude, but also of the phase. Two voltages can not simply be added, they must be added as *vectors*. Two voltages "added" are actually *subtracted* if they are 180° out of phase, and equal magnitudes then cancel. Each vector quantity must therefore be given with *two* (or more) numbers. We must introduce mathematical tools to deal with such double-numbered quantities.

A *vector* or a *complex number* is the answer, both are characterised by two numbers, e.g. by magnitude and direction.

Scalars and vectors
Mathematically, a vector in space is defined as a directed line segment (an arrow), with its initial point undefined. As long as it has the same length and is not rotated, the vector may be translated anywhere in space and it remains the same vector at all times. However, in physics we may impose restrictions on the initial point of the vector. A force may, for instance, be applied anywhere along its line of action to a rigid body: this is a *sliding* vector. The same force applied to an elastic body must be defined at a single initial point: it is a *fixed* or *bound* vector.

A vector is not necessarily referred to a Cartesian coordinate system, but may be referred, for example, to a neighbouring vector with respect to magnitude and direction. Rules about the addition, subtraction and multiplication (dot and cross products) of two vectors are a part of vector algebra.

Temperature, for instance, has no direction in space, so it is a *scalar*. However, a temperature variation in space (a *gradient*) has a direction, so it is a vector.

In addition to vectors in the *space* domain, vectors may also be defined in the *time* domain, in particular rotating vectors with a fixed initial point at the origin of a Cartesian coordinate system. A variable such as the electric field may be a vector

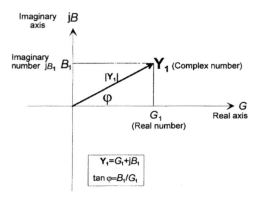

Figure 10.1. The complex plane (Wessel diagram).

both in time and in space. Every space vector may also be a time vector, and often it is not clear what sort of vector an author actually is dealing with. Vectors in the time domain are used for sine waves when the maxima do not occur simultaneously. These two-dimensional (planar) time vectors are more conveniently represented by complex numbers.

Complex numbers
Complex numbers and vectors are two ways of representing the same thing. The vector notation is often used in the space domain, the complex notation in the time domain. However, in contrast to a vector, a complex number is always referred to a two-dimensional Cartesian or polar coordinate system (Fig. 10.1).

A complex number is an ordered pair of real numbers, for instance G and B. Introducing the *imaginary unit* $\mathbf{j} = \sqrt{-1}$, the complex number $\mathbf{Y} = G + \mathbf{j}B$. G is the *real* part and can be written Y' or $\Re(\mathbf{Y})$, and B is the *imaginary* part written Y'' or $\Im(\mathbf{Y})$. $|\mathbf{Y}|$ is called the *absolute value, magnitude* or *modulus*, and the phase angle is $\varphi = \arctan(B/G)$.

A real number G can be regarded as a position on a number *line*. A complex number $\mathbf{Y}$ can be regarded as a point in the *plane* of a special Cartesian coordinate system: the complex plane, also called the Argand[32] or Wessel[33] diagram. G is an ordinary real number situated on the real x-axis. j (actually $\mathbf{j}$) simply indicates that B is to be situated on the imaginary y-axis. B is a real number; $\mathbf{j}B$ is an imaginary number; $\mathbf{Y}$ is a complex number. If the point representing $\mathbf{Y}$ is a function of, for example, frequency, the line described by all the points is called the *locus* of $\mathbf{Y}$.

Complex numbers like $\mathbf{Y}$ are printed in **bold** in this book. $\mathbf{Y}$ is represented by a point in the Wessel diagram determined by G and B, the *locus* of $\mathbf{Y}$ (cf. Fig. 10.1). $\mathbf{Y}^* = G - \mathbf{j}B$ is called the complex *conjugate* to $\mathbf{Y}$. Often the complex conjugate is

---

[32] Jean Robert Argand (1768–1822), French/Swiss mathematician. Proposed the complex plane presentation in 1806.
[33] Caspar Wessel (1745–1818), Norwegian surveyor. Proposed the complex plane presentation in 1797, nine years before Argand. Presentation in the imaginary plane is accordingly called a Wessel diagram in this book.

used in order to obtain positive values for the imaginary component in the Wessel diagram. Impedance loci, for instance, are usually plotted with the circular arcs upwards, so instead of $\mathbf{Z} = R + jX$, $\mathbf{Z}^* = R - jX$ is plotted. This is also the case for the permittivity $\varepsilon = \varepsilon' - j\varepsilon''$, usually ε^* is plotted.

Ohm's law in scalar form is written for ac (small i and v indicate sinusoidally varying quantities) as $i = vG$. There is no time lag in the circuit, e.g. only resistors. i, v and G are scalars and all maxima occur simultaneously. Under the condition that *all waveforms are sinusoidal*, Ohm's law is written

$$\mathbf{i} = v\mathbf{Y} = vG + vj\omega C \tag{10.1}$$

v is still a scalar, because it is chosen as the reference, the *independent* variable, e.g. coming from the signal generator. Phase shift is measured with reference to this sinusoidal voltage. $\mathbf{i}$ and $\mathbf{Y}$ are printed as bold characters, indicating that they are complex quantities. They are given by

$$\mathbf{i} = i_i + ji_q \tag{10.2}$$

i_i is the real part, the *in-phase* current component, meaning that this current is in phase with the imposed sine wave voltage. i_q is the imaginary part, the *quadrature* current component. The phase shift φ (relative to the voltage v) is given by

$$\varphi = \arctan(i_q/i_i) \tag{10.3}$$

$\mathbf{Y}$ is the complex *admittance*, composed of *conductance* G (in phase with the voltage; unit of siemens) and *susceptance* B (quadrature component; unit of siemens) in parallel:

$$\mathbf{Y} = G + jB = G + j\omega C \qquad |\mathbf{Y}| = \sqrt{G^2 + B^2} \tag{10.4}$$

For a capacitor $B = \omega C$, that is proportional to frequency. $f = 0$ gives $B = 0$: this is the dc case, with no influence from the capacitance. If $C = 0$ we have no capacitor, then $i_q = 0$, and $\varphi = 0°$: there is no phase shift, and the expressions reduce to real quantities only.

A complex number may also be given in *polar* form based upon *Euler's formula*:

$$e^{j\varphi} = \cos \varphi + j \sin \varphi \tag{10.5}$$

Then $\mathbf{Y} = Ye^{j\varphi} = Y\cos \varphi + jY\sin \varphi$. Since φ is the argument of sine and cosine functions, it is an angle, and therefore dimensionless.

Another important formula in polar form is that of *De Moivre*: $(\cos \phi + j \sin \phi)^\alpha = \cos \alpha\phi + j \sin \alpha\phi$. With $\phi = \pi/2$ we have the more specialised version used in the Cole equations:

$$j^\alpha = \cos(\alpha\pi/2) + j \sin(\alpha\pi/2) \tag{10.6}$$

The phasor, the sine wave and the operator $j\omega t$

A vector or complex number in the time domain is introduced under the assumption that the independent variable is a sinusoidal function of time.

What is so fundamental about the sine wave? It is the only signal containing only one frequency: it is the perfect and simplest form of *periodicity*. The sine wave is closely linked with the *circle*. A point on a circle rotating with constant frequency draws a sine wave if projected onto a paper moving with uniform speed. A vector or a complex number of unit magnitude rotating as a function of time around the origin is called a *phasor* (cf. Fig. 6.2). A phasor is best represented with polar coordinates: $Y = G + jB = Y(\cos\varphi + j\sin\varphi) = Ye^{j\varphi}$. If the angle φ is increasing uniformly with time, $\varphi = \omega t$ (radians). Then the phasor is $e^{j\omega t} = \cos\omega t + j\sin\omega t$. A clear distinction must be made between *frequency* and *angular frequency*. Both terms imply a dimension 1/second, but ωt is an angle. In the expression $\cos(ft)$, the product ft is not an angle and the expression is meaningless: in complex signal analysis the operator is $j\omega t$, not jft.

A more general mathematical treatment is possible by introducing the complex frequency $s = \sigma + j\omega$, allowing sine waves of variable amplitude, pulse waveforms, etc. Such Laplace analysis, however, is outside the scope of this book.

Implicit in the complex notation there is an important mathematical simplification. Because $\partial(e^{j\omega t})/\partial t = j\omega e^{j\omega t}$, the necessary differential equations are transformed to algebraic equations: the operator $\partial/\partial t$ becomes the factor $j\omega$.

Some algebraic rules for complex numbers presupposing sine waves

1. When a complex number is multiplied by j, the phase angle is increased by 90° ($\pi/2$).
2. When a complex number is divided by j, the phase angle is decreased by 90° ($\pi/2$).
3. $di/dt = ij\omega$ (i is complex sine wave)
4. $\int i \, dt = i/j\omega$ (i is complex sine wave)
5. $e^{j\varphi} = \cos\varphi + j\sin\varphi$ (Euler's formula)
6. $YY^* = |Y|^2 = G^2 + B^2$
7. $j^\alpha = \cos(\alpha\pi/2) + j\sin(\alpha\pi/2)$

The illustration of vector fields

A *field* is a function that describes a physical quantity at all points in space. An electric field is the distribution of field strength: the field strength and the flow of charges at each point are physical realities in the tissue volume, and as quantities they have both magnitude and direction in space. Such physical variables can be represented by vector fields, with an indication of the variable being a function of position in space, e.g. $E(x,y,z)$. The potential field is the distribution of the potential Φ at each point, but the potential has no direction so the potential field is a *scalar field*, the indication will be the same as for the vector field, e.g. $\Phi(x,y,z)$. The change of potential—the potential gradient—has a direction and is therefore a vector, the electric field $E = -\nabla\Phi$.

There are three different ways of making a two-dimensional illustration of a two- or three-dimensional vector field:

1. *Lines,* where the line direction is the direction of the vector and the *spacing* between lines is the local J value (the modulus of J) (Fig. 10.2a).

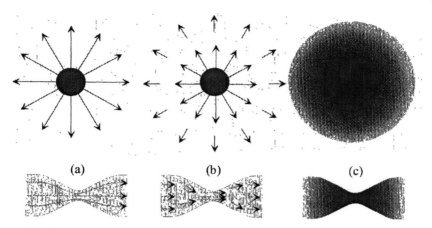

Figure 10.2. How to illustrate vector quantities. (a): Magnitude determined by distance between lines. (b): Magnitude determined by arrow length. (c): Magnitude determined by shading, no directional indication. Bottom: tube fluid flow analogy.

2. *Arrows*, each arrow indicating modulus and direction of **J** (Fig. 10.2b).
3. *Grey scale* (Fig. 10.2c) or colour scale or dot size scale. As this does not indicate direction, it is not a genuine vector presentation, but is well suited for magnitudes or scalars.

These are illustrated in Fig. 10.2 for the current density vectors caused by a current-carrying spherical electrode in a homogeneous conductive medium (Fig. 5.1). The same principles are illustrated underneath for water flow in a tube.

10.2.   Equivalent Circuit Equations

All the circuits in this section are constructed with ideal components, that is frequency-independent resistance, conductance and capacitance. Derived parameters, however, are often frequency dependent, e.g. C_{ext} in eq. (10.12). C_{pext} is the parallel capacitance as seen from the outside at the terminals. Two-component circuits are treated in Section 7.2.2.

10.2.1. Equations for two resistors + one capacitor circuits

2R–1C *series* circuit (Fig. 10.3)

$$\mathbf{Z} = R + (G - j\omega C)/(G^2 + \omega^2 C^2)$$
$$\mathbf{Z} = R + (1 - j\omega\tau_Z)/G(1 + (\omega\tau_Z)^2) \tag{10.7}$$

$$\mathbf{Z} = R + 1/G(1 + j\omega\tau_Z)$$
$$\tau_Z = C/G \tag{10.8}$$

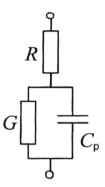

Figure 10.3. Two resistors + one capacitor series model, ideal components.

$$\varphi = \arctan\{\omega C/[G(1 + RG) + \omega^2 C^2 R]\}$$
$$\varphi = \arctan\{\omega C/G(1 + RG + \omega^2 \tau_Z \tau_2)\} \tag{10.9}$$

$$\mathbf{Y} = [G(1 + RG) + \omega^2 C^2 R + j\omega C]/[(1 + RG)^2 + \omega^2 C^2 R^2]$$
$$\mathbf{Y} = G(1 + RG + \omega^2 \tau_Z \tau_2 + j\omega \tau_Z)/[(1 + RG)^2 + (\omega \tau_2)^2] \tag{10.10}$$

$$\tau_2 = CR \tag{10.11}$$

$$C_{\text{pext}} = \varepsilon' = C/[(1 + RG)^2 + (\omega \tau_2)^2] \qquad \text{(unit cell)} \quad (10.12)$$

$$\varepsilon'' = G(1 + RG + \omega^2 \tau_Z \tau_2)/\omega[(1 + RG)^2 + (\omega \tau_2)^2] \qquad \text{(unit cell)} \quad (10.13)$$

Impedance is the preferred parameter characterising the two resistors–one capacitor series circuit, because it is defined by one unique time constant τ_Z (eq. 10.7). This time constant is independent of R, as if the circuit were current driven. The impedance parameter therefore has the advantage that the measured characteristic frequency determining τ_Z is directly related to the capacitance and parallel conductance (e.g. membrane effects in tissue), undisturbed by an access resistance. The same is not true for the admittance: the admittance is dependent on both τ_Z and τ_2, and therefore both on R and G.

when
$$\omega \to 0: \quad \mathbf{Y} \to G/(1 + RG)$$
$$B \to \omega C/(1 + RG)^2$$
$$C_{\text{pext}} \to C/(1 + RG)^2$$
$$\varphi \to 0°$$

when
$$\omega \to \infty: \mathbf{Y} \to 1/R$$
$$B \to \omega C/\omega^2 C^2 R^2 \to 0$$
$$C_{\text{pext}} \to C/\omega^2 C^2 R^2 \to 0 \tag{10.14}$$
$$\varphi \to 0°$$

Equation (10.14) is of particular interest. C_{pext} is the capacitance measured on the terminals, e.g. with a bridge or a lock-in amplifier. At high frequencies the susceptance part $B = Y''$ is small, and C_{pext} is strongly frequency dependent $(1/\omega^2)$. In this frequency range the strong capacitance increase with decreasing frequency is externally true as measured at the network port, but it does not reflect any frequency dependence of the internal capacitor component. It only reflects the simple fact that we do not have direct access to the capacitor, only through the (at high frequencies) dominating series resistance R.

Selectivity

It is important to analyse this circuit with respect to selectivity: let us assume that our black box contains the two resistors–one capacitor series circuit. Under what conditions will measured Y' be proportional to the unknown G and not be disturbed by variations in R and C? Correspondingly: under what conditions will measured Y'' be proportional to unknown C and not be disturbed by variations in R and G?

From eq. (10.10) we see that Y' is proportional to G only if the following three conditions are met:

(a) $RG \ll 1$
(b) $\omega^2 C^2 R^2 \ll 1$
(c) $\omega^2 C^2 R \ll G$

Y'' will be proportional to C only if the following two conditions are met:

(a) $RG \ll 1$
(b) $\omega^2 C^2 R^2 \ll 1$

If the conditions for Y'' are satisfied, then it will be possible to follow, for example, the unknown C directly by single frequency measurement of **Y**, without calculations based on results from measurements on several frequencies.

2R–1C *parallel* **circuit** (Fig. 10.4)

$$\mathbf{Y} = G + (\omega^2 C^2 R + j\omega C)/(1 + \omega^2 C^2 R^2)$$
$$\mathbf{Y} = G + (\omega^2 C\tau_Y + j\omega C)/[1 + (\omega\tau_Y)^2] \tag{10.15}$$

$$\tau_Y = CR \tag{10.16}$$

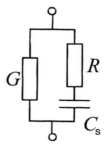

Figure 10.4. Two resistors + one capacitor parallel model, ideal components.

$$\varphi = \arctan\{\omega C/[G + \omega^2 C^2 R(1 + RG)]\}$$
$$\varphi = \arctan\{\omega C/G[1 + \omega^2 \tau_Y \tau_2 (1 + RG)]\} \tag{10.17}$$

$$Z = [G + \omega^2 C^2 R(1 + RG) - j\omega C]/[G^2 + \omega^2 C^2 (1 + RG)^2]$$
$$Z = [1 + \omega^2 \tau_Y \tau_2 (1 + RG) - j\omega \tau_2]/G[1 + (\omega \tau_2)^2 (1 + RG)^2] \tag{10.18}$$

$$\tau_2 = C/G$$
$$\varepsilon' = C_{\text{pext}} = C/(1 + \omega^2 C^2 R^2) \qquad \text{(unit cell)} \tag{10.19}$$
$$\varepsilon'' = G/\omega + \omega C^2 R/(1 + \omega^2 C^2 R^2) \qquad \text{(unit cell)}$$

The admittance time constant is uniquely defined by τ_Y, independent of G, as if the circuit were voltage driven. The admittance parameter therefore has the advantage that the measured characteristic frequency determining τ_Y is directly related to the capacitance (membrane effects) and series resistance in tissue. The same is not true for impedance: the impedance is defined by both τ_Y and τ_2.

When $\omega \to 0$: $Y \to G$
                        $B \to \omega C$
                        $\varphi \to 0°$
                        $\varepsilon' = C_{\text{pext}} = C$ (unit cell)
                        $\varepsilon'' \to \infty$ (unit cell)   NB! Loss per cycle diverges!

When $\omega \to \infty$: $Y \to G + 1/R$
                        $B \to \omega C/\omega^2 C^2 R^2$
                        $\varphi \to 0°$
                        $\varepsilon' = C_{\text{pext}} = C/\omega^2 C^2 R^2 \to 0$ (unit cell)
                        $\varepsilon'' \to 0$ (unit cell)

10.2.2. Equations for two capacitors + one resistor circuits

2C–1R *series* circuit (Fig. 10.5)

$$\varepsilon' = C_{\text{pext}} = C_s[G^2 + \omega^2 C_p (C_p + C_s)]/[G^2 + \omega^2(C_p + C_s)^2] \quad \text{(unit cell)}$$
$$\varepsilon' = C_{\text{pext}} = C_s(1 + \omega^2 \tau_2(\tau_1 + \tau_2))/(1 + \omega^2(\tau_1 + \tau_2)^2) \quad \text{(unit cell)}$$
$$\varepsilon'' = \omega\, C_s^2 G/[G^2 + \omega^2(C_p + C_s)^2] \quad \text{(unit cell)} \quad (10.20)$$
$$\varepsilon'' = C_s\, \omega \tau_1/(1 + \omega^2(\tau_1 + \tau_2)^2) \quad \text{(unit cell)}$$
$$\varepsilon = [C_s(1 + \omega^2 \tau_2(\tau_1 + \tau_2)) - j\omega \tau_1\, C_s]/(1 + \omega^2(\tau_1 + \tau_2)^2) \quad \text{(unit cell)}$$

$$\tau_1 = C_s/G \tag{10.21}$$

$$\tau_2 = C_p/G \tag{10.22}$$

$$Z = [\omega C_s G - j(\omega^2 C_s C_p + \omega^2 C_p^2 + G^2)]/[\omega C_s(\omega^2 C_p^2 + G^2)]$$
$$Z = [\omega \tau_1 - j(\omega^2 \tau_1 \tau_2 + (\omega \tau_2)^2 + 1)]/\omega C_s[(\omega \tau_2)^2 + 1] \tag{10.23}$$

$$\varphi = \arctan\{(\omega^2 C_p(C_p + C_s) + G^2)/\omega C_s G\}$$
$$\varphi = \arctan\{(\omega^2 \tau_2(\tau_1 + \tau_2) + 1)/\omega \tau_1\} \tag{10.24}$$

$$Y = [\omega^2 C_s^2 G + j\omega C_s(G^2 + \omega^2 C_p(C_p + C_s))]/[G^2 + \omega^2(C_p + C_s)^2]$$
$$Y = [\omega^2 C_s \tau_1 + j\omega C_s(1 + \omega^2 \tau_2(\tau_1 + \tau_2))]/[1 + (\omega(\tau_1 + \tau_2))^2] \tag{10.25}$$

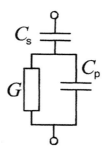

Figure 10.5. Two capacitors + one resistor series model, ideal components.

This is not a preferred one resistor–two capacitors circuit, because no parameters are uniquely defined with one unique time constant.

When $\omega \to 0$:
$$\varepsilon' \to C_s \qquad \text{(unit cell)}$$
$$\varepsilon'' \to 0$$
$$\mathbf{Y} \to 0$$
$$B \to \omega C_s$$
$$C_{\text{pext}} \to C_s$$
$$\varphi \to 90°$$

When $\omega \to \infty$:
$$\varepsilon' \to C_s \tau_2/(\tau_1 + \tau_2) \qquad \text{(unit cell)}$$
$$\varepsilon'' \to 0$$
$$\mathbf{Y} \to \infty$$
$$B \to \omega C_p C_s/(C_p + C_s)$$
$$C_{\text{pext}} \to C_p C_s/(C_p + C_s)$$
$$\varphi \to 90°$$

2C–1R *parallel* circuit (Fig. 10.6)

$$\varepsilon' = C_{\text{pext}} = C_p + C_s/[1 + (\omega\tau_Y)^2] \qquad \text{(unit cell)}$$
$$\varepsilon'' = C_s\omega\tau_Y/[1 + (\omega\tau_Y)^2] \qquad \text{(unit cell)} \qquad (10.26)$$
$$\varepsilon = C_p + C_s/(1 + j\omega\tau_Y) \qquad \text{(unit cell)}$$

$$\mathbf{Y} = j\omega C_p + (\omega^2 C_s^2 R + j\omega C_s)/(1 + \omega^2 C_s^2 R^2)$$
$$\mathbf{Y} = j\omega C_p + j\omega C_s/(1 + j\omega C_s R) \quad (10.27)$$
$$\mathbf{Y} = j\omega C_p + j\omega C_s/(1 + j\omega\tau_Y)$$

$$\tau_Y = C_s R \qquad (10.28)$$

$$\varphi = \arctan\{-(C_p + C_s + \omega^2 C_s^2 R^2 C_p)/\omega C_s^2 R\}$$
$$\varphi = \arctan\{-[C_p + C_s + (\omega\tau_Y)^2 C_p]/\omega\tau_Y C_s\} \qquad (10.29)$$

$$\mathbf{Z} = [\omega C_s^2 R - j(C_p + C_s + \omega^2 C_s^2 R^2 C_p)]/\omega[(\omega C_s C_p R)^2 + (C_p + C_s)^2]$$
$$\mathbf{Z} = [\omega\tau_Y C_s - j(C_p + C_s + (\omega\tau_Y)^2 C_p)]/\omega[(\omega\tau_Y C_p)^2 + (C_p + C_s)^2]$$
$$(10.30)$$

This is the preferred one resistor–two capacitors circuit, because all parameters are uniquely defined with one unique time constant.

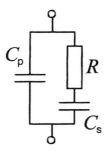

Figure 10.6. Two capacitors + one resistor parallel model, ideal components.

When $\omega \to 0$:
$$\varepsilon' = C_{\text{pext}} \to C_p + C_s \quad \text{(unit cell)}$$
$$\varepsilon'' \to 0 \quad \text{(unit cell)}$$
$$Y \to 0$$
$$B \to \omega(C_p + C_s)$$
$$C_{\text{pext}} \to (C_p + C_s)$$
$$G \to \omega^2 C_s^2 R \to 0$$
$$\varphi \to 90°$$

When $\omega \to \infty$:
$$\varepsilon' = C_{\text{pext}} \to C_p \quad \text{(unit cell)}$$
$$\varepsilon'' \to 0 \quad \text{(unit cell)}$$
$$Y \to \infty$$
$$B \to \omega C_p$$
$$G \to 1/R$$
$$\varphi \to 90°$$

10.2.3. Equations for four-component *series* circuit (simple Maxwell–Wagner model) (Fig. 10.7)

$$Y' = [(G_1 + G_2)(G_1 G_2 - \omega^2 C_1 C_2) - \omega^2(C_1 + C_2)(C_1 G_2 + C_2 G_1)]$$
$$\div [(G_1 + G_2)^2 + \omega^2(C_1 + C_2)^2]$$
$$Y'' = \omega[(G_1 + G_2)(C_1 G_2 + C_2 G_1) - (C_1 + C_2)(G_1 G_2 - \omega^2 C_1 C_2)] \quad (10.31)$$
$$\div [(G_1 + G_2)^2 + \omega^2(C_1 + C_2)^2]$$

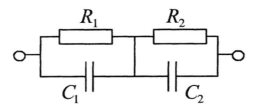

Figure 10.7. Four-component series circuit, ideal components.

When $\omega \to 0$:
$$Y' \to G_1 G_2/(G_1 + G_2)$$
$$Y'' \to 0$$
$$C_{\text{pext}} \to (C_1 G_2^2 + C_2 G_1^2)/(G_1 + G_2)^2$$
$$\varphi \to 0°$$

When $\omega \to \infty$:
$$Y' \to [C_1 C_2 (G_1 + G_2) - (C_1 + C_2)(C_1 G_2 + C_2 G_1)]/(C_1 + C_2)^2$$
$$Y'' \to \omega \, C_1 C_2/(C_1 + C_2)$$
$$C_{\text{pext}} \to C_1 C_2/(C_1 + C_2)$$
$$\varphi \to 90°$$

This is the Maxwell–Wagner model of a capacitor with two dielectric layers. Even with only two layers the equations are complicated, with three layers they become much worse.

10.3. Global Symbols (Table 10.1)

A *quantity* is usually defined as the product of a *numerical value* and a *unit*. A current is, say, 4 amperes: current is the quantity, 4 is the numerical value and ampere is the unit. In equations we use *symbols* for the quantities. For the units we also use symbols or abbreviations. The symbol for current is I, the abbreviation for ampere is A. The SI system used in this book defines seven base units (CRC 1998). The *dimension* of a derived unit is the product of the powers of the base units. A quantity with dimension = 1 is called *dimensionless*. A number, e.g. of electrons, is dimensionless.

Global symbols are symbols used in all the chapters, if not otherwise stated. Capital letters are often used for dc quantities, lower case letters for ac quantities (for example, dc voltage V, ac voltage v). The quantity called *voltage* is used in electrical circuits, and *potential* in space. Vectors and complex quantities are printed in **bold**. All space vectors may also be time vectors, and all scalar variables may be time vectors. The subscripts 0 and ∞ are used for frequency extremes, except for ε_0 where the subscript is for vacuum. The subscript s means static in the electrostatic meaning, which is without current flow and must therefore not be confused with dc conditions when current *is* flowing.

10.4. Physical Dimensions (Table 10.2)

The physical dimensions of a particle represent an important parameter: for example, the smaller a sphere, the larger the surface-to-volume ratio ($= 3/r$). The smaller a sphere, the more surface properties dominate. Table 10.2 shows some important dimensions of the components we are dealing with. The dimensions of most single atoms are of the same order of magnitude of 0.1 nm (nm = nanometer = 10^{-9} metre), but depend on measuring conditions and bonds (free, covalent, ionic). Because this is such an important practical dimension it was given a special unit, the angstrom (Å). In the internationally accepted SI-system the angstrom unit is not used, and 1 Å = 0.1 nm.

Table 10.1. Global symbols

Scalar only	Vector in time	Vector in space	Quantity	Unit (abbreviation) (dimension)
		r	Radius (variable)	metre (m). SI system base unit
		a	Radius (constant)	(m)
		L	Length	(m)
v			Volume; ac voltage	litre (L) (m^3); volt
		V	Velocity	m/s
		d	Diameter, thickness	(m)
		A	(Cross-sectional) Area	(m^2)
		f	Force	newton (N)
			Number of particles; relative mass	mol, dimensionless. Amount of substance of 6×10^{23} particles. SI system base unit
c			Concentration	(mol/L, %weight or volume of total)
η			Viscosity	(Pa s)
$q; F$			Charge; quantity of electricity	coulomb (C), (As) faraday = 96 500 coulomb/mol
q_v			Volume charge density	(C/m^3)
q_s			Surface charge density	(C/m^2)
e			Elementary charge	1.6×10^{-19} coulomb (C), + or −: proton or electron
z			Number of electrons for transfer (electrovalency)	dimensionless, + or −
n			Number of ions per volume	(1/L), + or −
N			Number of molecules per volume	(1/L)
μ			Ionic mobility	(m^2/Vs) ($\mu = v/E$)
	ε		Permittivity = $\varepsilon_r \varepsilon_0$	(F/m)
ε_0			Permittivity of vacuum	8.8×10^{-12} F/m
ε_s			Static permittivity	(F/m)
	ε_r		Relative permittivity (dielectric constant)	dimensionless
	M		Modulus function = $1/\varepsilon$	(m/F)
		E	Electric field strength	(volt/m)
		D	Electric flux density (displacement)	(C/m^2)
		P	Polarisation	(C m/m^3) or (C/m^2)
		p	Electric dipole moment	(C m) (debye (D) = 3.3×10^{-30} C m)
		m	Current dipole moment = iL	(A m)
	σ		Conductivity	(S/m)
	ρ		Resistivity	(Ω m)
C			Capacitance	farad (F)
C_p			Parallel capacitance	(F)
C_s			Series capacitance	(F)
t			Time	second (s). SI system base unit
τ			Time constant	(s)
f, v			Frequency	hertz (Hz, periods/s, 1/s)
ω			Angular frequency, $2\pi f$	(rad/s, 1/s)
m			Frequency exponent, f^m	dimensionless, $0 \le m \le 1$
φ			Phase angle	dimensionless, degrees (°)

(Continued)

Table 10.1. *Continued*

Symbol (in equations)				
Scalar only	Vector in time	Vector in space	Quantity	Unit (abbreviation) (dimension)
α			Phase angle coefficient and frequency exponent combined	dimensionless, $\alpha = \varphi/90°$, $0 \leq \alpha \leq 1$ For CPE: $\alpha = m$
δ			Loss angle	dimensionless, $\delta = 90° - \varphi$
G			Conductance	siemens (S), $1/\Omega$, mho, $\mho$
B			Susceptance	(S)
	Y		Admittance	(S)
R			Resistance (in series)	ohm (Ω)
X			Reactance	(Ω)
	Z		Impedance	(Ω)
I (dc)	i (ac)		Current	ampere (A). SI system base unit
		J	Current density	(A/m^2)
			Electromotive voltage (emv*)	(volt)
Φ			Potential (in space)	volt
V (dc)	v (ac)		Voltage (in circuit)	volt
$\hat{E}$			Energy, work	1 J = 1 watt-second (Ws); 1 eV = 1.6×10^{-19} J
W			Power = energy per second	watt = J/s
W_v			Volume power density	(watt/m^3)
T			Temperature difference from absolute zero	°C, K. 0 K = -273.16°C. SI system base unit
j			Imaginary unit	$\sqrt{-1}$
h			Planck's constant	6.6×10^{-34} J s
k			Boltzmann's constant	1.38×10^{-23} J/K

* Also called electromotance or electromotive *force* (emf). The use of emv in this book is limited to potentials caused by chemical or thermal energy, or induced in a circuit by a changing magnetic flux, causing a flow of current. Thus a potential across a resistor or a charged capacitor as ideal components is not termed emv. In the literature before 1940, emf was often used more generally for the voltage difference across a component. The use of the term "force" is related to eq. (3.2).

When the dimension is smaller than the order of 1 nm, the rules of quantum mechanics prevail. The concept of a single, well-defined particle must be abandoned and we know neither the exact dimension nor the position or velocity of each particle. Our picture of a single electron is more like an electron gas or cloud, with a certain probability of finding it at a given position. So it is also for the atoms of a gas. However, atoms in a solid evidently have more well-defined positions, e.g. in a crystal, but owing to thermal vibrations even here it is a question of the most probable position. Thermal (Brownian) movements are proportional to thermal energy and the Boltzmann factor kT, indicating an increasing uncertainty with temperature.

Our macro-world of experience, where the laws of Newtonian physics are valid, is based upon so many particles that the law of large numbers dominates. There is a gradual transition from single-particle or a few-particle quantum-mechanical laws to the classical laws governed by many particles in sum. *Tunnelling* is an example of a

Table 10.2. Dimensions of particles and other entities

Particle/entity	Dimensions	Remarks
Electron	10^{-15} ($= 1$ fm)	Gas-like probability
Nuclei, most atoms	10^{-14}	Gas-like probability
Proton	2×10^{-15}	Nucleus without electrons
Hydrogen atom	0.037 nm	Radius
Oxygen atom	0.066 nm	Radius
Nitrogen atom	0.07 nm	Radius
Ions: Na^+, O^{2-}, Cl^-	0.10; 0.13; 0.18 nm	Solid state
Water molecule, approx.	0.1 nm	
Electric double layer	0.1–10 nm	Strong/dilute solution
Amino acid	0.5 nm	Alanine
Small protein	1.5 nm	Myoglobin
Bilayer membrane, liposome	5 nm	
Membrane, human cell	7 nm	
Molecule, molecular weight 14 000	1.5 nm	Myoglobin
Molecule, molecular weight 500 000	50 nm	Myosin, length
Molecule, molecular weight 4 000 000	120 nm	DNA, double-helix length
Debye length	10 nm	Space charge region around an ion
Ribosome	15 nm	
Liposome sphere	25 nm	
Virus	20–200 nm	
Synaptic cleft	20 nm	
Mitochondria	0.1–5 μm	
Cell	2 μm	E. coli
Cell	20 μm	Liver
Cell	5 μm; 1 m	Nerve, diameter, length
Colloidal particles	1 nm–1 μm	Third dimension unlimited
Erythrocyte (red blood cell)	2; 8 μm	Thickness; diameter
Bacteria	0.2–2; 1–400 μm	Diameter; length
Electrode diffusion zone	0.01–0.5 mm	Stirred/unstirred

f (femto) $= 10^{-15}$; p (pico) $= 10^{-12}$; n (nano) $= 10^{-9}$; μ (micro) $= 10^{-6}$; m (milli) $= 10^{-3}$

quantum-mechanical effect violating macro-laws, where single electrons may cross an energy barrier over small distances of some hundred picometres, e.g. from a metal electrode to an ion in the solution or between redox centres on a protein molecule.

Defining the dimensions for a molecule may be difficult: is a whole crystal to be regarded as a macromolecule? Large organic molecules may form a helix or a double helix, with a total string length many times the external helix length. The dimensions may change as a function of, e.g., water content, and the whole molecule may be more or less rigid under stretching and torsion. The shape may vary from one moment to the next. Because molecules are continually rotating, their effective (apparent) volume is greater than their real volume. Two parameters are often used to specify dimensions: end-to-end distance and radius of gyration.

Of course *volume* is sometimes a more relevant quantity than a linear dimension. Weight and therefore density may also be of interest. In molecular biology the *dalton* (Da) unit is sometimes used: it is the same as the relative atomic weight (without unit), with the atomic weight of carbon-12 *defined* as 12 dalton.

References and Further Reading

Recommended books

Books, general topics

Gilbert W (1600/1958): *De magnete*. Dover, New York.

Maxwell JC (1873): *A treatise on electricity and magnetism*. 3rd edition (1891) Clarendon Press, Dover Publications (1954).

Kuo FF (1962): *Network analysis and synthesis*. Wiley International edition, New York.

Feynman RP, Leighton RB, Sands M (1965): *The Feynman lectures on physics*. Addison-Wesley, Reading, Massachusetts.

Castellan GW (1971): *Physical chemistry*. Addison-Wesley, Reading, Massachusetts.

Kreyszig E (1988): *Advanced engineering mathematics*. John Wiley, New York.

Lorrain P, Corson DP, Lorrain F (1988): *Electromagnetic fields and waves*. WH Freeman & Comp., New York.

CRC (1998): *Handbook of chemistry and physics*, 79th edn. CRC Press, Boca Raton, Florida.

Books, specialised topics

Debye P (1929/45): *Polar Molecules*. Dover, New York.

Rajewsky B (1938): *Ergebnisse der Biophysikalischen Forschung*. Georg Thieme, Leipzig.

Smyth CP (1955): *Dielectric behaviour and structure*. McGraw-Hill, New York.

Frölich H (1958): *Theory of dielectrics*. Oxford University Press, 2nd edn.

Plonsey R, Collin RE (1961): *Principles and applications of electromagnetic fields*. McGraw-Hill, New York.

Davis M (1965): *Some electrical and optical aspects of molecular behaviour*. Pergamon Press, Tarrytown, New York.

Tregear RT (1966): *Physical functions of skin*. Academic Press, New York.

Daniel VV (1967): *Dielectric relaxation*. Academic Press, New York.

Hill N, Vaughan WE, Price AH, Davies M (1969): *Dielectric properties and molecular behaviour*. Van Nostrand, New York.

Nyboer J (1970): *Electrical impedance plethysmography*. Charles Thomas.

Cole KS (1972): *Membranes, ions and impulses*. University of California Press, Berkeley.

Geddes LA (1972): *Electrodes and the measurement of bioelectric events*. Wiley-Interscience, New York.

Böttcher CJF (1973): *Theory of Electric Polarisation*. vol I: *Dielectrics in Static Fields*. 2nd edn. Elsevier, New York.

Hasted JB (1973): *Aqueous dielectrics*. Chapman and Hall, London.

Miller HA, Harrison DC (1974): *Biomedical electrode technology. Theory and practice*. Academic Press, New York.

Böttcher CJF, Bordewijk P (1978): *Theory of Electric Polarisation*. vol III: *Dielectrics in Time Dependent Fields*. 2nd edn. Elsevier, New York.

Gabler R (1978): *Electrical interactions in molecular biophysics: an introduction*. Academic Press, New York.

Grant EH, Sheppard RJ, South GP (1978): *Dielectric behaviour of biological molecules in solution*. Oxford University Press.

Schanne OF, Ruiz P, Ceretti E (1978): *Impedance measurements in biological cells*. John Wiley, New York.

Pethig R (1979): *Dielectric and electronic properties of biological materials*. John Wiley, Chichester.

Jonscher, AK (1983): *Dielectric relaxation in solids*. Chelsea Dielectrics Press.

Nordenstrøm, B (1983): *Biologically closed electric circuits*. Nordic Medical Publications, Stockholm.

Blaustein MP, Lieberman M (1984): *Electrogenic transport.* Raven Press, New York.
Polk C, Postow E (1986): *Handbook of biological effects of electromagnetic fields.* CRC Press, Boca Raton, Florida.
Macdonald JR (1987): *Impedance spectroscopy, emphasizing solid materials and systems.* John Wiley, New York.
Feder J (1988): *Fractals.* Plenum Press, New York.
Plonsey R, Barr RC (1988): *Bioelectricity. A quantitative approach.* Plenum Press, New York.
Geddes LA, Baker LE (1989): *Applied biomedical instrumentation.* Wiley Interscience, New York.
Neumann E, Sowers AE, Jordan CA (1989): *Electroporation and electrofusion in cell biology.* Plenum Press, New York.
Takashima S (1989): *Electrical properties of biopolymers and membranes.* Adam Hilger, Bristol.
Duck FA (1990): *Physical properties of tissue. A comprehensive reference book.* Academic Press, London.
Webster JG (1990): *Electrical impedance tomography.* Adam Hilger, Bristol.
Koryta J (1991): *Ions, electrodes and membranes.* John Wiley, Chichester.
Webster J (1992): *Medical instrumentation.* Houghton Mifflin, Dallas.
Bassingthwaighte JB, Liebovitch LS, West BJ (1994): *Fractal physiology.* Oxford University Press.
Elsner P, Berardesca E, Maibach H (1994): *Bioengineering of the skin: water and the stratum corneum.* CRC Press, Boca Raton, Florida.
Low J, Reed A (1994): *Electrotherapy explained. Principles and practice.* Butterworth-Heinemann Ltd, Oxford.
Craig DQM (1995): *Dielectric analysis of pharmaceutical systems.* Taylor & Francis, London.
Malmivuo J, Plonsey R (1995): *Bioelectromagnetism.* Oxford University Press, New York.
Thomasset A-L (1995): *Impédancemétrie bio-electrique. Principes et applications cliniques.* Meditions, Lyon.
Morucci J-P, Valentinuzzi ME, Rigaud B, Felice CJ, Chauveau N, Marsili P-M (1996): Bioelectrical impedance techniques in medicine. *Critical Rev in Biomed Eng*, **24**, 223–681. New York.
Akay M (1998): *Time frequency and wavelets in biomedical signal processing.* IEEE Press, New York.

References

Abramson HA, Gorin MH (1939): Skin reactions VII. Relationship of skin permeability to electrophoresis of biologically active materials into the living human skin. *J Phys Chem*, **43**, 335–346.
Abramson HA, Gorin MH (1940): Skin reactions IX. The electrophoretic demonstration of the patent pores of the living human skin; its relation to the charge of the skin. *J Phys Chem*, **44**, 1094–1102.
Ackmann JJ, Seitz MA (1984): Methods of complex impedance measurements in biological tissue. *CRC Crit Rev Biomed Eng*, 11(4), 281–311.
Akay M (1998): *Time frequency and wavelets in biomedical signal processing.* IEEE Press.
Alanen E, Lahtinen T, Nuutinen J (1998): Measurement of dielectric properties of subcutaneous fat with open-ended coaxial sensors. *Phys Med Biol*, **43**, 475–485.
Almasi JJ, Schmitt OH (1970): Systemic and random variations of ECG electrode system impedance. *Ann NY Acad Sci*, **170**, 509–519.
Amoussou-Guenou KM, Teyssier F, Squitiero B, Voutay M, Rusch Ph, Healy JC (1995): Second generation of cell transit analyzer. *Innov Technol Biol Med*, 16(5), 609–622.
Anfinsen O-G, Kongsgaard E, Foerster A, Aass H, Amlie JP (1998): Radio frequency current ablation of porcine right atrium: Increased lesion of bipolar two catheter technique compared to unipolar application in vitro and in vivo. *Pacing Clin Electrophys*, **21**, 69–78.
Arnold WM, Zimmerman U (1982): Rotating-field-induced rotation and measurement of the membrane capacitance of single mesophyll cells of *Avena sativa*. *Z Naturforsch*, **C37**, 908–915.
Baden HP (1970): The physical properties of nail. *J Invest Dermatol*, **55**, 115–122.
Barker AT, Jalinous R, Freeston IL (1985): Non invasive magnetic stimulation of the human motor cortex. *Lancet*, **1**: 1106–1107.
Barlow DJ, Thornton JM (1983): Ion pairs in proteins. *J Mol Biol*, **168**, 867–885.

Barnett A (1937): The basic factors involved in proposed electrical methods for measuring thyroid function. III The phase angle and the impedance of the skin. *Western J Surg Obstet Gynecol*, **45**, 540–554.

Bassingthwaighte JB, Liebovitch LS, West BJ (1994): *Fractal Physiology*. Oxford University Press.

Bauer HH (1964): Theory of Faradaic distortion: equation for the 2nd-harmonic current. *Australian J Chem*, **17**, 715.

Becker FF, Wang X-B, Huang Y, Pethig R, Vykoukal J, Gascoyne PRC (1994): The removal of human leukaemia cells from blood using interdigitated microelectrodes. *J Phys D: Appl Phys*, **27**, 2659–2662.

Becker FF, Wang X-B, Huang Y, Pethig R, Vykoukal J, Gascoyne PRC (1995): Separation of human breast cancer cells from blood by differential dielectric affinity. *Proc Natl Acad Sci USA*, **92**, 860–864.

Bickford RG, Fremming BD (1965): Neural stimulation by pulsed magnetic fields in animals and man. *Digest of the 6th International Conference on Medical Electronics and Biological Engineering, Tokyo*, Paper 7-6.

Blad B (1996): Clinical applications of characteristic frequency measurements: preliminary in vivo study. *Med Biol Eng Comput*, **34**(5), 362–365.

Block H, Hayes EF (1970): Dielectric behavior of stiff polymers in solution when subjected to high voltage gradients. *Trans Faraday Soc*, **66**, 2512–2525.

Bozler E, Cole KS (1935): Electric impedance and phase angle of muscle in rigor. *J Cell Comp Physiol*, **6**, 229–241.

Bragos R, Rosell J, Riu P (1994): A wide-band AC-coupled current source for electrical impedance tomography. *Physiol Meas*, **15**(suppl A), 91–99.

Brown BH, Barber DC, Wang W, Lu L, Leathard AD, Smallwood RH, Hampshire AR, Mackay R, Hatzigalanis K (1994a): Multi-frequency imaging and modelling of respiratory related electrical impedance changes. *Physiol Meas*, **15**(suppl), 1–12.

Brown BH, Barber DC, Leathard AD, Lu L, Wang W, Smallwood RH and Wilson AJ (1994b): High frequency EIT data collection and parametric imaging. *Innov Technol Biol Med*, **15**(1), 1–8.

Bull HB, Breese K (1969): Electrical conductance of protein solutions. *J Colloid Interface Sci*, **29**, 492.

Burt JPH, Pethig R, Talary MS (1998): Microelectrode devices for manipulating and analysing bioparticles. *Trans Inst Meas Control*, **20**(2), 82–90.

Casas O, Bragos R, Riu P, Rosell J, Tresanchez M, Warren M, Rodriguez-Sinovas A, Carreño A, Cinca J (1999): In-vivo and in-situ ischemic tissue characterisation using electrical impedance spectroscopy. *Ann NY Acad Sci*, **873**, 51–59.

Chilcott TC, Coster HGL (1991): AC impedance measurements on Chara Corallina. 1: Characterization of the static cytoplasm. *Australian J Plant Physiol*, **18**(2), 191–199.

Cole KS (1928a): Electrical impedance of suspension of spheres. *J Gen Physiol*, **12**, 29–36.

Cole KS (1928b): Electric impedance of suspensions of arbacia eggs. *J Gen Physiol*, **12**, 37–54.

Cole KS (1932): Electrical phase angle of cell membranes. *J Gen Physiol*, **15**, 641–649.

Cole KS (1934): Alternating current conductance and direct current excitation of nerve. *Science*, **79**, 164–165.

Cole KS (1940): Permeability and impermeability of cell membranes for ions. *Cold Spring Harbor Sympos Quant Biol*, **8**, 110–122.

Cole KS (1972): *Membranes, ions and impulses*. University of California Press.

Cole KS, Cole RH (1941): Dispersion and absorption in dielectrics. I. Alternating current characteristics. *J Chem Phys*, **9**, 341–351.

Cooper R (1946): The electrical properties of salt-water solutions over the frequency range 1–4000 Mc/s. *J Inst Electr Engineers*, **93**, 69–75.

Cornish BH, Thomas BJ, Ward LC (1993): Improved prediction of extracellular and total body water using impedance loci generated by multiple frequency bioelectrical impedance analysis. *Phys Med Biol*, **38**, 337.

Cornish BH, Jacobs A, Thomas BJ, Ward LC (1999): Optimising electrode sites for segmental bioimpedance measurements. *Physiol Meas* (in press).

CRC (1998): *Handbook of chemistry and physics*, 79th edn. CRC Press, Boca Raton, Florida.

Dalziel CF (1954): The threshold of perception currents. *AIEE Trans Power Appl Syst*, **73**, 990–996.

Dalziel CF (1972): Electric shock hazard. *IEEE Spectrum*, **9**, 41.

Daniel VV (1967): *Dielectric relaxation.* Academic Press, New York.

d'Arsoneval MA (1893): Production des courants de haute fréquence et de grand intensité; leurs effets physiologiques. *Comptes Rendus Soc de Biol,* **45,** 122–124.

Davidson DW, Cole RH (1951): Dispersion and absorption in dielectrics. *J Chem Phys,* **9,** 341–351.

Davis JM, Giddings JC (1986): Feasibility study of dielectric field-flow fractionation. *Sep Sci Technol,* **21,** 969–989.

Debye P (1929/45): *Polar molecules.* Dover, New York.

Dissado LA, Hill RM (1979): Non-exponential decay in dielectrics and dynamics of correlated systems. *Nature,* **279,** 685–689.

Duck FA (1990): *Physical properties of tissue. A comprehensive reference book.* Academic Press, New York.

Einolf CW, Carstensen EL (1973): Passive electrical properties of microorganisms. V Low frequency dielectric dispersion in bacteria. *Biophys J,* **13,** 8.

Emtestam L, Nyrén M (1997): Electrical impedance for quantification and classification of experimental skin reactions. *Am J Contact Dermatitis,* **8**(4), 202–206.

Emtestam L, Nicander I, Stenström M, Ollmar S (1998): Electrical impedance of nodular basal cell carcinoma: a pilot study. *Dermatology,* **197,** 313–316.

Epstein BR, Foster KR (1983): Anisotropy in the dielectric properties of skeletal muscles. *Med Biol Eng Comput,* **21,** 51.

Etter HS, Pudenz RH, Gersh I (1947): The effects of diathermy on tissues contiguous to implanted surgical materials. *Arch Phys Med Rehab,* **28,** 333–344.

Feder J (1988): *Fractals.* Plenum Press, New York.

Fink H-W, Schönenberger C (1999): Electrical conduction through DNA molecules. *Nature,* **398,** 407–410.

Forslind B (1970): Biophysical studies of the normal nail. *Acta Dermatol Venereol (Stockh),* **50,** 161–168.

Foster KR, Schwan HP (1986): Dielectric properties of tissue. In: Polk C, Postow E (eds) *CRC handbook of biological effects of electromagnetic fields.* Part I: Dielectric permittivity and electrical conductivity of biological materials. CRC Press, Boca Raton, Florida.

Foster KR, Schwan HP (1989): Dielectric properties of tissue. *CRC Crit Rev Biomed Eng,* **17,** 25–104.

Foster KR, Schepps JL, Stoy RD, Schwan HP (1979): Dielectric properties of brain tissue between 0.01 and 10 GHz. *Phys Med Biol,* **24,** 1177.

Foster KR, Epstein BR, Gealt MA (1987): "Resonances" in the dielectric absorption of DNA? *Biophys J,* **52,** 421–425.

Freiberger H (1933): Der elektrische Widerstand des menschlichen Körpers gegen technischen Gleich- und Wechselstrom. *Elektrizitätswissenshaft,* **32,** 373–375, 442–446.

Fricke H (1924): A mathematical treatment of the electrical conductivity and capacity of disperse systems. I. The electrical conductivity of a suspension of homogeneous spheroids. *Phys Rev,* **24,** 575–587.

Fricke H (1925): A mathematical treatment of the electrical conductivity and capacity of disperse systems. II. The capacity of a suspension of conducting spheroids surrounded by a non-conducting membrane for a current of low frequency. *Phys Rev,* **26,** 678–681.

Fricke H (1932): Theory of electrolytic polarisation. *Phil Mag,* **14,** 310–318.

Fricke H (1953): The Maxwell–Wagner dispersion in a suspension of ellipsoids. *J Phys Chem,* **57,** 934–937.

Fricke H (1955): The complex conductivity of a suspension of stratified particles of spherical cylindrical form. *J Phys Chem,* **59,** 168.

Fuhr G, Hagedorn R, Müller T, Benecke W, Wagner B, Gimsa J (1991): Asynchronous travelling-wave induced linear motion of living cells. *Stud Biophys,* **140,** 79–102.

Fuhr G, Zimmermann U, Shirley SG (1996): Cell motion in time-varying fields: principles and potential. In: Zimmermann U, Neil GA (eds) *Electromanipulation of Cells.* CRC Press, Boca Raton, Florida.

Gabor D (1946): Theory of communication. *JIEE,* **93,** 429–457.

Gabriel S, Lau RW, Gabriel C (1996): The dielectric properties of biological tissue: II. Measurements in the frequency range 10 Hz to 20 GHz. *Phys Med Biol,* **41**(11), 2251–2269.

Gabrielli C (1984): *Identification of electrochemical processes by frequency response analysis.* Solartron Technical report number 004/83.

Geddes LA (1972): *Electrodes and the measurement of bioelectric events.* Wiley-Interscience, New York.

Geddes LA, Baker LE (1966): The relationship between input impedance and electrode area in recording the ECG. *Med Biol Eng,* **4,** 439–450.

Geddes LA, Baker LE (1967): The specific resistance of biological material—A compendium of data for the biomedical engineer and physiologist. *Med Biol Eng,* **5,** 271–293.

Geddes LA, Baker LE (1989): *Applied Biomedical Instrumentation.* Wiley Interscience, New York.

Geddes LA, Valentinuzzi ME (1973): Temporal changes in electrode impedance while recording the electrocardiogram with "dry" electrodes. *Ann Biomed Eng,* **1,** 356–367.

Geddes LA, Tacker A, Cabler B, Kidder H, Gothard R (1975a): The impedance of electrodes used for ventricular defibrillation. *Med Instrum,* **9,** 177–178.

Geddes LA, Tacker A, Cabler B, Chapman R, Rivera R, Kidder H (1975b): The decrease in tranthoracic impedance during successive ventricular defibrillation trials. *Med Instrum,* **9,** 179–180.

Geddes LA, Tacker A, Schoenlein W, Minton M, Grubbs S, Wilcox P (1976): The prediction of the impedance of the thorax to defibrillating current. *Med Instrum,* **10,** 159–162.

Gencer NG, Ider YZ (1994): A comparative study of several exciting magnetic fields for induced current EIT. *Physiol Meas,* **15**(suppl 2A), 51–57.

Gersing E (1991): Measurement of electrical impedance in organs—measuring equipment for research and clinical applications. *Biomedizinische Technik,* **36**(1–2), 6–11.

Gersing E (1998): Impedance spectroscopy on living tissue for determination of the state of organs. *Bioelectrochem Bioenergetics,* **45**(2), 145–149.

Gersing E, Krüger W, Osypka M, Vanupel P (1995): Problems involved in temperature measurements using EIT. *Physiol Meas,* **16**(suppl 3A), 153–160.

Gheorghiu E (1996): Measuring living cells using dielectric spectroscopy. *Bioelectrochem Bioenergetics,* **40**(2), 133–139.

Gholizadeh G (1998): Human skin perception in open MR and perception related to electrostatic discharge. MSc thesis, Department of Physics, University of Oslo, Norway.

Giaever I, Keese CR (1993): A morphological biosensor for mammalian cells. *Nature,* **366,** 591–592.

Gibson LE, Cooke RE (1959): A test for concentration of electrolytes in sweat in cystic fibrosis of the pancreas utilising pilocarpine by iontophoresis. *Pediatrics,* **23,** 545–549.

Gordon DH (1975): Triboelectric interference in the ECG. *IEEE Trans Biomed Eng,* 252–255.

Gougerot L, Fourchet M (1972): La membrane de l'hematie est-elle un dielectrique parfait? *Ann Phys Biol Med,* **6,** 17–42.

Grahame DC (1952): Mathematical theory of the faradaic admittance. *J Electrochem Soc,* **99,** 370c–385c.

Grant EH (1965): The structure of water neighbouring proteins, peptides and amino acids as deduced from dielectric measurements. *Ann NY Acad Sci,* **125,** 418–427.

Greatbatch W (1967): Electrochemical polarization of physiological electrodes. *Med Res Eng,* **6,** 13–17.

Griffiths H, Ahmed A (1987): Applied potential tomography for non-invasive temperature mapping in hyperthermia. *Clin Phys Physiol Meas,* **8**(suppl A), 147–153.

Grimnes S (1982): Psychogalvanic reflex and changes in electrical parameters of dry skin. *Med Biol Eng Comput,* **20,** 734–740.

Grimnes S (1983a): Impedance measurement of individual skin surface electrodes. *Med Biol Eng Comput,* **21,** 750–755.

Grimnes S (1983b): Skin impedance and electro-osmosis in the human epidermis. *Med Biol Eng Comput,* **21,** 739–749.

Grimnes S (1983c): Dielectric breakdown of human skin in vivo. *Med Biol Eng Comput,* **21,** 379–381.

Grimnes S (1983d): Electrovibration, cutaneous sensation of microampere current. *Acta Physiol Scand,* **118,** 19–25.

Grimnes S (1984): Pathways of ionic flow through human skin in vivo. *Acta Dermatol Venereol (Stockh),* **64,** 93–98.

Grimnes S, Piltan H, Martinsen ØG, Gholizadeh G (1998): Threshold of perception of dc current in human skin—a function of current or current density? *Proc 10th Int Conf Electrical Bio-Impedance,* Barcelona. 139–142.

Grosse C, Foster KR (1987): Permittivity of a suspension of charged spherical particles in electrolyte solution. *J Phys Chem*, **91**, 3073.

Guy AW, Davidow S, Yang GY, Chou CK (1982): Determination of electric current distribution in animals and humans exposed to a uniform 60 Hz high intensity electric field. *Bioelectromagnetics*, **3**, 47.

Hacking I (1983): *Representing and intervening*. Cambridge University Press.

Halldorsson H, Ollmar S (1998): Signal analysis of non-invasive impedance spectra of transplanted kidneys in vivo. *Proc 10th Int Conf Electrical Bio-Impedance, Barcelona*, 351–354.

Hanai T (1960): Theory of the dielectric dispersion due to the interfacial polarization and its application to emulsion. *Kolloid Z*, **171**, 23–31.

Harris ND, Suggett AJ, Barber DC, Brown BH (1987): Applications of applied potential tomography (APT) in respiratory medicine. *Clin Phys Physiol Meas*, **8**(suppl A), 155–165.

Hasted JB (1973): *Aqueous dielectrics*. Chapman and Hall, London.

Havriliak S, Negami S (1966): A complex plane analysis of α-dispersions in some polymer systems. *J Polym Sci: Part C*, **14**, 99–117.

Hayakawa R, Kanda H, Sakamoto M, Wada Y (1975): New apparatus for measuring the complex dielectric constant of a highly conductive material. *Jpn J Appl Phys*, **14**, 2039–2052.

Hodgkin AL, Huxley AF (1952): A quantitative description of membrane current and its application to conductance and excitation in nerve. *J Physiol*, **117**, 500–544.

Hodgkin AL, Katz B (1949): The effect of sodium ions on the electrical activity of the giant axon of squid. *J Physiol*, **108**, 37–77.

Hoenig SA, Gildenberg PL, Murthy K (1978): Generation of permanent, dry, electric contacts by tattooing carbon into skin tissue. *IEEE Trans Biomed Eng*, **25**, 380–382.

Holder DS (1992): Detection of cortical spreading depression in the anesthetised rat by impedance measurement with scalp electrodes—implications for noninvasive imaging of the brain with electrical impedance tomography. *Clin Phys Physiol Meas*, **13**(1), 77–86.

Holder DS (1998): Electrical impedance tomography in epilepsy. *Electron Eng*, **70**(859), 69–70.

IEC-60601 (1988): *Medical electrical equipment. General requirements for safety.*

IEC-60601-1-1 (1992): *General requirements for safety. Safety requirements for medical electrical systems.*

Jaron D, Schwan HP, Geselowitz DB (1968): A mathematical model for the polarisation impedance of cardiac pacemaker electrodes. *Med Biol Eng*, **6**, 579–594.

Jaron D, Briller A, Schwan HP, Geselowitz DB (1969): Nonlinearity of cardiac pacemaker electrodes. *IEEE Trans Biomed Eng*, **16**, 132–138.

Jones P (1979): High electric field dielectric studies of aqueous myoglobin solutions. *Biophys Chem*, **9**, 91–95.

Jonscher AK (1974): Hopping losses in polarisable dielectric media. *Nature*, **250**, 191–193.

Jonscher AK (1977): The "universal" dielectric response. *Nature*, **267**, 673–679.

Jonscher AK (1983): *Dielectric relaxation in solids*. Chelsea Dielectrics Press.

Jossinet J (1996): Variability of impedivity in normal and pathological breast tissue. *Med Biol Eng Comput*, **34**(5), 346–350.

Jossinet J, McAdams ET (1991): The skin/electrode interface impedance. *Innov Tech Biol Med*, **12**(1), 22–31.

Jossinet J, Schmitt M (1998): Alternative parameters for the characterisation of breast tissue. *Proc 10th Int Conf Electrical Bio-Impedance, Barcelona*, 45–48.

Kirkwood JG (1939): The dielectric polarization of polar liquids. *J Chem Phys*, **7**, 911.

Knudsen V (1999): *Verification and use of a numerical computer program for simulations in bioimpedance*. MSc thesis, Dept of Physics, University of Oslo, Norway (in Norwegian).

Ko WH, Hynecek J (1974): Dry electrodes and electrode amplifiers. In: Miller HA, Harrison DC (eds) *Biomedical electrode technology. Theory and practice*. Academic Press, New York.

Kontturi K, Murtomäki L, Hirvonen J, Paronen P, Urtti A (1993): Electrochemical characterization of human skin by impedance spectroscopy. The effect of penetration enhancers. *Pharm Res*, **10**(3), 381–385.

Korjenevsky AV, Cherepenin VA (1998): Measuring system for induction tomography. *Proc 10th Int Conf Electrical Bio-Impedance, Barcelona*, 365–368.

Kramers HA (1926): Theory of dispersion in the X-ray region. *Physik Z*, **30**, 52.

Kronig RdeL (1929): The theory of dispersion of X-rays. *J Opt Soc Am*, **12**, 547.

Lahtinen T, Nuutinen J, Alanen E, Turunen M, Nuortio L, Usenius T, Hopewell JW (1999): Quantitative assessment of protein content in irradiated human skin. *Int J Radiat Oncol Biol Phys*, **43**, 635–638.

Lindholm-Sethson B, Han S, Ollmar S, Nicander I, Jonsson G, Lithner F, Bertheim U, Geladi P (1998): Multivariate analysis of skin impedance data in long term type 1 diabetic patients. *Chemometrics Intell Lab Syst*, **44**(1–2), 381–394.

Lindsey CP, Patterson GD (1980): Detailed comparison of the William–Watts and Cole–Davidson functions. *J Chem Phys*, **73**, 3348–3357.

Lionheart WRB (1997): Conformal uniqueness results in anisotropic electrical impedance imaging. *Inverse Problems*, **13**(1), 125–134.

Lozano A, Rosell J, Pallas-Areny R (1995): A multifrequency multichannel electrical impedance data aquisition system for body fluid monitoring. *Physiol Meas*, **16**, 227–237.

Lumry R, Yue RHS (1965): Dielectric dispersion of protein solutions containing small zwitterions. *J Phys Chem*, **69**, 1162–1174.

Macdonald JR (1987): *Impedance spectroscopy, emphasizing solid materials and systems*. John Wiley, Chichester.

Maleev VT, Kashpur VA, Glibitski GM, Krasnitskaya AA, Veretelnik YV (1987): Does DNA absorb microwave energy? *Biopolymers*, **26**, 1965–1970.

Malmivuo J, Plonsey R (1995): *Bioelectromagnetism*. Oxford University Press.

Mandel M (1977): Dielectric properties of charged linear macromolecules with particular reference to DNA. *Ann NY Acad Sci*, **303**, 74–87.

Mandel M, Jenard A (1963): Dielectric behavior of aqueous polyelectrolyte solution. *Trans Faraday Soc*, **59**, 2158–2177.

Mandel, M, Odijk, T (1984): Dielectric properties of polyelectrolyte solutions. *Annu Rev Phys Chem*, **35**, 75–108.

Mangnall YF, Baxter AJ, Avill R, Bird NC, Brown BH, Barber DC, Seagar AD, Johnson AG, Read NW (1987): Applied potential tomography: a new non-invasive technique for assessing gastric function. *Clin Phys Physiol Meas*, **8**(suppl A), 119–130.

Markx GH, Talary M, Pethig R (1994): Separation of viable and non-viable yeast using dielectrophoresis. *J Biotechnol*, **32**, 29–37.

Markx GH, Dyda PA, Pethig R (1996): Dielectrophoretic separation of bacteria using a conductivity gradient. *J Biotechnol*, **51**, 175–180.

Martinsen ØG, Grimnes S, Henriksen I, Karlsen J (1996): Measurement of the effect of topical liposome preparations by low frequency electrical susceptance. *Innov Technol Biol Med*, **17**(3), 217–222.

Martinsen ØG, Grimnes S, Sveen O (1997a): Dielectric properties of some keratinised tissues. Part 1: Stratum corneum and nail in situ. *Med Biol Eng Comput*, **35**, 172–176.

Martinsen ØG, Grimnes S, Nilsen S (1997c): Absolute water content and electrical admittance of human nail. *Exp Dermatol*, **6**(5), 264.

Martinsen ØG, Grimnes S, Karlsen J (1998): Low frequency dielectric dispersion of microporous membranes in electrolyte solution. *J Colloid Interface Sci*, **199**, 107–110.

Martinsen ØG, Grimnes S, Haug E (1999): Measuring depth depends on frequency in electrical skin impedance measurements. *Skin Res Technol*, **5**, 179–181.

Martinsen ØG, Grimnes S, Kongshaug ES (2000): Dielectric properties of some keratinised tissues. Part 2: Human hair. *Med Biol Eng Comput*, **35**, 177–180.

Martinsen ØG, Grimnes S, Mirtaheri P (2000): Non-invasive measurements of post mortem changes in dielectric properties of haddock muscle—a pilot study. *J Food Eng*, **43**, 189–192.

Masuda S, Washizu M, Iwadare M (1987): Separation of small particles suspended in liquid by nonuniform travelling field. *IEEE Trans Ind Appl*, **23**, 474–480.

Maxwell JC (1873): *Treatise on electricity and magnetism*. Oxford University Press.

McAdams ET, Jossinet J (1991a): DC nonlinearity of the solid electrode-electrolyte interface impedance. *Innov Technol Biol Med*, **12**, 330–343.

McAdams ET, Jossinet J (1991b): The impedance of electrode-skin impedance in high resolution electrocardiography. *Automedica*, **13**, 187–208.

McAdams ET, Jossinet J (1994): The detection of the onset of electrode-electrolyte interphase impedance nonlinearity: a theoretical study. *IEEE Trans Biomed Eng*, **41**(5), 498–500.

Metherall P, Barber DC, Smallwood RH, Brown BH (1996): Three dimensional electrical impedance tomography. *Nature*, **380**(6574), 509–512.

Min M, Parve T (1996): A current mode signal processing in lock-in instruments for bio-impedance measurement. *Med Biol Eng Comput*, **34**(suppl 1, part 2), 167–178.

Min M, Parve T (1997): A current signal processing as a challenge for improvement of lock-in measurement instruments. *Proc XIV IMEKO World Congress*, Vol. VII, 186–191.

Mørkrid L, Qiao ZG (1988): Continuous estimation of parameters in skin electrical admittance from simultaneous measurements at two different frequencies. *Med Biol Eng Comput*, **26**, 633–640.

Mørkrid L, Ohm O-J, Hammer E (1980): Signal source impedance of implanted pacemaker electrodes estimated from spectral ratio between loaded and unloaded electrograms in man. *Med Biol Eng Comput*, **18**, 223–232.

Morucci JP, Marsili PM, Granie M, Shi Y, Lei M, Dai WW (1994): A direct sensitivity matrix approach for fast reconstruction in electrical impedance tomography. *Physiol Meas*, **15**(suppl 2A), 107–114.

Moussavi M, Schwan HP, Sun HH (1994): Harmonic distortion caused by electrode polarisation. *Med Biol Eng Comput*, **32**, 121–125.

Munk H (1873): Über die galvanische Einführung differenter Flüssigkeiten in der unversehrten lebenden Organismus. *Arch Anat Physiol Wissens Med*, 505–516.

Murphy D, Burton P, Coombs R, Tarassenko L, Rolfe P (1987): Impedance imaging in the newborn. *Clin Phys Physiol Meas*, **8**(suppl A), 131–140.

Neuman RN (1992): Biopotential electrodes. In: Webster JG (ed.) *Medical instrumentation*. Houghton Mifflin, Boston.

Neumann E, Katchalsky A (1972): Long lived conformation changes induced by electric impulses in biopolymers. *Proc Natl Acad Sci USA*, **69**, 993–997.

Newell JC, Peng Y, Edic PM, Blue RS, Jain H, Newell RT (1998): Effect of electrode size on impedance images of two- and three-dimensional objects. *IEEE Trans Biomed Eng*, **45**(4), 531–534.

Nicander I (1998): *Electrical impedance related to experimentally induced changes of human skin and oral mucosa*. PhD thesis, Karolinska Institute, Stockholm.

Nicander I, Ollmar S, Eek A, Lundh Rozell B, Emtestam L (1996): Correlation of impedance response patterns to histological findings in irritant skin reactions induced by various surfactants. *Br J Dermatol*, **134**, 221–228.

NIH (1994): Bioelectrical impedance analysis in body composition measurements. *Nat Inst Health: Technol Assessment Conf Statement*, 1–37.

Nordenstrøm, Bjørn (1983): *Biologically closed electric circuits*. Nordic Medical Publications, Chaska, Minnesota.

Norlén L, Nicander I, Lundh Rozell B, Ollmar S, Forslind B (1999): Differences in human stratum corneum lipid content related to physical parameters of skin barrier function in vivo. *J Invest Dermatol*, **112**, 72–77.

Nuutinen J, Lahtinen T, Turunen M, Alanen E, Tenhunen M, Usenius T, Kolle R (1998): A dielectric method for measuring early and late reactions in irradiated human skin. *Radiother Oncol*, **47**, 249–254.

Øberg PÅ (1973): Magnetic stimulation of nerve tissue. *Med Biol Eng Comput*, **11**, 55–64.

Ollmar S (1997): Noninvasive monitoring of transplanted kidneys by impedance spectroscopy—a pilot study. *Med Biol Eng Comput*, **35**(suppl, Part 1), 336.

Ollmar S (1998): Methods for information extraction from impedance spectra of biological tissue, in particular skin and oral mucosa—a critical review and suggestions for the future. *Bioelectrochem Bioenergetics*, **45**, 157–160.

Ollmar S, Nicander I (1995): Information in multi frequency measurement of intact skin. *Innov Technol Biol Med*, **6**, 745–751.

Ollmar S, Eek A, Sundström F, Emtestam L (1995): Electrical impedance for estimation of irritation in oral mucosa and skin. *Med Prog Technol*, **21**, 29–37.

Onaral B, Schwan HP (1982): Linear and nonlinear properties of platinum electrode polarisation. Part I: Frequency dependence at very low frequencies. *Med Biol Eng Comput*, **20**, 299–306.

Onsager L (1934): Deviations from Ohm's law in weak electrolytes. *J Chem Phys*, **2**, 599–615.

Page CC, Moser CC, Chen X, Dutton L (1999): Natural engineering principles of electron tunnelling in biological oxidation reduction. *Nature*, **402**, 47–52.

Pauly H, Schwan HP (1959): Über die Impedanz einer Suspension von Kugelformigen Teilchen mit einer Schale. *Z Naturforsch*, **14b**, 125–131.

Pauly H, Schwan HP (1966): Dielectric properties and ion mobility in erythrocytes. *Biophys J*, **6**, 621.

Pethig R (1979): *Dielectric and electronic properties of biological materials*. John Wiley, Chichester.

Pethig R, Kell DB (1989): The passive electrical properties of biological systems: their significance in biology, biophysics and biotechnology. *Phys Med Biol*, **32**, 933–970.

Pliquett U, Weaver JC (1996): Electroporation of human skin: simultaneous measurement of changes in the transport of two fluorescent molecules and in the passive electrical properties. *Bioelectrochem Bioenergetics*, **39**, 1–12.

Plonsey R, Barr R (1982): The four-electrode resistivity technique as applied to cardiac muscle. *IEEE Trans Biomed Eng*, **29**(7), 541–546.

Pohl HA (1958): Some effects of nonuniform fields on dielectrics. *J Appl Phys*, **29**, 1182–1188.

Polk C, Postow E (1986): *Handbook of biological effects of electromagnetic fields*. CRC Press, Boca Raton, Florida.

Qiao ZG, Mørkrid L, Grimnes S (1987): Three-electrode method to study event-related responses in skin electrical potential, admittance and blood flow. *Med Biol Eng Comput*, **25**, 567–572.

Qu M, Zhang Y, Webster JG, Tompkins WJ (1986): Motion artefact from spot and band electrodes during impedance cardiography. *IEEE Trans Biomed Eng*, **33**(11), 1029–1036.

Riu P, Rosell J, Lozano A, Pallas-Areny R (1992): A broadband system for static imaging in electrical impedance tomography. *Clin Phys Physiol Meas*, **13**(suppl A), 61–66.

Riu P, Rosell J, Lozano A, Pallas-Areny R (1995): Multifrequency static imaging in electrical impedance tomography. Part 1: Instrumentation requirements. *Med Biol Eng Comput*, **33**, 784–792.

Robbins CR (1979): *Chemical and physical behaviour of human hair*. Van Nostrand Reinhold, New York.

Rosell J, Riu P (1992): Common-mode feedback in electrical impedance tomography. *Clin Phys Physiol Meas*, **13**(suppl A), 11–14.

Rosell J, Colominas J, Riu P, Pallas-Areny R, Webster JG (1988a): Skin impedance from 1 Hz to 1 MHz. *IEEE Trans Biomed Eng*, **35**(8), 649–651.

Rosell J, Murphy D, Pallas-Areny R, Rolfe P (1988b): Analysis and assessment of errors in a parallel data acquisition system for electrical impedance tomography. *Clin Phys Physiol Meas*, **9**(suppl A), 93–100.

Rosen D (1963): Dielectric properties of protein powders with adsorbed water. *Trans Faraday Soc*, **59**, 2178–2191.

Rosendal T (1940): *The conducting properties of the human organism to alternating currents*. Thesis, Munksgaard, Copenhagen.

Salter DC (1979): Quantifying skin disease and healing in vivo using electrical impedance measurements. In: Rolfe P (ed.) *Non-invasive physiological measurements*, vol. 1. Academic Press, New York.

Salter DC (1981): *Studies in the measurement, form and interpretation of some electrical properties of normal and pathological skin in vivo*. PhD thesis, University of Oxford.

Salter DC (1998): Examination of stratum corneum hydration state by electrical methods. In: Elsner *et al.* (eds) *Skin bioengineering. Techniques and applications in dermatology and cosmetology*. Karger, Farmington, Connecticut.

Scaife JM, Tozer RC, Freeston IL (1994): Conductivity and permittivity images from an induced current electrical impedance tomography system. *IEE Proc—Sci Meas Technol*, **141**(5), 356–362.

Schäfer M, Schlegel C, Kirlum H-J, Gersing E, Gebhard MM (1998): Monitoring of damage to skeletal muscle tissue caused by ischemia. *Bioelectrochem Bioenergetics*, **45**, 151–155.

Scharfetter H, Hartinger P, Hinghofer Szalkay H, Hutten H (1998): A model of artefacts produced by stray capacitance during whole body or segmental bioimpedance spectroscopy. *Physiol Meas*, **19**(2), 247–261.

Schnelle T, Mueller T, Fiedler S, Shirley SG, Ludwig K, Herrmann A, Fuhr G, Wagner B, Zimmermann U (1996): Trapping of viruses in high-frequency electric field cages. *Naturwissenschaften*, **83**, 172–176.

Schwan HP (1954): Die elektrischen Eigenschaften von Muskelgewebe bei Niederfrequenz. *Z Naturforsch*, **9b**, 245.

Schwan HP (1957): Electrical properties of tissue and cell suspensions. In: Lawrence JH, Tobias CA (eds) *Advances in biological and medical physics*, Vol V, 147–209. Academic Press, New York.

Schwan HP (1963): Determination of biological impedances. In: Nastuk WL (ed.) *Physical techniques in biological research*, Vol 6, 323–406. Academic Press, New York.

Schwan HP (1982): Nonthermal cellular effects of electromagnetic fields: ac-field induced ponderomotoric forces. *Br J Cancer*, **43**(suppl 5), 220–224.

Schwan HP (1985): Dielectric properties of the cell surface and biological systems. *Studia Biophysica*, **110**, 13–18.

Schwan HP (1992a): Early history of bioelectromagnetics. *Bioelectromagnetics J*, **13**, 453–467.

Schwan HP (1992b): Linear and non-linear electrode polarisation and biological materials. *Ann Biomed Eng*, **20**, 269–288.

Schwan HP (1993): Early organizations of biomedical engineering in the US. *IEEE Eng in Med Biol Mag*, Sept, 25–29.

Schwan HP, Ferris CD (1968): Four-electrode null techniques for impedance measurement with high resolution. *Rev Sci Instrum*, **39**(4), 481–485.

Schwan HP, Foster KR (1980): RF-field interactions with biological systems: Electrical properties and biophysical mechanism. *Proc IEEE*, **68**(1), 104–113.

Schwan HP, Kay CF (1957): The conductivity of living tissue. *Ann NY Acad Sci*, **65**, 1007.

Schwan HP, Morowitz HJ (1962): Electrical properties of the membranes of the pleuro-pneumonia-like organism A5969. *Biophys J*, **2**, 295.

Schwan HP, Sher LD (1969): Electrostatic fields induced forces and their biological implications. In: Pohl HA, Pickard WF (eds) *Dielectrophoretic and electrophoretic deposition*, 107–126, The Electrochemical Society, Pennington, New Jersey.

Schwan HP, Schwartz G, Maczuk J, Pauly H (1962): On the low frequency dielectric dispersion of colloidal particles in electrolyte solution. *J Phys Chem*, **66**, 2626–2636.

Schwan HP, Takashima S, Miyamoto VK, Stoeckenius W (1970): Electrical properties of phospholipid vesicles. *Biophys J*, **10**, 1102.

Schwartz G (1962): A theory of the low frequency dispersion of colloidal particles in electrolyte solution. *J Phys Chem*, **66**, 2636.

Schwartz G (1967): On dielectric relaxation due to chemical rate processes. *J Phys Chem*, **71**, 4021–4030.

Schwartz G (1972): Dielectric relaxation of biopolymers in solution. *Adv Mol Relaxation Processes*, **3**, 281.

Shankar TMR, Webster JG, Shao S-Y (1985): The contribution of vessel volume change and blood resistivity change to the electrical impedance pulse. *IEEE Trans Biomed Eng*, **32**(3), 192–198.

Slager CJ, Schuurbiers JCH, Oomen JAF, Bom N (1993): Electrical nerve and muscle stimulation by radio frequency surgery: role of direct current loops around the active electrode. *IEEE Trans Biomed Eng*, **40**, 182–187.

Smallwood RH, Mangnall YF, Leathard AD (1994): Transport of gastric contents. *Physiol Meas*, **15**(suppl 2A), 175–188.

Smith SR, Foster KR (1985): Dielectric properties of low-water content tissues. *Phys Med Biol*, **30**, 965.

South GP, Grant EH (1973): The contribution of proton fluctuation to dielectric relaxation in protein solutions. *Biopolymers*, **12**, 1937–1944.

Stoy RD, Foster KR, Schwan HP (1982): Dielectric properties of mammalian tissues from 0.1 to 100 MHz. A summary of recent data. *Phys Medl Biol*, **27**, 501–513.

Stuchly MA, Stuchly SS (1990): Electrical properties of biological substances. In: Gandhi OP (ed.) *Biological effects and medical applications of electromagnetic energy*. Prentice Hall, Englewood Cliffs, New Jersey.

Swanson DK, Webster JG (1983): Errors in four-electrode impedance plethysmography. *Med Biol Eng Comput*, **21**, 674–680.

Takashima S (1962): Dielectric dispersion of protein solutions in viscous solvents. *J Polymer Sci*, **56**, 257–265.

Takashima S (1967): Effect of ions on the dielectric relaxation of DNA. *Biopolymers*, **5**, 899–913.

Takashima S (1989): *Electrical properties of biopolymers and membranes*. Adam Hilger.

Takashima S, Schwan HP (1965): Dielectric dispersion of crystalline powders of amino acids, peptides and proteins. *J Phys Chem*, **69**, 4176–4182.

Takashima S, Schwan HP (1974): Passive electrical properties of the squid axon membrane. *J Membr Biol*, **17**, 51–68.

Takashima S, Yantorno RE (1977): Investigation of the voltage dependent membrane capacity of squid axon. *Ann NY Acad Sci*, **303**, 306–321.

Talary MS, Burt JPH, Tame JA, Pethig R (1996): Electromanipulation and separation of cells using travelling electric fields. *J Phys D: Appl Phys*, **29**, 2198–2203.

Therkildsen P, Hædersdal M, Lock-Andersen J, Olivarisu FdF, Poulsen T, Wulf HC (1998): Epidermal thickness measured by light microscopy: a methodological study. *Skin Res Technol*, **4**, 174–179.

Thomasset AL (1965): Mesure du volume des liquides extra-cellulaires par la methode electro-chimique signification biophysique de l'impedance a 1 kilocycle du corps humain. *Lyon Med*, **214**, 131–143.

Tozer JC, Ireland EH, Barber DC, Barker AT (1998): Magnetic impedance tomography. *Proc 10 Int Conf Electrical Bio-Impedance, Barcelona*, 369–372.

Tregear RT (1965): Interpretation of skin impedance measurements. *Nature*, **205**, 600–601.

Tregear RT (1966): *Physical functions of skin*. Academic Press, New York.

Ugland OM (1967): Electrical burns. *Scand J Plastic Reconstruct Surg*, (suppl 2), 1–74.

Venables PH, Christie MJ (1980): Electrodermal activity. In: Martin I, Venables PH (eds) *Techniques in psychophysiology*. John Wiley, Chichester.

Vistnes AI, Wormald I, Isachsen S, Schmalbein D (1984): An efficient digital phase-sensitive detector for use in electron-spin-resonance spectroscopy. *Rev Sci Instrum*, **55**, 527–532.

Wada A, Nakamura H (1981): Nature of charge distribution in proteins. *Nature*, **293**, 757–758.

Wagner W (1914): Explanation of the dielectric fatigue phenomena on the basis of Maxwell's concept. *Arch ElektroTechnol (Berlin)*, **2**, 271.

Wang X-B, Huang Y, Hölzel R, Burt JPM, Pethig R (1993): Theoretical and experimental investigations of the interdependence of the dielectric, dielectrophoretic and electrorotational behaviour of colloidal particles. *J Phys D: Appl Phys*, **26**, 312–322.

Warburg E (1899): Über das Verhalten sogenannte unpolarisierbare Elektroden gegen Wechselstrom. *Ann Phys Chem*, **67**, 493–499.

Washizu M, Kurosawa O (1990): Electrostatic manipulation of DNA in microfabricated structures. *IEEE Trans Ind Appl*, **26**, 11657–11672.

Washizu M, Suzuki S, Kurosawa O, Nishizaka T, Shinohara T (1994): Molecular dielectrophoresis of biopolymers. *IEEE Trans Ind Appl*, **30**, 835–843.

Webster J (1992): *Medical instrumentation*. Houghton Mifflin, Boston.

Wien M (1928): Über die Abweichungen der Elektrolyte vom Ohmischen Gesetz. *Phys Z*, **29**, 751–755.

Wien M (1931): Über Leitfähigkeit und Dielektrizität Konstante von Elektrolyten bei Hochfrequenz. *Phys Z*, **32**, 545–547.

Williams G, Watts DC (1970): Non-symmetrical dielectric relaxation behavior arising from a simple empirical decay function. *Trans Faraday Soc*, **66**, 80–85.

Yamamoto T, Yamamoto Y (1976): Electrical properties of the epidermal stratum corneum. *Med Biol Eng*, **14**, 592–594.

Yamamoto T, Yamamoto Y (1981): Non-linear electrical properties of the skin in the low frequency range. *Med Biol Eng Comput*, **19**, 302–310.

Yamamoto Y, Yamamoto T, Ozawa T (1986): Characteristics of skin admittance for dry electrodes and the measurement of skin moisturisation. *Med Biol Eng Comput*, **24**, 71–77.

Yelamos D, Casas O, Bragos R, Rosell J (1999): Improvement of a front end for bioimpedance spectroscopy. *Ann NY Acad Sci*, **873**, 306–312.

Zhang MIN, Repo T, Willison JHM, Sutinen S (1995): Electrical impedance analysis in plant tissues: on the biological meaning of Cole–Cole α in Scots pine needles. *Eur Biophys J*, **24**, 99–106.

Zhou X-F, Markx GH, Pethig R, Eastwood IM (1995): Differentiation of viable and non-viable bacterial biofilms using electrorotation. *Biochim Biophys Acta*, **1245**, 85–93.

Zoll PM, Linenthal AJ (1964): External electrical stimulation of the heart. *Ann NY Acad Sci*, Vol 111, 932–937.

Index

α parameter 223–224
α-dispersion 83
β-dispersion 83
γ-dispersion 83
δ-dispersion 83

3D EIT 150

ablation 290
ac (perception) 302–303
ac 323
access resistance 218
accommodation effect 280
acne vulgaris 112
action potential 96–97, 101–103
activation energy 44
active electrode 128
activity coefficient 12
acupuncture 277–278
adiabatic conditions 73, 131
adipose tissue 103–104
admittance
admittance xiii, 59, 153–154, 233, 323
    specific 63
    transfer 155
admittivity 63
adsorbed counterions 77–79
adsorption isotherm 28
adsorption, specific 30
after-field effect 292
Ag/AgCl electrode 8–11, 41, 250
AgCl wet gel electrode 259–261
aliasing 173, 192
amino acids 89–93
ampereometry 40
amplifier
    instrumentational 176–177, 180
    operational 176–178
    transconductance 192
    transresistance 155, 176–177
amplitude 160
amplitude modulation 188
amplitude spectrum 184
amplitude-modulated signal 165
anaesthetic agents 279
analogue lock-in amplifier 190–192

analyser
    Fourier 193
    frequency response 192
    impedance 192–193
    network 193
    spectrum 193
anatomical imaging 149
anelectrotonus 279
angular frequency 160, 324
anion 8, 12
anisotropy 99–101, 120
anode 8, 37
aperiodic waveform 167–169
applications 241–312
applied part 179, 309
Argand diagram xiii, 322
Argand, Jean Robert 322
Arrhenius, Svante August 11
Arrhenius theory of dissociation 11
associative memory network 183
atom relative mass 14
atomic radius 4
attractor
    fractal 182
    strange 182
auto-balancing bridge 192
autocorrelation 173
averaging 173

baby wavelet 186
back-projection 151
back-propagation algorithm 183
band electrode 263
bandwidth 171, 189
basic electrolytic experiment 7–11
basic membrane experiment 79–81
basic suspension experiment 81–83
beat frequency 163, 282
bilayer lipid membrane (BLM) 94–95, 290
bioelectricity 1
bioimmittance 1
bioimpedance 1
biomaterials 87
Biot, Jean-Baptiste 150
Biot–Savart law 150
bipolar 128

bispectral analysis 174
black box 153–159, 175–176
  non-linear 157
Blackman function 172
blood 104–105
Bode diagram 206–207, 235
Bode, Hendrik W. 235
body
  (type B applied part) 309–312
  composition 282–284
  floating (BF) 179, 309–312
  fluid balance 282
  fluids 87–88
  forces 32
  liquid electrolytes 19
  position 283
  segment 115–116
  water 282
Boltzmann equation 27
Boltzmann factor 38
bolus 273
bonds
  covalent 6, 89
  ionic 6
  metallic 6
  molecular 6
  Van der Waals 6
bone fracture growth 122
bone tissue 103–104
bootstrapping 177
bound charge 202
bound vector 321
bound water 88
boundary value problem 147
brain electroconvulsion 285–286
brain tissue 101–104, 116
breakdown 122–124
bridge 187–188
  auto-balancing 192
Brownian motion 38, 182
Butler–Volmer equation 47

calculated measured data 232
capacitive coupling 150
capacitor 154
carbohydrate 93
carbon electrode 8–11, 251
cardiac floating (CF) 179, 309–312
cardiac pacing 284–285
cardiac tissue 116
carrier 165
catelectrotonus 279
cathode 8, 37
cation 8, 12

Caton Richard 315
causal network 156
cell 94–99
  attachment 293–294
  characterisation 292–293
  constant 267
  electrolytic 7
  membrane 94–98
  membrane (excitable) 96–99
  sorting 292–293
  suspension 77, 99, 290–294
  swelling 116
cell galvanic 7
cellular patch clamp 266
cellular spin resonance 34
CGS units xiii, 202
chaos theory 182–183
characteristic frequency 66, 69, 236
charge
  bound 51, 202
  free 51, 202
chemisorption 30
chord resistance 96
chronaxie 280
circular arc 230
class I equipment 179, 309
class II equipment 179, 309
Clausius–Mossotti equation 56
CMRR 180
coagulation 31
coaxial probe 297
Cole
  brothers 318
  compatibility 239
  complete parallel system 220
  complete series system 218
  equations 213–214, 216–221
  parallel element 218
  series element 220
  system 83, 227–232
  systems, multiple 227–232
  Kenneth S. 316–319
  Robert H. 318–319
Cole–Cole
  equation 221–222
  system 83
colloidal electrolyte 18
colloidal particle 18
common mode rejection ratio (CMRR) 180
common mode voltage 178–182
common mode feedback 151
complex conductivity 61
complex number 321–325
complex permittivity 59–63

complex resistivity 61
concentration, electrolytic 14
concentration wave 48
concentric ring electrode 264–265
conductance 11–22, 323
  equivalent 13
  membrane 97
conducting polymer 251
conduction band 20
conductivity 11–22
  complex 61
  ionic 12–15
  molar 13–14
  of weak acids 16
  temperature dependence 17
  tissue 119
conductors, mixed 19
conformal mapping 53
constant phase element (CPE) 209–212
constrictional resistance 115
constrictional zone 131
contact electrolyte 111, 252–254
contact medium 262
continuous wavelet transform 185
convolution 173
corneo-retinal potentials 276
corona discharge 243
correlation 173
Coulomb's law 53, 202
coulombic forces 6
Coulter counter 294
counterion 91–92
  diffusion 78–79
  effect 84
  polarisation 29
  adsorbed 77–79
covalent bond length 7
covalent bond 6–7, 89
cross-correlation 173
Curie–von Schweider law 216
curl 201
current clamp 174
current density under disk electrode 133
current
  constrictional zone 131
  diadynamic 281
  diffusion controlled 41
  displacement 202, 242
  eddy 248–249
  electronic 3
  faradaic 9
  faradic 281
  galvanic 159
  induced 248–249

ionic 3
  paths 100
current mode lock-in amplifier 191–192
current-carrying electrode 127–128
current-injecting electrode 127–128
cut mode (electrosurgery) 287
cyclic voltammetry 42–43

data 195–239
  presentation 232–239
Davidson–Cole model 118
dc 120, 159, 254–257, 278–280, 300
  ablation 279
  conductance 11–22
  conductivity electronic 21
  conductivity ionic 21
  hazards 304
  perception 301–302
  shock 279
De Moivre's formula 323
death process 115–117
Debye
  equation 56
  length 11, 27
  Peter Joseph 11, 317
  relaxation model 64–70
  unit 54
Debye–Falkenhagen effect 17
Debye–Hückel
  approximation 27
  theory 49
defibrillation 49, 123, 285–287
demand pacemaker 284
denaturation 91
dental galvanism 121
derived data 232
dermatitis 296–297
descriptive model 195–201
diadynamic current 281
diagnostic methods 241
diagnostic radiography 247
diagram
  Argand xiii, 322
  Wessel xiii, 67, 206–208, 322
dialysis 31
diathermy 245–246, 287–290
dielectric constant 54, 59
dielectric decrement 59
dielectric increment 90
dielectric loss 59, 71
dielectric spectroscopy 83–85
dielectrics 51–85
dielectrophoresis 32–33, 292–293
  travelling wave 32 34

differential input 180
diffuse electric layer 26
diffusion 24–25, 47
    controlled current 41
    ion 77–79
    counterion 78–79
digital lock-in amplifier 188–190
dipolar 128
    ion 89
dipole 11, 136–143
    equipotential lines 137
    induced 55
    moment 53–56
    source 129
discharge, electrostatic 124–125
discrete parameter wavelet transform 185
discrete spectrum 165
discrete time wavelet transform 185
discrete wavelet transform 186
disk electrode 131
    current density 131
dispersion 63–64, 83–85, 229
    α, β, γ and δ 83
    magnitudes 101
dispersive electrode 128
displacement 54
displacement current 202, 242
dissipation factor 61
dissociation 6, 11
distribution of relaxation times (DRT) 223
divergence 201
DNA molecule 92, 291
Donnan potential difference 39
double insulation 179
double layer
    thickness 27
    electric 25
driving electrode 127–128
driving point immittance 127, 154
Du Bois-Raymond 315
dynamic cylinder model (plethysmography)
    273

ECG 268–270, 305
ECT 287
eddy current 248–249
EDR 121, 297–299
EEG 275
effective electrode area 258, 263
EGG 276
Einthoven
    triangle 269
    Willem 268, 315
EIT 149–152

electret 55
electric
    arc 254
    double-layer 25
    fence 309
electric field 242
    rotational 55
    static 53
electric flux density 54
electrical impedance tomography (EIT) 149–
    152
electrical safety of electromedical equipment
    309–312
electroacupuncture 278
electrocardiography (ECG) 268–270, 305
electrocautery 287
electrochemical membrane 95
electroconversion 285
electroconvulsion 285–286
electroconvulsive therapy (ECT) 287
electrocorticography 275
electrocution 309
electrode
    active 128
    Ag/AgCl 8–11, 41, 250
    AgCl wet gel 259–261
    area 132, 258
    band 263
    carbon 8–11, 251
    concentric ring 264–265
    current density under disk 131
    current-carrying 127–128
    current-injecting 127–128
    design 242–268
    disk 131
    dispersive 128
    driving 127–128
    equivalent circuit 44–45
    gel 111
    half-sphere 130
    hydrogel/aluminium 261–262
    indifferent 128
    insulated 244–245
    invasive 121
    invasive needle 265
    lead 127–128
    linear region 49
    micro 265
    multiple-point 264
    needle 135
    neutral 128
    nickel 251
    non-polarisable 41
    passive 128

pH 18
pick-up 127–128
platinum 8–11, 40, 250–251
platinum black 251, 258–259
platinum/hydrogen 256
polarisable 43
polarisation 45, 48
polarisation immittance 48
polarisation impedance 129, 257
processes 49–50
properties 242–268
reaction 36–37, 46
receiving 127–128
recessed 145–146
recording 127–128
reference 44, 256–257
registering 127–128
series resistance 262
silent 128
silver 250
skin surface 262–263
source 127–128
stainless steel 251
stimulating 127–128
titanium 251
VHF/UHF 268
working 128
electrodermal response (EDR) 121, 297–299
electrodialysis 31
electrodics 35–50
electrodiffusion 23
electroencephalography (EEG) 275
electrofusion 290–292
electrogastrography (EGG) 276
electrogenic pump 95–96
electrogenic transport 23
electrogram 128
electrogustometry 302
electrokinesis 31–35
electrolithotrity 279
electrolysis 4
electrolyte 3, 18
    colloidal 18
    contact 111, 252–254
    fused 18
    indifferent 45
    solid 18
electrolytes, body liquid 19
electrolytic
    cell 7
    components xiii
    concentration 14
    effects 307

electrolytics 3–51
    non-linear properties 48–50
electromagnetic field effects 304
electromagnetic wave 242, 246
    penetration depth 247
electromedical equipment, safety 179, 309–312
electromotive voltage (emv) 176
electromyography (EMG) 276–277
electron
    affinity 6
    shell configuration 4
    transfer 25, 44
    transfer resistance 46
    valence 5
    migration velocity 3
electronegativity 5
electroneurography (ENeG) 277
electroneutrality 13
electronic
    current 3
    dc conductivity 21
    polarisation 56
electronystagmography (ENG) 276
electro-oculography (EOG) 121, 276
electro-osmosis 31–32, 300
electroparacentesis 280
electrophoresis 31–32
electroporation 290–292
electroretinography (ERG) 276
electrorotation 32, 34, 292–293
electroshock 285–287
electrostatic
    discharge 124–125, 303
    theory 136
electrostriction 124
electrosurgery 124, 287–290
electrotherapy 277–282
electrotonus 279, 281
electrovalency 4–5
electrovibration 302
EMG 276–277
endoelectrogenic source 121
endogenic source 127
endorphins 277
endosomatic measurements 298
ENeG 277
energy activation 44
energy barrier 20
ENG 276
enhancers 199
EOG 121, 276
epicardial mapping 268
epithelia 105

equations 201–232
equilibrium potential (electrode) 36
equipotential line 128
equipotential lines of dipole 137
equivalent
    circuit equations 325–331
    circuit 153, 195–232
    conductance 13
    model, parallel 61
    model, series 61
ERG 276
erythrocytes 100, 104
    lysed 99, 104
Euler's formula 323
event related signal 174
evoked response 174, 298
excitable cell membrane 96–99
excitation 305
    step 157, 168
excitatory effect 279
excited nerves and muscles 122
exogenic source 127
exosomatic measurements 298
explanatory model 195–201

far field 203
faradaic
    current 9, 23
    impedance 46
Faraday
    cage 178
    Michael 22, 314
    stimulation 315
    law of electrolysis 22
    law of induction 202, 248
faradic current 23, 281
faradisation 281
far-field 243, 246
fast Fourier transform (FFT) 170
fat mass 282
feedforward network 183
Fermi–Dirac statistics 21
fibrosis 296–297
Fick's first law 23
Fick's second law 24
filed-flow fractionation 293
filter, high/low-pass 177–178
finite differences 147
finite element method (FEM) 105, 120, 147–
    149
fish muscle 117
fitted arc 230
fixed vector 321
floating applied part 179

flow-generated potential 32–33
fluid balance 282
flux density 12
    electric 54
flux
    scalar 12
    vector 12
force
    body 32
    coulombic 6
    ponderomotive 32
forward problem 129
four-component model 214–216
four-electrode system 145–147
Fourier
    analyser 193
    series 160–167
    transform 170–174
    Joseph 170
fractal 182–183
    attractor 182
    dimension 182
Franklin
    currents 314
    Benjamin 313
free charge 202
frequency 160
    domain 170
    filtering 177–178
    folding 173
    response analyser 192
    spectrum 162
Fricke
    compatibility 222–223, 234, 239
    Hugo 317
Fricke's law 211–213
frog sartorius muscle 116
fulguration 287
functional
    examination 241
    grounding 179
fused electrolyte 18

gain-phase measurements 192
Galvani, Luigi 314
galvanic
    cell 7
    contact 242–250
    current 159
    separation 179–180, 310
    skin response (GSR) 121, 298
galvanisation 278
galvanometer 315
gap junctions 105, 116

gas
    bubbles 8–11
    noble 4
gating theory 277
Gauss law 202
geometrical
    analysis 127–152
    scattering 247
Gilbert, William 313
Gildemeister 316
global symbols 332
glycocalyx 95
Gouy layer 28
Gouy–Chapman's model 26
gradient 201, 321
Grotthuss mechanism 15
ground 178–182
grounding
    functional 179
    safety 179
GSR 121, 298
guard ring 267
Guericke, Otto von 313
Guldberg, Cato Maximilian 16
Guldberg–Waage law 16
gyromagnetic ratio 70

haemoglobin 104
hair 112–114
    follicles 112
half-cell electrode potential 7
half-sphere electrode 130
Hanai's equation 76
Hanning function 172
harmonic analysis 50
harmonics 160
Hauksbee, Francis 313
Havriliak–Negami 118
hazards 301–312
    threshold levels 307
heart vector 269
heating effect 306–307
Helmholtz
    layer 26, 28
    model 26
    Hermann von 315
Henderson's equation 39
heterodyning 193
high-pass filter 177–178
His
    Jr., Wilhelm 270
    bundle 268
history of our field 313–319
Hoeber, Rudolf 316

hopping mechanism 15
Hurst
    exponent 182
    rescaled range analysis 182
hydration 11–12, 30
    number 11
    stratum corneum 294–296
    primary sphere 30
    secondary sphere 30
hydrogel 252
hydrogel/aluminium electrode 261–262
hydrogen production 280
hyperhidrosis 299–300

ice 88
ICG 274–275
IEC-60601 179, 193, 310
immittance xiii, 153–154
    distribution 149
    driving point 127, 154
    plethysmography 120, 152, 270–275
    polarisation 257–262
    specific 63
immittivity 63
    surface 62
impedance xiii, 59, 153–154, 233
    analyser 192–193
    cardiographic curve 274–275
    cardiography (ICG) 274–275
    plethysmography 270–275
    specific 63
    tomography 120
    transfer 155
impedivity 63
implants 289–290
incremental resistance 41
index 283, 233, 297
indifferent
    electrode 128
    electrolyte 23, 45
induced current 248–249
inductive field 242
inhibitory effect 279
initial value problem 147
injury potentials 122
in-phase 58, 323
instrumentalism 201
instrumentation 153–193
instrumentational amplifier 176–177, 180
insulated electrodes 244–245
insulation 179
insulator 51
integration, number of cycles 189
interface vs. interphase 23

interfacial polarisation 73–79
interferential currents 281–282
interphase
    phenomena 22–35
    vs. interface 23
interstitial liquid 98
invasive
    electrode 121
    needle electrode 265
inverse problem 129
ion
    diffusion 77–79
    dipolar 89
    exchanger 31
    migration velocity 3
    pump 96
    radius 4
ionic
    atmosphere 11
    bonds 6
    channels 95–98
    conductivity 12–15
    current 3
    dc conductivity 21
    liquid 18
    pairs 14
    polarisation 56
ionisation 4–6
    of neutral species 8
    potential 4
ionising radiation 247
iontophoresis 124, 279, 299–301
irreversible process 42
irreversible reaction 38
ischemia 116–117
isoelectric 128
    point 31, 89
iterative process 151

Joule effect 71–73

keratin 19
    α and β 113
keratinised tissue 105–115, 294–301
Kohlrausch, Friedrich Willhelm Georg 12
Kramers–Kronig transform 67, 158, 211

Langmuir adsorption isotherm 28
Laplace
    equation 202
    transform 156
    Pierre Simon de 156

Larmor equation 70
latency 299
layer
    diffuse electric 26
    Gouy 28
    Helmholtz 26, 28
    Stern 27
LCR-meter 192–193
lead
    sensitivity 140–143
    vector 140–143
    vector field 142
leads (in ECG) 268
leakage 172
    current 310
let-go current 306
levitation 32, 34
Leyden jar 315
lie detector 298
lightning 308
line spectrum 165
linear
    network 156
    region (electrode) 49
lipid 93
liposome 94
liquid
    junction potential 39, 255
    water 88
local volume sensitivity 142
lock-in amplifier
    analogue 190–192
    current mode 191–192
    digital 188–190
loss 61
    angle 61, 224
    dielectric 71
    factor 61
    tangent 61
low-pass filter 178
lysed erythrocytes 99, 104
lysis 290–291

macroshock 305
magnetic
    coil 150
    excitation 70
    field 242
    field coupling 248–250
    resonance imaging (MRI) 250
    stimulation 249–250
mammalian tissue 118
many-body interactions 223

mass action law 16
Mateucci, Carlo 315
Maxwell, James Clerk 202, 315
    equations 201, 248
    spherical particles mixture equation
        76
Maxwell–Fricke equation 76
Maxwell–Wagner effects 74–77, 84
measuring
    cell 267
    depth 134, 295–297
membrane 301
    cell 94–98
    conductance 97
    electrochemical 95
metallic bonds
metals 7
microelectrode 265
micromotion detection 293–294
micropipette 265
microshock 305
migration 23, 78
    velocity electrons 3
    velocity ions 3
minimum phase shift 158
mitosis 88
mixed conductors 19
model 153, 195–239
    descriptive 195–201
    explanatory 195–201
    four-component 214–216
    Gouy–Chapman's 26
    Helmholtz 26
    parallel 203–208
    series 205–208
    Stern 28
    three-component 214–216
    two-component 203–209
modulation 165
modulus function 61
molar conductivity 13–14
molecular
    bonds 6–7
    interactions 57
monomer 91
monopolar 128
monopole source 129
mother wavelet 185
motion sensing (cells) 293–294
MRI 250
mucosa 297
Müller, Hermann 315
multifrequency EIT 149
multiple-point electrode 264

muscle
    mass 282
    tissue 101–103, 115–117
nail 112–115
near-field 203, 243
needle electrode 135
Nernst equation 37
nerve
    stimulation 49
    tissue 101–103
net charge of particle 31
network
    analyser 193
    associative memory 183
    causal 156
    feedforward 183
    linear 156
    neural 183–184
    non-linear 157
    passive 156
    reciprocal 156
    recurrent 183
    self-organising 183
    theory 153–159
neural network 183–184
neuron 183
neutral electrode 128
nickel electrode 251
NMR 70
noble gas 4
node 183
noise 172, 188, 254–257
    power line 180–182
    reduction 180–182
    white 169
Nollet, Jean-Antoine 314
non-conductor 51
non-galvanic contact 242–250
non-linear
    black box 157
    network 157
    properties of electrolytes 48–50
    tissue parameters 122–124
non-polarisable electrode 41 44
non-symmetrical distribution of relaxation
        times 226–227
non-union 279
nuclear magnetic resonance (NMR) 70
Nyquist criterion 173
nystagmus 276

Ohm's law 12
one-compartment model (plethysmography)
        271–272

Onsager
  Lars 15
  theory 15
operational amplifier 176–178
oral mucosa 297
organ state 116
organelle 95
orientational polarisation 56
osmosis 31, 116
overvoltage 41–42, 45
oxidation 10
oxonium 15

pacemaker 284–285
  catheter stimulation 49
parallel
  circuit 325–331
  Cole element 218
  equivalent model 61
  model 203–208
Parseval
  des Chênes 171
  theorem 171
particle
  colloidal 18
  net charge 31
passive
  electrode 128
  network 156
patch clamp 266
Pauling
  Linus 5
  scale of electronegativity 5
Pauly–Schwan equations 77
pearl chain formation 32 35
penetration enhancers 199
peptide 91
perception 301–309
  threshold levels 307
period 160
periodic waveform 160–167
permittivity 54
  complex 59–63
Pflüger's law 280
pH electrode 18
phase
  angle 61, 160, 224
  sensitive detector (PSD) 190
  space 182
  locked loop 192
phasor 323
Philippson 316
phospholipid 94

photo-plethysmography 270
physical dimensions 334
pick-up electrode 127–128
piezoelectricity 124
Plancheral's theorem 171
plant tissue 118–119
plasma 104, 254
platinum
  black electrode 251, 258–259
  electrode 8–11, 40, 250–251
  hydrogen electrode 256
plethysmography 120, 152, 270–275
  dynamic cylinder model 272
  one-compartment model 271–272
  two-compartment model 272
Poisson equation 27, 38, 202
polar substances 18
polarisability 55
polarisable electrode 43
polarisation 52–57, 63–73
  counterion 29
  electrode 45
  electronic 56
  immittance 257–262
  interfacial 73–79
  ionic 56
  orientational 56
polarography 40
polyelectrolyte 91
polymer 91
  conducting 251
polysaccaride 93
ponderomotive effects 33–35
porcine liver 116
pore formation 290–292
pores 78
post mortem 115–117
post-excision changes 115–117
potential 53
  equilibrium (electrode) 36
  evoked 174
  flow-generated 32–33
  half-cell 7
  ionisation 4
  liquid junction 39, 255
  resting 98
  sedimentation 31–32
  streaming 31, 33
  zeta 28
  injury 122
potentiometry 36
potentiostat 144
power 72
  absorption 247

density 130
density spectrum 172
line noise 180–182
spectrum 172
pre-amplifier 192
precipitation 31
presentation of data 232–239
primary hydration sphere 30
probability density function 174
product of two sine waves 161–165
proportionality 156
protective earth 310
protein 19, 88–93, 95
proton hopping 15
prototype 196
proximal zone 131
pseudomaterial constants 63

quadrature 58, 323
quadropolar 128
quasi bipolar system 264
quasi-circular arcs 226
quasi-dipoles 145

radiation 203, 242
radiography 247
Ranvier nodes 308
raw data 232
Rayleigh scattering 247
reactive power 72
receiving
    coil 250
    electrode 127–128
recessed electrode 145–146
reciprocal
    excitation 142
    network 156
reciprocity theorem 141, 156
recording
    electrode 127–128
    leads 138–143
recovery time 299
rectifier, synchronous 190
recurrent network 183
redox
    process 10
    system 36–37
reduction 9–10
reference 178–182
    electrode 44, 256–257
refractory period 96
registering electrode 127–128
Rein, Herman 316

relative
    atomic mass 14
    permittivity 54, 59
relaxation 63–73
    characteristic frequency 66, 69
    Debye 64–70
relaxation time 11
    distribution (DRT) 223
    non-symmetrical distribution 226–227
    symmetrical distribution 224–226
REM sleep 276
resistance
    access 218
    chord 96
    electron transfer 46
    incremental 41
resistivity, complex 61
resistor 154
resonance 70
resting potential 98
reversible
    process 42
    reaction 37
RF coil 250
rheobase 278, 280
rigor 117
rise time 299
root-mean-square (rms) 72, 161
rotational E-field 55

saccaride 93
safety 179, 309–312
    grounding 179
salt bridge 39, 255–256
sample-and-hold 145
Savart, Félix 150
scalar 321–325
    field 324
    flux 12
scattering 247
Schwan
    Herman P. 319
    law of non-linearity 50
Schwarz theory 29
sebaceous glands 112
sebum 112
secondary hydration sphere 30
sedimentation 82
    potential 31–32
self-affinity 182
self-organising network 183
self-similarity 182
semiconductor theory 20

sensation 301–309
sensitivity 133
   lead 140–143
   matrix 151
series
   circuit 325–331
   Cole element 220
   equivalent model 61
   model 205–208
shadowing effect 262, 267
Sheffield Mark III system 150
short time Fourier transform (STFT) 168,
   184
short-wave diathermy 245–246
Siemens, Werner von 316
signal
   averaging 173
   generator 174–175
silent electrode 128
silver electrode 251
sine waves
   product 161–165
   sum 161–165
single-ended input 180
single-pass process 151
SI system of units xiii
skeletal muscle 117
skin 105–112, 294–301
   series resistance 110
   abrasion 254
   depth 249
   drilling 110
   effect 249
   impedance 253
   irritation 296–297
   site dependence of impedance 109
   surface electrode 262–263
sliding vector 321
solid electrolyte 18
sorption surface 46
source electrode 127–128
space domain 321
specific
   absorption rate (SAR) 248
   admittance 63
   adsorption 30
   immittance 63
   impedance 63
spectral
   density 168
   distribution 171
spectroscopy, dielectric 83–85
spectrum analysis 170–174, 193
spontaneous activity 298

spread coagulation mode (electrosurgery) 287
SQUID 248
St. Elmo's fire 313
stainless steel electrode 251
state space 182
static electric fields 53
step
   excitation 157
   function 168
   voltage 308
Stern
   layer 27
   model 28
stimulating electrode 127–128
stirring 255
Stoke's law 13
strange attractor 182
stratum corneum 105–112
   hydration 294–296
   thickness 107
streaming potential 31, 33, 78
substitute 196
sum of two sine waves 161–165
superconducting quantum interference
   device (SQUID) 248
superposition 156
surface immittivity 62
surface sorption 46
susceptance 323
susceptibility 55
susceptivity 55
sweat duct 110
symmetrical distribution of relaxation times
   224–226
synchronous rectifier 190

tattooing 254
temperature
   coefficient 101
   dependence of conductivity 17
   rise 71–73, 131, 309
TENS 277–278, 308
terminal 153
Tesla, Nikola 315
tetanic muscle contractions 281
tetrapolar 128
   system 145–147
therapeutic methods 241
thickness of double layer 27
three-component model 214–216
three-electrode system 143–145
threshold levels for perception and hazard
   307
threshold of perception 301–309

tight junctions 105
time
    constant 69, 157–158, 204, 206
    domain 170, 321
tissue
    adipose 103–104
    blood 104–105
    bone 103–104
    brain 101–104, 116
    cardiac 116
    conductivity 119
    electrical properties 87–125
    hair 112–114
    keratinised 105–115
    muscle 101–103, 115–117
    nail 112–115
    nerve 101–103
    plant 118–119
    mammalian 118
titanium electrode 251
tomography 120
transconductance 155
    amplifier 192
transcutaneous electrical nerve stimulation
        (TENS) 277–278, 308
transdermal drug delivery 300–301
transfer
    admittance 155
    coefficient 47
    function 155
    impedance 155
    sensitivity 128
transference number 13
transmittance 127–128, 154
transmitting coil 250
transport electrogenic 23
transresistance amplifier 155, 176–177
travelling wave dielectrophoresis 32 34
triboelectric series 124
triboelectricity 124–125
tripolar 128
tunnelling 20, 333
two-compartment model (plethysmography)
        272
two-component model 203–209

u/v-plot 223, 239
unipolar 128
    augmented leads 268
unit
    charge 22
    impulse 168–169

valence
    band 20
    electrons 5
Van der Waals bonds 6
vasodilatation 278
vector 321–325
    field 324
    flux 12
vectorcardiography 270
VHF/UHF electrode 268
Vinci, Leonardo da 313
Volta, Alessandro 314
voltage
    clamp 144, 174
    common mode (CMV) 178–182
    follower 177
    step 157
voltammetry 40
    cyclic 42
volume sensitivity 128, 131, 142

Waage Peter 116
Warburg immittance 47
water
    bound 88
    liquid 88
wave 203
waveform synthesis 165
waveguide 268
wavelet
    analysis 184–186
    prototype 185
    series 185
Wessel
    diagram xiii, 67, 206–208, 322
    Caspar 322
white noise 169
whole body impedance 114–116
Wien
    effect 49
    Wilhelm Karl Werner 49
working electrode 128
wound healing 279

x-ray absorption 270

zeta potential 28

Ørsted, Hans Christian 314

Printed in the United Kingdom
by Lightning Source UK Ltd.
121322UK00001B/49/A

9 780123 032607